To Mike,

Hope you enjoy
thanks for coming

C000155309

28/03/2018 - ST. PETER'S SCHOOL, SOUTHBOURNE

SEE IT/SHOOT IT

SEE IT/SHOOT IT

The Secret History
of the CIA's Lethal
Drone Program

Christopher J. Fuller

Yale UNIVERSITY PRESS
New Haven and London

Published with assistance from the foundation established in memory of Calvin Chapin of the Class of 1788, Yale College.

Yale University Press books may be purchased in quantity for educational, business, or promotional use. For information, please e-mail sales.press@yale.edu (U.S. office) or sales@yaleup.co.uk (U.K. office).

Set in Galliard type by IDS Infotech, Ltd. Printed in the United States of America.

Library of Congress Control Number: 2016954414
ISBN 978-0-300-21854-1 (cloth : alk. paper)

A catalogue record for this book is available from the British Library.

This paper meets the requirements of ANSI/NISO Z39.48-1992 (Permanence of Paper).

10 9 8 7 6 5 4 3 2 1

CONTENTS

Acknowledgments   vii

A Note on Language   x

List of Abbreviations   xii

Prologue: "Let Him Sleep"   1

Introduction: "The True Watersheds in Human Affairs Are Seldom Spotted"   9

1 "The Hamlet of Nations": The Rhetoric and Reality of the Reagan Administration's Counterterrorism Policy, 1980–1985   22

2 "Let's Find a Way to Go after Them": The Fadlallah Affair and the *Achille Lauro* Hijacking, 1985   63

3 "We Have to Find a Better Way to Send a Message": The CIA's Counterterrorism Center, from the Eagle Program to the Predator Drone, 1986–2001   88

4 "Talking about Capturing bin Laden": The Clinton Administration and the Legal Architecture of Lethal Force in Counterterrorism, 1993–2000   129

5 "Ninja Guys in Black Suits": Alternative Counterterrorism Tools, 1993–2008   178

6 "The Only Game in Town": The Strategy and Effectiveness of the CIA's Lethal Drone Campaign, 2009–2012   209

Conclusion: "It Sends Its Bloodhounds Everywhere"   247

Notes   255

Index   333

ACKNOWLEDGMENTS

I would like to express my deepest appreciation to Kendrick Oliver for his invaluable support in the writing of this book. From his tutelage and professional support during the research phase of the project, to the timely, clear, and constructive comments during the book's writing up, there is no doubt that this project would not have come to fruition without his help. Additionally, I benefited hugely from the input of Mick Jardine, whose careful reading of the book's first draft provided a number of nuanced contributions on the contentious language of terrorism and counterterrorism. My thanks also go to Helen Spurling, who has been both a supportive colleague and an invaluable friend throughout this project, offering limitless enthusiasm and encouragement as well as a generous supply of champagne at every milestone.

Early drafts of chapters from the project's manuscript were presented as papers at a number of conferences, including HOTCUS in both Northumbria and Reading, BAAS in Exeter, and two thought-provoking sessions at Surrey's Center for International Intervention. The astute questions, perceptive comments and helpful recommendations that resulted from these meetings helped to sharpen the arguments made in a number of the book's chapters, for which I am thankful to my colleagues and peers. I am also extremely grateful to my editor, Erica Hanson, and her colleagues at Yale University Press, in particular Otto Bohlmann, whose diligence and guidance helped ensure this book maintained the lucidity and coherence necessary to make this complex subject accessible. Furthermore, I would like to thank the anonymous reviewers, whose learned recommendations ensured the book met the rigorous standards such a contentions topic demands. It goes without saying that any faults that remain in the book are my responsibility alone.

Closer to home, I would like to thank my mother, Joan, and father, Charles, for instilling an appreciation of the benefits of hard work and delayed gratification—something that has come in incredibly handy during the lengthy process of bringing this research project to fruition. I am also grateful to my sister, Jennie, for helping sharpen my rhetorical arguments through our exchanges as kids. Most important of all, I am completely and utterly indebted to my wonderful wife, Carys. She has shown remarkable patience throughout this project—a virtue those who know her well will be aware does not come naturally to her—as long hours often kept me away from home or buried away in our office. During this time Carys has provided invaluable support. When I needed to talk ideas through, she was my intellectual sparring partner, playing devil's advocate for the more contentious arguments. I am particularly indebted to Carys for her contribution to the book's Prologue, where her superior storytelling skills helped refine my clumsy attempt to unleash my inner novelist. Her willingness to tolerate my lengthy discussions on counterterrorism, her good humor, and her shared interest in spiced rum have all been indispensable. This fortitude has been made all the more remarkable since January 2014, with the arrival of our adopted son, Leo. The fact that I was able to complete this book within the original time frame despite the addition of a toddler in our lives is entirely due to Carys. Leo's arrival transformed both our lives, but it was my wife who had to make the most significant changes to raise our boy, something for which I will be eternally grateful, and enamored. She has been a remarkable partner and mother.

Finally, I owe a debt to my young son, Leo, just for being his charming and fun-loving self. Studying terrorism, its causes and consequences, and the efforts of those who face the daunting task of trying to prevent it can be draining. The extent of the violence, brutality, and sheer ruthlessness those who seek to use fear and intimidation as a form of power can erode one's faith in humanity. Following long days of exposure to the sights and descriptions of such brutality, it has been life-affirming to have the joy of sharing a home with a young child whose innocence and good nature restore that lost faith. Such an environment serves to remind me of what it is those engaged in the challenging and often thankless task of counterterrorism are dedicated to protecting. It is my hope that this book, through its critical appraisal of counterterrorism methods past and present, can help in some small way to ensure that we do all we can to protect our societies, while at the same time not being forced to sacrifice the natural human love and kindness Leo so wonderfully demonstrates.

A NOTE ON LANGUAGE

This book uses a number of terms that hold serious and at times contentious connotations. The use of these terms has been carefully considered, and an explanation follows of why each is used.

AfPak—The term "AfPak," popularized by the Obama administration, is used throughout the book to reflect America's strategic thinking on the Afghanistan/Pakistan border regions, rather than as an acceptance of the term in a geopolitical sense, inferring that the border regions are a single, homogenous bloc. It is fully acknowledged that the tribal regions which make up the Federally Administrated Tribal Areas (FATA), the North-West Frontier Province (NWFP), and the border provinces of Afghanistan are all distinct regions, each with its own unique identity, tribes, culture, social groupings, and history.

Collateral damage—This military term is used to denote civilian casualties. It is not intended to sanitize the tragic losses it signifies but rather to ensure the language used in official sources is consistent with that of the book.

Drone—The term "drone" is unpopular with pilots due to connotations of a mindless or autonomous activity. So too are the terms "Unmanned Aerial Vehicle" (UAV) and "Unmanned Aerial System" (UAS), as both suggest aircraft that fly without pilots. The preferred term among operators is "Remotely Piloted Aircraft" (RPA). Due to its common usage in both academia and the media, however, the term "drone" will be used throughout the book to cover all variants of RPAs. It is used with a full acknowledgment that at no time is a drone flying without a pilot or the support of a sensor operator and ground crew.

Neutralization—The term "neutralize" is used in reference to the counterterrorism objectives of the United States. It is drawn from the Reagan administration's first counterterrorism document, National Security Decision Directive 138, and is used to collectively denote the most proactive counterterrorism interventions, such as killing, rendition, and imprisonment, as opposed to defensive measures or efforts to tackle the root causes of terrorism.

Targeted killing—This term is used to denote the killing of a known or suspected terrorist/militant in a drone strike. It is used in place of assassination in acknowledgment of the fact that within its own domestic law the United States government considers itself to be legally at war with al-Qaeda, its affiliates, and their Taliban allies.

Terrorist/Terrorism—Who is a terrorist and what constitutes terrorism is an essentially contested concept. As this book is primarily concerned with the decisions and actions of U.S. policy makers, the use of these terms throughout are reflections of the U.S. government's own definition, contained in Title 22 of the U.S. Code, Section 2656f(d):

- The term "terrorism" means premeditated, politically motivated violence perpetrated against noncombatant targets by subnational groups or clandestine agents
- The term "international terrorism" means terrorism involving the territory or the citizens of more than one country
- The term "terrorist group" means any group that practices, or has significant subgroups that practice, international terrorism

With regard to which groups this term is applied to, those officially designated as foreign terrorist organizations by the U.S. Department of State are referred to in the same manner throughout the book. It should be noted that this decision has been taken in order to employ terminology that is consistent with official sources, and does not indicate a rejection of other available usage of the same contested terms.

ABBREVIATIONS

ACC	Air Combat Command
AFPAK	Afghanistan/Pakistan border region
AIRRA	Aryana Institute for Regional Research and Advocacy (Pakistan)
ANSF	Afghan National Security Force
APNSA	Assistant to the President for National Security Affairs
AQAP	Al-Qaeda in the Arabian Peninsula (Yemen)
AQI	Al-Qaeda in Iraq
AUMF	Authorization for Use of Military Force
BIJ	Bureau of Investigative Journalism
BoC	Bureau of Counterterrorism (State Department)
CIA	Central Intelligence Agency
CINC	Commander in Chief (Regional, Department of Defense)
COTS	Commercial off-the-shelf
CTC	CIA's Counterterrorism Center (Note: the name of the CTC was changed from the Counterterrorist Center to the Counterterrorism Center in 2005)
DARO	Defense Airborne Reconnaissance Office
DARPA	Defense Advanced Research Projects Agency
DIA	Defense Intelligence Agency (Pentagon)
DCI	Director of Central Intelligence (1945–2005)
D/CIA	Director of Central Intelligence Agency (from 2005)

DI	Directorate of Intelligence (CIA)
DO	Directorate of Operations (CIA)
DoD	Department of Defense
DoS	Department of State
DoJ	Department of Justice
DSS	Diplomatic Security Service
DST	Directorate of Science and Technology (CIA)
EIT	Enhanced Interrogation Techniques
EO	Executive Order
EOB	Executive Office Building (White House staff)
FATA	Federally Administered Tribal Areas (Pakistan)
FLIR	Forward-looking infrared
GA	General Atomics
GA-ASI	General Atomics Aeronautical Systems Incorporated
GCHQ	Government Communications Headquarters
HUMINT	Human intelligence
ICC	International Criminal Court
INF	Intermediate-Range Nuclear Forces Treaty
IOB	Intelligence Oversight Board
IS	Islamic State
ISI	Inter-Services Intelligence
ISI	Islamic State of Iraq
ISIS	Islamic State of Iraq and al-Sham (previously ISI, aka ISIL)
JCS	Joint Chiefs of Staff
JITF-CT	Joint Intelligence Task Force for Combating Terrorism
JAN	Jabhat al-Nusra (al-Qaeda affiliate in Syria)
JSOC	Joint Special Operations Command
JTTF	Joint Terrorism Task Force
LSI	Leading Systems Incorporated
MAE	Medium altitude endurance
MNNA	Major Non-NATO Ally
MON	Memorandum of Notification
NAF	New American Foundation
NATO	North Atlantic Treaty Organization

NCC	National Counterterrorism Center
NGA	National Geospatial Agency
NGO	Nongovernmental Organization
NWFP	North-West Frontier Province
NSA	National Security Agency
NSC	National Security Council
NSDD	National Security Decision Directive (Reagan)
NSPG	National Security Planning Group
OMB	Office of Management and Budget
OSS	Office of Strategic Services
PAD	Pakistan Afghanistan Department (CIA)
PDD	Presidential Decision Directive (Clinton)
PLF	Palestine Liberation Front
PLO	Palestine Liberation Organization
QRF	Quick Reaction Force
SEAL	Sea, Air, Land
SIGINT	Signals intelligence
SNIE	Special National Intelligence Estimate
SOG	Special Operations Group (CIA)
SSCI	Senate Select Committee on Intelligence
TTIC	Terrorist Threat Integration Center
UAV	Unmanned aerial vehicle
U.N.	United Nations
USAF	United States Air Force
WMD	Weapons of mass destruction
WTC	World Trade Center

SEE IT/SHOOT IT

PROLOGUE: "LET HIM SLEEP"

North Waziristan, Pakistan, Friday, May 21, 2010

As Mustafa Abu al-Yazid's vehicle buffeted down the dirt track connecting the main town of Miranshah with the village of Boyya, he would have known he was taking a risk. After all, they were anticipating strikes.[1] It had been just two days since the Bagram operation, and while the dozen Taliban fighters had not succeeded in penetrating the American base's perimeter, the attack had still featured prominently in the Western media.[2] The veteran Egyptian jihadist was accustomed to American retaliation against such operations, but the unexpected news that his wife and children were close by in a loyal supporter's home had prompted his uncharacteristically spontaneous decision to make contact. Using one of the half dozen cell phones and prepaid SIM cards he carried in order to mask his identity from his hunters, al-Yazid had broken the very protocols he compelled his fellow jihadists to follow, and had called the safe house's owner to arrange his arrival.[3]

Fort Meade, Maryland, United States of America, Saturday, May 22, 2010

The sound of an alarm alerted the analysts on duty—the automated algorithm that scanned the huge electronic dragnet the National Security Agency had thrown up over the Pakistani tribal areas had registered a number of interest. The cell from which the call had originated was not logged on the Geolocation watch list, but the number the phone had connected to was. The algorithm instantly linked the digits to the owner of an al-Qaeda safe house in the North Waziristan village of Boyya.[4] A second program instantaneously sifted through the mass of data stored in the Disposition Matrix, comparing the voice pattern of the caller with the thousands held on record linked to known and suspected extremists.[5] Within seconds the system registered a match, linking the voice on the end of the unknown cell to a video statement issued on May 26, 2007, by Sheikh Sa'id, also logged under the name Mustafa abu-al-Yazid. Data on al-Yazid streamed onto the analyst's screen. An original member of al-Qaeda's Shura leadership council, his file identified the Egyptian as the organization's chief financial manager and, according to the 2007 statement, the network's overall commander in charge of operations in Afghanistan. This was, in the parlance of the targeted killing campaign, a jackpot target.[6] The National Security Agency (NSA) analyst quickly passed the lead up the chain of command. The voice recognition software was reliable, but if this really was al-Yazid, a strike would certainly follow. It was American protocol that humans had to stay in the loop—the decision to recommend lethal action could not be left up to a machine. Within minutes, two NSA linguists were replaying the intercepted call, comparing the phonetic features with those of the stored statement.[7] Confident they had just located al-Qaeda's most senior field commander in Afghanistan, the analysts forwarded the details to the Counterterrorism Center. A brief window of opportunity had opened up.

Counterterrorism Center (CTC), Langley, Virginia, Saturday, May 22, 2010

The arrival of the intelligence on al-Yazid sent a charge through the CTC. Located on Langley's sixth floor, the center's Pakistan-Afghanistan Department (PAD) had more resemblance to a war room than to the offices of a spy agency. Operating in a warren of cubicles and offices, and abuzz with a frenetic energy, the CTC's targeters oversaw the orbits of dozens of Predator and Reaper drones operating out of a constellation of secret bases across Pakistan and Afghanistan, their locations marked on maps mounted on the department's walls. A photographic tribute hung by the PAD's entrance, honoring seven colleagues killed six months earlier at one of these bases—Camp Chapman in the Afghan province of

Khost—by an al-Qaeda suicide bomber.[8] The attack had significantly set back the PAD's operations and was the single most devastating assault upon the Central Intelligence Agency (CIA) in twenty-five years. The personnel's killer, Humam Khalil Abu-Mulal al-Balawi, had been a triple agent who played his handlers in order to gain access to those who were coordinating the CTC's drone operations from the ground in Afghanistan. Following the attack, al-Yazid had praised al-Balawi, declaring that the former Jordanian doctor had avenged the jihadist's prime martyrs.[9] Now the CTC had a lead on this man, less than forty miles from where the bombing had taken place. As the targeters prepared the briefing package for the air force pilots who would guide the drone to al-Yazid's location, it was clear that this would be more than just another counterinsurgency airstrike for the staff of the CTC. This was a chance at revenge.

Bayya Village, North Waziristan, Pakistan, Friday, May 21, 2010

Al-Yazid had stayed longer than he had intended. The pull of his family had proven too much, overpowering his usual strict discipline and the well-developed sense of self-preservation that had kept the veteran alive through more than twenty years of guerrilla warfare against two super-powers. By the time his daughters were asleep, the sun had long since set behind the nearby mountains of the Hindu Kush. The Egyptian was aware he should leave, but the relentless nature of the American's pursuit had taken its toll on the fifty-four year old, both mentally and physically. With a woozy combination of fatigue and satisfaction, the doting father lay on the cushions that were scattered over the intricately patterned rug next to his sleeping girls, closed his eyes, and let sleep wash over his exhausted body.[10] The serenity of the family scene was disturbed shortly after 22:00 by the arrival of al-Yazid's son Ya'qub, who had been dispatched from Miranshah by the other members of the Shura council to retrieve the wayward commander. But as the young fighter entered the house, he was hushed by his mother, Umm-al Shaymah, who was sitting with her resting husband, cradling their sleeping granddaughter. "He is very tired," she told her son, "let him sleep." Respectful of his mother's wishes, Ya'qub left his father to rest. He would return to the safe house early the next morning to collect his father.[11]

Creech Air Force Base, Nevada, United States of America,
Saturday, May 22, 2010

The sensor operator gently tilted her joystick, instinctively compensating for the latency of the signal traveling to an aircraft seventy-five hundred miles away. The camera panned, tracking the oblivious figure as he left the house and walked back to the car he had arrived in just minutes earlier. As the vehicle proceeded back up the dirt track, the mission's CIA commander, watching via live feed at Langley, ordered the airman to refocus the unblinking stare of the drone's sensor ball onto the house below. Judging from the size of the heat signature picked up by the aircraft's infrared camera, there were a number of occupants in the house, all gathered in the same room. Al-Yazid was a high-value target, which gave the drone's air force pilots considerable leeway to act, but the presence of unidentified individuals in the same building complicated the strike. The pilot confirmed the GPS coordinates and locked his aircraft into a steady, preset circular orbit at ten thousand feet above the house as they waited for the collateral damage estimate to be completed.

Confirmation that the strike was within the rules of engagement arrived via the Ground Control Station's chat room screen from the air force commander supervising the operation from the nearby Wing Operations Center.[12] With the military's legal assessments completed, the mission's civilian commander at Langley granted the crew permission to fire. The pilot ran through the prelaunch checklist, powering up an M-Model Hellfire. The missile had been specifically manufactured by Lockheed Martin for exactly this kind of strike, its unique characteristics revealing much about the nature of the war the CIA had come to be waging. A nine-millisecond delay fuse had been fitted in order to ensure the munition would explode only after bursting through the roof of the target building. The intention was twofold; the blast, its force fully concentrated inside the target building, was particularly devastating for those in the immediate vicinity. At the same time, the delay was also intended to limit the damage to surrounding structures and people, revealing the difficulty associated with engaging an enemy who sheltered among the flimsy civilian compounds of Pakistan's tribal areas.[13]

The sensor operator sparkled the target with the drone's laser designator. A box appeared on the heads-up display, locking on to the pixels of the house's roof while the drone continued its slow orbit.[14] As the pilot's right index finger began to tighten on the control stick, a third air force officer hurried into the trailer to act as the safety observer for the shot, positioning himself behind the pilot's seat, his gaze fixed firmly on the monitoring screen.[15] Concentrating on the crosshairs superimposed over the house, the pilot drew a deep breath before beginning his countdown—"Three . . . two . . . one . . . rifle." A small electronic pulse exited the shipping-container-sized metal box in which the drone's crew sat, traveling by fiber-optic cable across the Nevada desert and continental United States, under the Atlantic Ocean and over Western Europe to a relay station at Ramstein Air Base in Germany. From there the tiny packet of data was broadcast to a satellite orbiting twenty-two thousand miles above Earth, and bounced back down to the receiver located in the bulbous nose of the twenty-seven-foot-long Predator drone.[16] Two seconds later the image on the screen pixelated as the aircraft yawed from the asymmetric thrust of the Hellfire's launch. As the image was restored, the sensor operator care-fully guided the missile to the target with the laser designator's beam. It took just thirty seconds for the Hellfire to reach its target. "Splash," declared the operator as the infrared display washed out from the heat of the explosion. As the digital bloom dissipated, the camera's live feed revealed the devastation below. The immense shock wave of the contained blast had cracked the walls of the house, causing the partial collapse of the building onto its occupants, while the thermobaric pressure had ignited the highly combustible furnishings upon which the occupants had rested.[17] The twisted and contorted outlines of bodies and body parts, both adult and infant, glowed from the heat of the explosion amid the scorched rubble.[18]

The Fallout of the Strike

News of the Boyya strike was broken in the Pakistani media that evening by Geo TV. The report made no mention of al-Yazid's death but disclosed that ten people, including women and a female minor, had been killed in the U.S. attack.[19] Later that night, *Dawn,* Pakistan's oldest and most widely read English-language newspaper, reported that five civilians, including women and children, were among at least ten casualties. The article went on to quote undisclosed sources as identifying five alleged militants among the dead, with a further seven civilians injured.[20] An hour later Reuters quoted an anonymous Pakistani intelligence source who claimed six militants had been killed. The report also cited a conflicting account from local North Waziristan residents who alleged the only victims from a volley of five American missiles had been twelve innocent civilians from the same family, with the body count including four women and two children.[21] By the time Pakistan's only internationally affiliated paper, the *International New York Times*–linked *Express Tribune,* posted its report of the strike, an additional witness account claimed that a further six women and two more minors had been hospitalized.[22] Despite the suggestion of high civilian casualties, the drone strike barely registered in the mainstream American media—earning a seventy-seven-word entry on CNN's blog four hours after the event.[23] The U.S. government, as had become common practice, refused to acknowledge the action.

It was ten days later that news of Mustafa Abu al-Yazid's death was announced by al-Qaeda's media wing, As Sahab—a deliberate delay by al-Qaeda to deny the Americans the opportunity to shift the news cycle from the Bagram attack. Through a statement released to the jihadist Al Ansar forum, al-Qaeda eulogized its former fundraiser, describing him as the "prince of financial princes" and commending his work as the commander in chief of al-Qaeda in Afghanistan.[24] News of the death of a senior al-Qaeda commander saw the previously overlooked strike reported on in the mainstream Western media, as anonymous government sources confirmed the attack.[25] The coverage focused predominantly upon the extent to which the loss of al-Yazid was a blow to al-Qaeda, with little mention of the previously reported civilian casualties. Some months later, more reflective reporting on the collateral costs of al-Yazid's death began to emerge. In the December 2010 edition of the *Atlantic* Peter Bergen and Katherine Tiedemann used the strike to debate the benefits and drawbacks of America's escalating use of remote targeted killing as a counterterrorism method.[26] A 2011 op-ed in the *New York Times* by the director of Reprive—an organization that advocates, among other causes, on behalf of drone strike victims—compared the retrieval of Hellfire missile fragments from the site of the strike to the recovery of a murder weapon. Most hard hitting was a 2012 piece in *Salon* that combined an image of al-Yazid, the shrapnel from the Hellfire that killed him, and a picture of one of the children also killed in the strike to give a face to the collateral damage of the CIA's campaign.[27] It was not until the release of documents seized by U.S. Navy SEALs during the raid on bin Laden's compound that the full details of the casualties of the May 21 strike became clear. An internal al-Qaeda communication, published among legal papers for a counterterrorism trial in March 2015, confirmed that al-Yazid's wife, three daughters, and granddaughter had all died in the attack, along with another Arab supporter, his young son, and four local militants.[28]

Yet despite the increasingly reflective nature of the reporting on strike victims, rising concerns regarding the legality, morality, and efficacy of the CIA's drone campaign, and the gradual drip-feeding of information on the conduct of the agency's covert operation, the most obvious question remained unasked—its irregularity obscuring it from observers, its sheer abnormality serving to hide it in plain sight. Even as the full details of the strike were revealed, commentators of the drone campaign were not prompted to try to provide an answer to the most fundamental question concerning the CIA's drone campaign: How was it that civilian staff at America's primary intelligence agency came to be giving U.S. Air Force personnel an order to undertake a lethal strike, against an Egyptian militant and his family, in the tribal region of Pakistan, utilizing a remotely piloted aircraft from seventy-five hundred miles away? What sequence of events, policy decisions, and legal conclusions could possibly have led to this abnormality having become the mainstream counterterrorism policy of the United States? This book answers that question, placing that strike, and the thousands of others like it, into a historical context that explains the true origins of the CIA's development of what Richard Clarke dubbed the "see it/shoot it option," and the lethal drone campaign it enabled.

INTRODUCTION: "THE TRUE WATERSHEDS IN HUMAN AFFAIRS ARE SELDOM SPOTTED"

The American historian Daniel J. Boorstin once stated, "The true watersheds in human affairs are seldom spotted amid the tumult of headlines broadcast on the hour."[1] Few events in recent history have generated as many headlines as the United States' controversial War on Terror. The culpability of the federal government in failing to stop the September 11 attacks; the challenge to domestic civil liberties through the PATRIOT Act and to international human rights through extraordinary rendition, torture, and the use of the Guantánamo Bay detention facility; the invasion of Afghanistan and the nation-building mission that followed; the challenge to the U.N. Charter through the policy of preemption, enabling the invasion of Iraq and the subsequent scandal of the absent weapons of mass destruction (WMD); the failure to eliminate bin Laden in the Tora Bora caves, and his eventual killing in a U.S. Navy SEAL raid in Abbottabad, undertaken without Pakistan's consent; and the emergence of the Islamic State, a by-product of a decade's upheaval and conflict throughout the Middle East. Despite the magnitude of these issues, it was another development in the ongoing conflict that came to dominate the headlines during Barack Obama's two terms in office—the CIA's use of remotely piloted aircraft (RPAs), or "drones" in common parlance, to launch lethal strikes upon al-Qaeda militants and their affiliates.

The U.S. government runs three drone programs. The first, managed by the U.S. Air Force (USAF), is publicly acknowledged and has operated in the recognized war zones of Afghanistan and Iraq, in Libya as part of the 2011 NATO intervention against Mu'ammar Gaddafi, and over Syria as part of the Pentagon's Operation Inherent Resolve. The second, executed by the Department of Defense's Joint Special Operations Command (JSOC), is a significantly more secretive and irregular program involving Special Forces operators who have been granted extensive leeway to "find, fix, and finish" those labeled terrorists by the U.S. government, predominantly in Somalia and Yemen, and increasingly in Syria and Iraq too.[2] The third—and the primary focus of this book—is the sustained campaign run by the CIA. The agency has refused to publicly discuss this campaign, classified as covert action, or to provide any details of how it operates, other than the fact that it is aimed at al-Qaeda and its affiliates, and has principally been focused upon those operating within the Afghanistan/Pakistan (AfPak) border region. In 2015, the CIA's drone warfare remit expanded to Syria, where it operates a hybrid program, combined with the operators of JSOC to target the senior leadership of the Islamic State.[3]

Due in equal parts to the secrecy surrounding their use, the technological novelty of their unmanned operation, and concerns over the agency's suitability to undertake lethal operations, the CIA's use of armed drones has garnered increased attention from academia and investigative journalists, in particular those working in the foreign policy, defense, and legal fields. While disagreements over the putative military benefits, ethical downsides, and legal complexities of the CIA's campaign are common, a number of persistent themes in media and scholarly discussions have emerged over recent years, crystalizing into a dominant set of commonly held views about the agency's execution of drone warfare, many of which are challenged in this book.

The first, and arguably most common, perception of the CIA's drone campaign is that it represents an escalation of the counterterrorism policies introduced after the events of 9/11. This argument, advanced by former president Jimmy Carter in a 2012 *New York Times* opinion piece, states that America's current counterterrorism practice "began after the terrorist attacks of September 11, 2001, and has been sanctioned and escalated by bipartisan executive and legislative actions."[4] "Many," explains the investigative journalist Chris Woods, have come to view the torture and extraordinary rendition used under George W. Bush as having been "replaced with industrial-scale extrajudicial execution by his successor," as part of a continued, militarized response to the 9/11 attacks.[5] This viewpoint earned President Obama a number of titles in the media, such as the "Drone President" and "drone warrior"; the neoconservative commentator Charles Krauthammer sarcastically referred to him as "Zeus the Avenger, smiting by lightning strike." Reporters and academics alike have argued that Obama has introduced a "drone doctrine," and that the use of these remotely piloted aircraft will be "Obama's most lasting legacy."[6] "No president," argued the *Washington Post*'s Greg Miller, "has ever relied so extensively on the secret killing of individuals to advance the nation's security goals."[7]

Yet true to Boorstin's adage, the real watershed of this drone campaign has been missed amid the tumult of articles and opinions. Contrary to the views discussed above, the drone campaign is not unprecedented, nor is it solely the product of the War on Terror policies pursued by the George W. Bush administration. "History," Boorstin observed, "rarely shows its hand swiftly," and the drone program is no exception. As chapter 1 of this book demonstrates, the concept of waging war against terrorists did not emerge from the smoking ruins of the World Trade Center but can in fact be traced back to a small group of counterterrorism hardliners within the Reagan administration. This cadre, consisting of Secretary of State George Shultz, Director of the CIA (DCI) William Casey, and the National Security Council (NSC) member responsible for low-intensity warfare, Lt. Col. Oliver North, pushed for the United States to adopt a policy of preemptive force and lethal retaliation as measures of self-defense against the emerging threat posed to U.S. citizens by increasingly well organized and motivated terrorist groups, in what they described as an undeclared war with terror. Though their calls for an aggressive American stance were never fully adopted, the philosophy of these hardliners, enshrined in two National Security Decision Directives and a presidential Finding, prompted the establishment of the CIA's Counterterrorist Center (CTC), the very institution that would eventually take on the hunt for Osama bin Laden and his affiliates, five years before the 9/11 attacks, and the body that now orchestrates the agency's drone campaign, executed under the principles of preemption and lethal retaliation for self-defense.

The second, closely related theme associated with the CIA's drone war revolves around the concern that the rapid expansion of the drone program has blurred long-standing boundaries between the CIA and the military.[8] While some, such as an anonymous former CIA official, have argued this is a positive change, transforming the CIA from "an agency that was chugging along" into "one hell of a killing machine," other commentators have shown greater concern.[9] "The current campaign can't be reconciled with the agreed premises for the separation of military and intelligence community activities in the National Security Act of 1947," the human rights lawyer Scott Horton has argued.[10] That act stipulated that the agency should be essentially civilian, and though it was granted the freedom to undertake "such other functions and duties related to intelligence affecting the national security as the President or the Director of National Intelligence may direct," its primary function was intended to be intelligence gathering and analysis, not waging covert wars.[11] Commentators such as the Council on Foreign Relations' Micah Zenko have called for the CIA's intelligence operations to be distinguished from military operations, which, as Title 10 of the U.S. Code identifies, should be in the hands of the Department of Defense.[12] According to Horton, the drone campaign marked the first time in U.S. history that a state-of-the-art, cutting-edge weapons system was placed in the hands of the CIA. This, critics argue, marks the continued evolution of the CIA as a paramilitary force, equipped with advanced tactical weaponry.[13]

Chapter 2 challenges this perception that the CIA has become militarized as a result of the War on Terror, instead revealing that the agency's role as the United States' vanguard in its violent confrontation with terrorism began as early as 1985, when it was actively involved in a paramilitary plot against a suspected terrorist leader, Ayatollah Mohammed Hussayn Fadlallah. With pressure growing on the Reagan administration to act against increasingly aggressive terrorist elements, but Secretary of Defense Caspar Weinberger determined to keep the Pentagon away from the messy business of counterterrorism lest it derail the administration's ambitious Cold War military buildup, it was DCI Casey who opted to put his agency on the front line. Though that plot was unsuccessful in its first effort, its fallout was to shape the way the CIA would conduct its future counterterrorist operations, evidence of which, as the chapter reveals, can be seen in the way the agency runs its drone campaign today.

Under the leadership of its first director, Duane Clarridge, the CIA's CTC—discussed in detail in chapter 3—was established to function as a war room against terrorists, supported by a Finding authorizing it to undertake a global campaign of preemptive neutralization against terrorists. Though the consequences of the Iran-Contra affair initially tempered the CTC's aggression, the foundations of the agency's post-9/11 role in the War on Terror were laid during this decade, and its aggressive pursuit of al-Qaeda today owes much to the approach Clarridge sought to instill in his new department. The origins of the agency's involvement with drones as tools of counterterrorism are also traced back to this period, with chapter 3 establishing that as early as 1986, through its classified Eagle program, the CIA first explored the concept of using unmanned aircraft for both intelligence gathering—seeking information on the location of the American hostages held in Beirut—and to conduct lethal precision strikes against targets such as the Libyan dictator and state sponsor of terror, Mu'ammar Gaddafi. The chapter tracks the evolution of the unmanned technology from this prototype stage, through its early joint use by the CIA and air force over the Balkans as part of NATO operations against the Serbs, to its eventual deployment over Afghanistan in the CTC's hunt for bin Laden. In establishing this timeline, the book further challenges the commonly held belief that the armed drone is a product of the pursuit of bin Laden and the War on Terror, instead revealing that the Reagan administration's clashes with the state-sponsored terrorist groups and the pursuit of better intelligence for air force airstrikes during Clinton's presidency served as the key drivers of drone technology.

Despite the Obama administration's abandonment of what were seen as the most internationally damaging polices of the Bush administration, questions surrounding the legality and accountability of the CIA's strikes make up the third commonly discussed theme related to drones. If the CIA is "acting like a military organization, shouldn't it have the (relative) transparency and accountability of a military organization?" Spencer Ackerman asked in 2011 as the agency's campaign increased in intensity.[14] The fact that most members of Congress lack the appropriate security clearances necessary to know anything about CIA operations has raised questions in the media about the lack of effective oversight for the drone campaign.[15] With the exception of Republican senator Rand Paul's filibuster over drone policy—which was focused upon their use against American citizens—Congress remained remarkably quiet on the topic despite Obama authorizing more strikes in his first year in the Oval Office than his prede-cessor had launched in the eight years prior.[16] Jane Mayer of the *New Yorker* remarked, "It's easy to understand the appeal of a 'push-button' approach to fighting Al Qaeda," but she extended Ackerman's concerns about lack of oversight by observing that the campaign "has occurred with remark-ably little public discussion, given that it represents a radically new and geographically unbounded use of state-sanctioned lethal force."[17] Others have argued that the accountability for the drone campaign goes beyond Congress and the American public, and that in setting new international precedents for the use of lethal force, the United States owes the world a clear accounting for its use of drones as a vehicle for targeted killings.[18] The law and philosophy scholar David Luban has gone as far as to argue that "the opacity and unaccountability of the program are, in and of them-selves, threats to the rule of law."[19]

On a domestic level, the legal debate is centered upon whether targeted killings undertaken with drones constitute assassination, which has been illegal in U.S. law since Executive Order 11905, signed by President Ford in 1976 as a direct result of the Church Committee's report on the CIA's previous assassination campaigns.[20] Unsurprisingly, this question has trig-gered debate among an array of legal scholars. Some, such as Mary Ellen O'Connell, Richard Murphy, and Afsheen John Radsan, former assistant general counsel to the CIA from 2002 to 2004, have condemned targeted killings as "extra-judicial, premeditated killing by a state of a specifically identified person not in its custody" and argue that therefore they are assassinations.[21] Other lawyers, such as former Justice Department official John Yoo and Andrew C. Orr, have instead agreed with the position put forward by administration officials and President Obama himself that targeted killing through drones is not assassination, and that the use of the devices rests on solid legal foundations.[22]

Despite a number of high-profile speeches on the subject, the forced publication of a Justice Department memo on the legal rationale employed for preemptive strikes against suspected terrorists, and a "Policy Standards and Procedures" factsheet released by the White House in May 2013, critics have argued that the details of the campaign remain concealed by ambiguity, with officials refusing to disclose precise details of how the decision to target specific individuals is made, to describe the safeguards in place to ensure that all targeted killings are legal and accurate, or to provide account-ability mechanisms for violations.[23] Most troubling to the critics of the campaign, through what is deemed an abuse of the CIA's authority to act covertly under Title 50 of the U.S. Code, has been the governments refusal to disclose who was killed and for what reason, and what the collateral consequences of such killings were.[24] This secrecy has resulted in more than twenty separate calls for greater transparency, from bodies such as the United Nations, Columbia Law School, and New York Bar Association to nongovernmental organizations such as Reprieve, Amnesty International, and Human Rights Watch, as well as politicians from both sides of the political aisle, such as Republican congressman Walter Jones and Californian Democrat Adam Schiff.[25] There have also been several unsuccessful legal cases aimed at challenging the campaign's secrecy. The most prominent of these thus far was brought by the American Civil Liberties Union (ACLU) and the Center for Constitutional Rights (CCR) in 2012, which argued that "the government must be held to account when it carries out such killings in violation of the Constitution and international law."[26]

Contrary to these arguments, chapter 4 of the book challenges the notion that the CIA's drone campaign lacks appropriate congressional oversight and domestic legal authorization, instead revealing that the agency sought extensive legal cover from both the executive branch and Congress before undertaking its role as aerial executioner. Introducing the concept of the covert action pendulum, the chapter argues that a historic cycle of agency excess followed by a backlash of congressional investigation and subse-quently increased oversight became established. This dysfunctional dynamic initially served to undermine the CIA's efficacy as a counterterrorism force by stripping out talented staff, sowing doubt among senior managers, and creating a culture of risk aversion within the agency charged with pursuing the newly emergent nonstate threat of al-Qaeda. The chapter examines how the hard-learned lessons from these pendulum swings prompted Langley's cautious managers to insist upon the creation of the complex legal architecture that now underwrites the United States' drone campaign. This architecture, though bolstered by post-9/11 legislation such as Congress's Authorization to Use Military Force against Terrorists (AUMF) and the dubious legal category of "illegal enemy combatant," has evolved over three decades, with the George W. Bush–era Findings adopted by the

Obama administration being based upon preexisting Memorandums of Notification, signed by President Clinton, which in turn were drawn from the authorizations to hunt terrorists first granted to the CTC by the Reagan administration. This legislation, introduced by the White House with the bipartisan support of the majority of the Congress, produced a remarkably stable legal architecture through the first decade of the War on Terror, suggesting the CIA is operating within a much more considered and secure legal remit than critics previously believed.

On an international level, the legal criticism revolves around the extent to which the drone campaign violates state sovereignty and the human rights of those targeted. The former Swedish foreign minister Anna Lindh was one of the first international critics of the campaign when she described a November 2002 Predator strike in Yemen as "a summary execution that violate[d] human rights."[27] In May 2010, Philip Alston, the U.N.'s special rapporteur for extrajudicial, summary, or arbitrary executions, added his voice to the debate when he criticized the United States for the lack of international legal justification for the drone program, describing it as "a vaguely defined license to kill." While not going as far as declaring the drone campaign illegal, Alston criticized the blurring of the boundaries of human rights law, the laws of war, and the law applicable to the use of interstate force.[28] Reflecting upon this exploitative interpretation of international law, Hina Shamsi, the director of the ACLU's National Security, project warned in the pages of the *Guardian* that the continued use of targeted killings put the United States at risk of being perceived as a legal pariah.[29] The killing of Anwar al-Awlaki, a U.S. citizen, in a drone strike in Yemen, raised further protests, both because of his citizenship and because of the evidence it provided of the expanding frontiers of America's drone campaign beyond traditional warzones.[30]

Chapter 4's examination of America's legal rationale for drone strikes in international law concludes that a judgment on its justification rests upon two debatable factors: first, one must accept that it is possible—as the United States claims—for an ongoing state of war to exist between the United States and al-Qaeda, thus justifying strikes as acts of self-defense under Article 51 of the U.N. Charter. Second, the chapter discusses the legal necessity for the campaign to be both precise and proportionate in order to meet the *jus ad bellum* requirements of a "just war," a claim made by U.S. officials but impossible to corroborate on account of the lack of transparency relating to drone casualties.

Just who the victims of drone strikes are represents the fifth and arguably most hotly contested theme regularly discussed in drone-related literature and reporting. There is fierce debate over the number of civilian casualties caused by the CIA's drone strikes. This dispute is made all the more intense by the agency's refusal, on grounds of national security, to release data on its strikes. Despite the impression the CIA's noncompliance gives of an agency with something to hide, the U.S. government has gone to considerable lengths to present the drone campaign as carefully executed, precise, and proportional. In a live Web chat on Google+ in January 2012, President Obama responded to what he described as a "perception that we're just sending in a whole bunch of strikes willy-nilly" by stating that the drone strikes were "very precise, precision strikes against al-Qaeda and their affiliates," which were part of "a targeted, focused effort [against] people who are on a list of active terrorists."[31] The strongest avowal of the accuracy of drone strikes came from John Brennan, then serving as deputy national security adviser for homeland security and counterterrorism, who, in June 2011, went as far as to state, "There hasn't been a single collateral death [as a result of drone strikes] because of the exceptional proficiency, precision of the capabilities we've been able to develop."[32]

Those seeking to challenge Brennan's assertion of accuracy have highlighted the *New York Times*'s May 2012 revelation that the White House classes "all military-age males [killed] in a strike zone to be combatants . . . unless there is explicit intelligence posthumously proving them innocent."[33] Rejecting this methodology, the Bureau of Investigative Journalism (BIJ) published the evidence of its own investigations into civilian deaths in the Federally Administered Tribal Area (FATA) region of Pakistan in June 2012. The bureau concluded that in the 330 drone strikes (278 under Obama) that had occurred in the region since 2004, more than 2,500 people had been killed, including at least 482 who could be credibly reported as civilian casualties.[34] The New American Foundation (NAF) also published the results of its own study into casualties. While NAF's figures were slightly lower than those of the BIJ, they still contested Brennan's claims, reporting that between 152 and 191 civilians were killed by drone strikes between 2004 and October 2012, with a further 130 to

268 victims unable to be credibly identified as either militants or civilians.[35] Investigations such as these have fueled the perception that the U.S. drone campaign is not proportional and does not take account of civilian casualties.

Such controversy, with its potential damage to America's interests and reputation, raises the question of why the Obama administration adopted drone strikes as America's primary counterterrorism tool in the first place. By examining the methods applied by Obama's predecessors, such as the deployment of proxy agents; cruise missile strikes; rendition, imprisonment, and interrogation; and Special Forces raids, chapter 5 places this decision in context. In doing so, it demonstrates the extent to which the adoption of different counterterrorism policies from president to president has as much to do with the successes and failures of their predecessors' efforts as it does the ideological outlook of the policy makers themselves. In the case of the Obama administration, the chapter reveals how the hugely negative international response to the United States' association with torture though its use of rendition and the damaging existence of the Guantánamo Bay detention facility caused this approach to be rejected entirely. This chapter also explores the impact of wider factors on influencing policy decisions; for example, the American public's fatigue over foreign occupations, the financial limitations imposed by a decade of nation building and the economic ravages of the 2008 economic crash, and the advent of new technological innovations, such as the deployment of the more capable Reaper drone and the enhancements made in the National Security Agency's electronic surveillance techniques.

The final theme of frequent contention and debate relating to the CIA's drone war concerns the efficacy of the program itself, in particular the risk of potential blowback the frequent strikes pose. Since the campaign's inception, a range of observers have argued that far from making the United States safer, the increased dependency upon lethal drone strikes has actually stirred anti-Americanism and resulted in greater hostility toward the United States, ultimately reducing the nation's security. "The use of drone aircraft," argued David Cortright, director of policy studies at the Kroc Institute for International Peace Studies at the University of Notre Dame, "perpetuates the illusion that military force is an effective means of countering terrorism. . . . We should know better by now."[36] Former Australian military officer and Pentagon adviser David Kilcullen agreed with this position, testifying before Congress in March 2009 that drone strikes aroused feelings of anger that risked coalescing local populations around the targeted extremists, who somewhat ironically come to represent a form of resistance against the very attacks they attracted.[37] Kilcullen also published a widely discussed opinion piece in the *New York Times* describing the drone campaign as a strategic error, which had distracted from the more important job of isolating extremists and hindered the vital task of building local partnerships by angering Pakistani locals.[38]

Two years after Kilcullen's warnings, Leila Hudson, Colin Owens, and Matt Flannes came to a similar conclusion in *Middle East Review*, arguing that "rather than calming the region through the precise elimination of terrorist leaders, . . . the accelerating counterterrorism program has compounded violence and instability [in Pakistan]."[39] Noah Shachtman, contributing editor to *Wired*, agreed that the use of armed drones is breeding anger and frustration and leading to a backlash against the Pakistani and U.S. governments.[40] Supporting this view, Peter Bergen and Katherine Tiederman have reported that two successful terrorist attacks and one failed one have been carried out in direct response to drone attacks, and Hillary Clinton faced several noisy protests against the drone strikes when she visited Pakistan as secretary of state in September 2009.[41] Even the 2012 Republican presidential candidate Mitt Romney argued in a debate with the president that "America cannot kill its way out" of its conflict with al-Qaeda, while the *New Yorker*'s Jane Mayer warned of Americans' apparent lack of regard for the acute anger their country's employment of drone warfare in Pakistan had created.[42]

Despite the numerous warnings of potential blowback, U.S. policy makers have shown virtually no reservations about the continued use of drones in Pakistan. Peter W. Singer, a Brookings Institute fellow and author of *Wired for War*, argued that this support might have more to do with the seductive nature of the technology, which has created the perception that the military response is costless.[43] This in turn, argues Cortright, "reduces the political inhibitions against the use of deadly violence," creating what former U.N. special rapporteur Philip Alston described as "a PlayStation mentality to killing," which may induce public callousness.[44]

Chapter 6 of the book addresses these notions of accuracy, collateral damage, and blowback, assessing from what evidence is available just how accurate the CIA's program is, and the extent to which it has aided the United States in its efforts to end its ongoing War on Terror. Ultimately, the chapter concludes that, if one is to measure the campaign by its primary goals—the decapitation the al-Qaeda leadership; the denial of safe haven in the AfPak region; and the undermining of the Taliban's insurgency against the U.S.-backed Afghan government in Kabul—the campaign has been a success. Al-Qaeda's core was devastated, and its remaining senior figures were prevented from playing an influential role in the shifting jihadist scene following the events of the Arab Spring and the eruption of civil war in Syria. The group's limited time and resources had to be diverted from its core business of organizing anti-American plots, recruitment, and training, and marketing its cause, to simply trying to avoid detection and stay alive. Meanwhile the Taliban, though not defeated, were degraded, introducing at least the possibility of a negotiated peace with Afghanistan's current government. Yet, while the drone's ability to consistently loiter over the mountains and valleys of the AfPak region may have succeeded in cutting off al-Qaeda's leadership and reducing the threat posed by that particular group, the chapter argues that the CTC's drone campaign also played a key role in creating the power vacuum into which the Islamic State was able to step, arguably just replacing the clandestine, nonstate threat of al-Qaeda with the larger, wealthier state-based danger posed by the self-declared caliphate.

In its concluding chapter, the book explores the likely legacy of the CIA's drone war on U.S. counterterrorism, wider U.S. national security policy, and the conduct of America's rivals, both nation-states and terrorist groups. It contemplates the nature of technological progress, judging that innovations always introduce potential threats and opportunities in equal measure, and hypothesizing that while it is almost inevitable that terrorist groups will exploit drone technology for heinous ends, the technology also offers wider commercial and civilian society exceptional opportunities, just as previous transformative technologies, first developed for the purpose of taking lives, eventually came to transform them in positive ways.

Ultimately, what this book demonstrates is that the CIA's drone campaign, though unique in scale, is not unprecedented and actually marks a return to, rather than a further departure from, counterterrorism methods and strategy developed in the decades preceding 9/11. Thus, the drone campaign should not be regarded as a post-9/11 policy innovation or the result of the seductive nature of remote control warfare. Instead, the use of drones to neutralize terrorists is best understood as the embodiment of America's long-term counterterrorism goal—what Richard Clarke dubbed the "see it/shoot it option"—made possible by incremental advancements in both technology and the willingness of the U.S. government to clearly authorize the CIA to undertake the lethal counterterrorist actions it first set out to execute three decades before the drone campaign even began.[45]

1

"THE HAMLET OF NATIONS": THE RHETORIC AND REALITY OF THE REAGAN ADMINISTRATION'S COUNTERTERRORISM POLICY, 1980–1985

We cannot allow ourselves to become the Hamlet of nations, worrying endlessly over whether and how to respond.

—*Secretary of State George Shultz*

As the threat of international terrorism against the United States grew during the early years of the Reagan administration, a small group of influential policy makers began to coalesce around a shared hardline position on counterterrorism. Secretary of State George Shultz, Director of Central Intelligence (DCI) William Casey, and the National Security Council (NSC) staff member responsible for low-intensity warfare, Lt. Col. Oliver North, collectively came to believe that the most effective way to meet the emerging threat was to deploy lethal force, preemptively if possible, against terrorist operatives and their leaders. These hardliners encountered considerable resistance to their ideas from within their own administration, with the likes of Secretary of Defense Caspar Weinberger publicly criticizing their hardline agenda. Over the past three decades, however, this view has gradually moved from the hardline fringes of U.S. counterterrorism to being accepted as the mainstream policy approach adopted by successive American administrations. In the wake of the 9/11 attacks the focus upon Bush's declaration of a War on Terror has meant that the impact of these officials' ideas and actions has been overlooked. Yet their legacy is evident in the counterterrorism policies adopted by the Clinton, George W. Bush, and Obama administrations, with the eventual emergence of the CIA's drone campaign and the practice of targeted killing that it embodies owing more to these officials than to any other policy makers.

Reagan's Counterterrorism in Rhetoric

Reagan came to power at a time of heightened public interest in, and fear of, acts of terrorism. Globally, terrorist attacks had been increasing at an alarming rate over the previous decade. In 1970, there had been a total of 643 terrorist incidents worldwide. In 1980, this figure had increased to 2,621 for the year.[1] Americans themselves had a particular sense of their vulnerability to this tactic due to the humbling ordeal of fifty-two American citizens held hostage in Iran for 444 days after the successful overthrow of the American-backed shah by Ayatollah Khomeini's Islamic revolution. President Carter had attempted a military rescue operation, but its failure had left him looking enfeebled and the United States humiliated. The mission, code-named Operation Eagle Claw, was conducted by the U.S. Army's elite new counterterrorism unit, Delta Force, on April 24, 1980. The effort ended in debacle when one of the team's helicopters collided with a fuel-laden C-130 in the Iranian desert, killing eight Americans. The bodies were discovered by the Iranian authorities and pictures of the charred corpses were used as a symbol of American failure.[2] Not only had this ordeal demonstrated defenselessness on the part of the superpower, it also revealed that terrorism was no longer just a foreign policy issue but instead had the power to cross into domestic politics. Acts of terror—and the federal government's response to them—affected the public mood and carried with them electoral ramifications for presidents.

Many Americans had complained about Carter's conduct during the ordeal of the Iranian hostage crisis, and Reagan had been able to exploit this during his presidential challenge. In a television campaign speech a month before the election, he spoke of the need for "the United States to assume a leadership role in curbing the spread of international terrorism." He went on: "I will direct the resources of my administration against this scourge of civilization and toward the expansion of our cooperation with other nations combatting terrorism in its main forms."[3] Nine days later in the candidates' only debate, a question was asked about international terrorism, indicating its relevance as a political issue. Again Reagan raised leadership, implying Carter had not been doing enough to combat the threat and declaring that it was "high time that the civilized countries of the world made it plain that there is no room worldwide for terrorism."[4] So great was the impact of the hostage crisis on Carter's campaign and so acute the sense of weakness, one of his political advisers later lamented that "the President's chances of re-election probably died in the desert of Iran with the eight brave soldiers who gave their lives trying to free the American hostages."[5] "The greatest casualty of Operation Eagle Claw," reflects Richard Immerman, "was Jimmy Carter."[6] The failure did not just have an impact on the Carter presidency, it also left a lasting legacy for Reagan and in particular the man who would become his secretary of defense, Caspar Weinberger. Weinberger sought to ensure that no such disaster would befall the U.S. military during his tenure at the Pentagon by doing all he could to keep its forces out of high-risk counterterrorism operations.

From the moment he took office, Reagan's rhetoric regarding terrorism reflected his campaign promise to provide strong, decisive leadership in the face of terror threats. During his inaugural address, the new president warned "those who practice terrorism and prey upon their neighbors" that the "will and courage of free men and women" could never be shaken by their violent methods.[7] Minutes after the new president completed his address, it appeared that his forceful stance was already paying dividends as Iran finally released the Americans hostages. The convenient timing of the liberation of the U.S. detainees gave rise to the allegation that members of Reagan's campaign team conspired with the Iranian authorities to ensure Carter could not pull off an October surprise and boost his reelection chances by resolving the crisis before Americans went to the polls in November. In return for the Iranians dragging the crisis out it is alleged that members of the Reagan camp promised to lift trade sanctions and release billions of dollars of Iranian gold and assets frozen after the embassy seizure.[8] Two congressional inquiries and a number of media investigations failed to provide any solid evidence that the conspiracy was true, but it has persisted, in no small part due to the fact that a number of notable individuals who were involved in the hostage negotiations at the time, including former Iranian president Abulhassan Banisadr, Israel's former prime minister Yitzhak Shamir (serving as foreign minister at the time), and Gary Sick, a Carter NSC member, have all stood by the allegations.[9] The exposure of the Reagan administration's later collusion to supply arms to Iran in return for U.S. hostages in Beirut lends further credence to this theory.[10]

Whatever the truth regarding the hostages' release, Reagan used their return to stake out his administration's first tangible policy departure from the Carter administration, vowing to "take an unrelentingly tough line against any such future acts of terrorism."[11] For the Reagan administration, in public at least, there was to be no flexibility in dealing with the perpetrators and sponsors of terrorist acts. Responding to a spate of kidnappings of American citizens in Lebanon in 1984, Reagan warned: "Yielding to violence and terrorism today may seem to provide temporary relief, but such a course is sure to lead to a more dangerous and less manageable future crisis."[12] Building upon this theme of resistance in the face of terrorist threats, members of Reagan's government criticized the Carter administration for opening diplomatic channels with Iranian officials in his efforts to free the hostages. "The present administration would not have negotiated with Iran for the release of the hostages," said William Dyess, a State Department spokesman.[13] Instead of diplomacy, acts of terrorism against the United States would now be met with a policy of "swift and effective retribution," promised the president.[14] During a press conference relating to the American hostages held in Lebanon, the president promised: "Every effort will be made to find the criminals responsible for this act of terrorism so this despicable act will not go unpunished."[15] As the anti-American terrorist threat escalated from the Iranian-backed militia in Lebanon to the state-sponsored violence conducted by the Libyan dictator Mu'ammar Gaddafi's agents, so too did Reagan's rhetoric. "When our citizens are abused or attacked anywhere in the world on the direct orders of a hostile regime," warned the president during an address following military strikes against Libyan targets in 1986, "we will respond so long as I'm in this Oval Office."[16]

The bellicose rhetoric used by the president and other members of his administration perpetuated an image of a government that was tough on terrorism. The members of Reagan's team presented themselves as offering a new, swift, and decisive approach to counterterrorism, which put "those who would instigate acts of terrorism against U.S. citizens or property on notice" that the United States would "vigorously confront" them and bring them to justice, whether that be through diplomatic and economic sanctions, their arrest for trial in U.S. courts, or the full-scale deployment of American military force.[17] The reality behind the rhetoric, however, was significantly different to the image it projected. The Reagan administration was deeply divided on how best to deal with the threat of international terrorism and its state sponsors. "The executive branch was itself so fragmented," reflected George Shultz, "that it was impossible to orchestrate all counterterrorist efforts effectively or even to get agreement that there should be a specific counterterrorist effort."[18] This chapter examines the various viewpoints and policy proposals put forward by key members of Reagan's administration on counterterrorism. It then provides an assessment of why Reagan was unable to forge these viewpoints into a coherent counterterrorism policy, and why it was the CIA that ended up at the forefront of the U.S. counterterrorist mission. The remainder of the chapter demonstrates how these policies, though not successfully enacted at the time, would eventually form the strategic backbone of the CIA's drone campaign.

The Key Architects of U.S. Counterterrorism Policy in the Reagan Administration

Alexander M. Haig, Secretary of State, 1981–1982

Though Haig was not in his post long enough to make a significant impact upon the administration's counterterrorism policies, he was the first cabinet member to set out a tangible policy regarding terrorism.[19] During his service as supreme allied commander in Europe in 1979, he narrowly survived an assassination attempt, which West German intelligence had concluded was perpetrated by the Baader-Meinhof Gang. Such groups, Haig was told, operated with significant Eastern bloc assistance.[20] This personal brush with Soviet-sponsored terrorism had left Haig convinced that the United States had to make fighting international terrorism a central pillar of Cold War foreign policy. He used his first press conference to set out a hardline, anti-Soviet position he believed to be in step with the tough rhetoric Reagan had used during the election campaign. He argued that terrorism was not a regional issue but rather a Cold War issue, with terror groups coordinated globally by the Soviets as part of a covert war against Western democracies.[21] Taking a swipe at the Carter administration, he declared: "International terrorism will take the place of human rights" in the Reagan administration. "The greatest problem to me in the human-rights area today," the secretary of state continued, "is the area of rampant international terrorism." Haig went on to place the blame for this international terrorism squarely on the Soviet Union for its role in the "training, funding and equipping" of the world's most significant terrorist groups.[22] This declaration was not well received by many within the Reagan administration, especially James Baker, the new White House chief of staff. Baker berated Haig for upstaging the president on a strategic policy matter and signaling to the Soviet Union that the United States might hold it culpable for any future terrorist attacks against it and its allies, anywhere in the world.

Despite Baker's criticism, Haig's views received support from the American journalist Claire Sterling in her 1981 book, *The Terror Network*, in which she argued that the KGB's role in international terrorism was not a matter of guesswork but documented fact.[23] The Soviet Union, Sterling argued, coordinated with terrorist groups to attack strategic Cold War targets: "The strongest and weakest links in the Western democratic chain . . . had plainly been singled out as front-line states by the terrorists and the Kremlin both. In neither of these countries did the evidence of Soviet complicity depend merely on connections two or three times removed. The KGB was involved directly, in tandem with the security services of Russia's East European satellites."[24]

Experts across the U.S. government and commentators in the media were divided over the role played by the Soviets in the escalating number of international terror acts.[25] The CIA produced a draft Special National Intelligence Estimate that rejected Sterling's argument, concluding that the Soviets were largely uninvolved, while the Pentagon's Defense Intelligence Agency (DIA) produced a contradictory draft that asserted Soviet guilt. Eventually the standoff was ended by an independent review of both reports and their sources undertaken by Lincoln Gordon, a former president of Johns Hopkins University. Gordon's conclusion was that, though it may have been possible to argue that the Soviets had contributed indirectly to international terrorism, they were not using it as a strategic tool to destabilize Western democracies and Third World nations.[26] The Soviets were not the hidden hand behind international terrorism. It was a total rejection of both Haig's argument and Sterling's thesis, and left the struggling secretary of state's intended policies on counterterrorism in tatters, hastening his early exit from the cabinet.

With hindsight, the views expressed by Haig and Sterling were not as far from the truth as Gordon's analysis suggested. While the Soviet Union was not the perpetrator behind *all* international terrorism, Soviet archives have since revealed the USSR and its satellites were far from innocent bystanders. Details of close links between the KGB and significant terrorist groups have emerged.[27] Furthermore, the statistics relating to international terrorism reveal a substantial reduction in terrorist attacks following the fall of the Soviet Union, further supporting Haig's theory. In the 1980s there had been 5,431 international terrorist incidents, in which 4,684 people died. In the 1990s these figures dropped to 3,824 incidents, which resulted in 2,468 deaths. While improved counterterrorism efforts played a credible part in this reduction, the demise of leftist terror groups such as the Red Brigades in Italy, the Red Army Faction in Germany, Direct Action in France, and the Communist Combatant Cells in Belgium was a key factor. As terrorism expert Paul Pillar reflects, "The West's winning [of the Cold War] sealed their fate. The groups lost both a credible ideology and the material support they had received from communist states in the East."[28]

George W. Shultz, Secretary of State, 1982–1989

Shultz was the administration's most ardent supporter of the use of force to fight terrorism and is arguably the forefather of the concept of the War on Terror. He pushed a number of significant policy positions that can be recognized in the CIA's conduct of its drone campaign. He held a firm view that by 1980 state-sponsored terrorism "had become a weapon of unconventional war against the democracies of the West, taking advantage of their openness and building on political hostility toward them."[29] By this rationale, when terrorists targeted American citizens or U.S. interests, they were actively "waging war" against the United States. In order to win this war, Shultz believed that the United States had to show terrorists that they would "pay a price for their crimes" through the use of lethal force. "Every nation has the right under international law to take defensive action," explained the secretary of state in his autobiography, three chapters of which are dedicated to his efforts to shape counterterrorism policy. "Part of that defense," he continued, "is to be prepared to take the offensive when the proper occasion arises."[30]

Shultz first publicly raised the issue of offensive action against terrorists in a statement to the Trilateral Commission on April 3, 1984. The choice of a meeting with the commission as the forum in which to deliver his statement—a body created by David Rockefeller to encourage dialogue between public and private leaders in Europe, North America, and Asia on the most pressing problems of the time—reflected the seriousness with which Shultz regarded the issue. "It is increasingly doubtful," warned the outspoken secretary, "that a purely passive strategy can even begin to cope with the problem." His address went on to consider the issues that effectively countering terrorism posed for Western liberal democratic societies. In what circumstances and with what means should the United States respond to a terrorist threat, Shultz asked his audience. Following up on this line of inquiry he posed further questions to the assembled members of the commission: "When—and how—should we take preventive or preemptive action against known terrorist groups? What evidence do we insist upon before taking such steps?"[31]

The secretary of state's questions remained rhetorical for only three weeks, after which Shultz returned to the theme in a speech delivered at the Jonathan Institute in Washington, D.C., on June 24, 1984. The choice of venue was deliberate on his part. The Jonathan Institute was founded in 1979 by Benjamin Netanyahu and was named after his late brother, Lt. Col. Jonathan Netanyahu, an Israeli officer who was killed during a rescue mission to free predominantly Israeli hostages from an aircraft hijacked by terrorists in Entebbe, Uganda.[32] By delivering his address from an institution named in honor of a man who gave his life fighting terrorism, Shultz was making clear his belief that U.S. counterterrorism should draw inspiration from Israel's more aggressive approach. "It is time to think long, hard, and seriously about more active means of defense," he declaimed. He went on to define what he meant by a more "active" defense by calling for the development of the "appropriate preventative or preemptive actions against terrorists before they strike."[33] The secretary's speech marked an extremely significant moment in the evolution of U.S. counterterrorism. Never before had a U.S. official publicly called for the United States to take action against known or suspected terrorists *before* an attack was launched.

Shultz continued to build his case for preemptive action, delivering his most impassioned promotion of the approach four months later at the Park Avenue Synagogue in New York. He publicly criticized what he regarded as the Reagan administration's lack of decisive action against the growing terrorist threat. "We cannot allow ourselves to become the Hamlet of nations," Shultz cautioned, "worrying endlessly over whether and how to respond." As a great power with global responsibilities, he explained, the United States could not afford to be "hamstrung by confusion and indecisiveness." He recognized the concerns of those who were hesitant to commit the United States to a forceful counterterrorist approach, acknowledging that fighting terrorism would be neither a "clean or a pleasant contest" but insisted that doing so was a necessity in order to prevent and deter future terrorist acts—something he firmly believed was the duty of the U.S. government. In order for this policy of prevention and deterrence to be successful, he continued to push the view that the United States would need to be willing to use lethal force "before each and every fact is known" regarding an imminent terrorist threat, and that such decisions could not be "tied to opinion polls."[34]

By raising the issue of opinion polls, Shultz was acknowledging that without the clarity of an attack being undertaken in retaliation for an act of terrorism, it could be difficult for the government to justify to the public why lethal force had been deployed. The American public were used to terrorism being treated as a criminal act, for which perpetrators would be arrested and put on trial in the federal legal system. The concept of preemptively using lethal force against possible terrorists was an extremely hardline position, one that posed a number of legal, practical, and ethical challenges. Over time, however, this approach would become the default policy of the federal government in its war against al-Qaeda, and a primary element of the CIA's drone campaign. In this chicken-and-egg scenario, then, the policy preceded the technology, not the other way around.

One of the main factors motivating Shultz to push for a more proactive and forceful counterterrorism strategy was his certainty that terrorism was being shown to be a viable strategy for successfully challenging the United States. The bombing of the U.S. Marine barracks in Beirut on October 23, 1983, was a case in point. A truck bomb delivered by the Islamic Jihad militant group (later known to be Hezbollah) killed 241 and injured a further sixty American service members. Minutes later a second, coordinated bombing killed fifty-eight French paratroopers. The soldiers were stationed in Beirut as part of a multinational force charged with a poorly thought out peacekeeping role during the Lebanese civil war. Four months after the attack the United States withdrew its forces, ending the peacekeeping mission without having achieved its objectives. Far from removing the threat of further American losses, Shultz was convinced the rapid retreat sent a clear message to the United States' enemies that terrorism works.[35] "Once it becomes established that terrorism works—that it achieves its political objectives—its practitioners will be bolder, and the threat to us will be greater," the secretary argued.[36] Terrorism expert Brian Jenkins agreed, stating in his Rand study *New Modes of Conflict*, published shortly after the barracks bombings, that state-sponsored terrorist attacks had become a form of covert or surrogate warfare through which smaller and weaker states could confront larger, more powerful rivals without the risk of retribution.[37] Later statements from Osama bin Laden suggest that Shultz's analysis that a lack of clear deterrence would serve to make America's enemies bolder was also accurate. Citing both the events in Beirut and a later withdrawal of U.S. forces from Somalia, bin Laden described America as a "paper tiger" that was "unable to endure the strikes that were dealt to [its] army," instead fleeing in the face of violence and casualties.[38] This perspective served as a significant motivation behind al-Qaeda's decision to focus upon the "far enemy" of the United States.[39]

Shultz put the administration's unwillingness to use military force against terrorists down to three things. First, he blamed the "Vietnam Syndrome." He believed that the senior officials in the Department of Defense (DoD) and Secretary of Defense Caspar Weinberger were so concerned about avoiding another Vietnam scenario that they were "abdicating the duties of leadership."[40] He reasoned that while lessons of the past had taught the United States to avoid "no-win situations," this should not extend to a refusal to engage in "hard-to-win situations," where action would be prudent. For the secretary of state, fighting terrorism was certainly a challenge, but not an unwinnable one. By opting out of the contest, Shultz believed the DoD was allowing the world's future to be "determined by others—most likely by those who are the most brutal, the most unscrupulous, and the most hostile to our deeply held principles."[41] Furthermore, as a man with prior military experience himself, he believed Weinberger's strategy to deter the Soviet Union from initiating a conflict by building the U.S. military into an unchallengeable conventional force severely limited America's strategic options. Not unlike Eisenhower's policy of massive retaliation, it was an approach that presented the president with just two options to respond to threats—do nothing, or initiate all-out warfare. As a realist, Shultz believed diplomacy was at its most effective when the credible threat of force was part of the equation. He held that by presenting military action as exclusively a measure of last resort, Weinberger was emboldening terrorists. Despite huge military expenditure, the DoD lacked the sort of flexible units that could be deployed against the small, mobile militias and militant groups that hid in safe havens around the world, granting terrorists a sense of security.[42]

Second, Shultz blamed those on the political left whom he regarded as apologists for terrorist's actions through their calls for a greater focus upon the "root causes" of terrorism. Former assistant secretary of state George Ball advanced such an argument in an op-ed piece in the *New York Times* on December 16, 1984, in which he criticized Shultz's State Department for its failure to investigate the persistent causes of terrorist violence coming from the same source, specifically a given population over the span of several years, such as the Palestinians. Articulating the view of many who opposed Shultz's call for military force as the solution to terrorism, Ball went on to argue that the United States' prime objective should be to understand and then try to correct, or at least mitigate, the "fundamental grievances that nourish terrorism rather than engage in pre-emptive and retaliatory killing of those affected by such grievances."[43] While Shultz agreed that it was important for the United States should work toward social betterment, the advancing of human rights and seeking diplomatic solutions to conflicts, he also argued that he felt it was folly for the United States ever to be willing to excuse terrorism as the consequence of other factors, such as poverty, oppression, anger at American policies, or abject desperation. "If we got ourselves in the frame of mind that these terrorist acts could be justified and legitimized—and that somehow *we* were to blame," Shultz wrote, "then we would have lost the battle."[44] In the secretary's view, it was exactly this sort of self-defeating mind-set that was handing terrorists the upper hand against the United States.

Finally, Shultz blamed what he saw as overcautiousness within the Reagan administration to support the necessary legislation to enable offensive actions against terrorist groups. For example, when National Security Decision Directive 138 (NSDD 138), introduced on April 3, 1984, called for a "shift [of] policy focus from passive to active defense measures" against terrorism, he felt it lacked the political backing to cure the administration's paralysis.[45] Challenging this view, Ball argued that the United States should not "through panic and anger, . . . embrace counter-terrorism and international lynch law and thus reduce our nation's conduct to the squalid level of the terrorists." The choice of the word "anger" is particularly important here, as Ball went on to suggest that Shultz had lost perspective due to his "obsession with terrorism," which was distorting his "normally judicious view of the world."[46] This was a reference to the fact that Shultz, a former U.S. Marine, had a deep emotional connection to what had happened in Beirut. Furthermore, it was the State Department that was bearing the brunt of terrorist attacks. Diplomacy was becoming an increasingly dangerous business—twenty-eight Foreign Service staff members were killed by terrorist actions between 1976 and 1988. This included seventeen Americans killed in the bombing of the U.S. embassy in Beirut on April 18, 1983, the deadliest attack on an American diplomatic mission up to that time.[47] Among the casualties of that bombing was the CIA's Robert Ames, a Middle East expert. Ames had become Shultz's reader of intelligence and handler when it came to the Middle East, and the loss of such a trusted adviser was a significant professional and personal blow for the secretary of state.[48] All this meant that Shultz was personally invested in counterterrorism, and arguably his anger and desire for revenge were causing him to push America's counterterrorism policy toward forceful retaliation. Shultz, of course, saw it differently. To him, terrorism was a genuine threat to the United States and to global order, one that could only be stopped effectively through the preemptive application of force.

William J. Casey, Director of Central Intelligence, 1981–1987

Next to Shultz, DCI William Casey was the Reagan administration's second most ardent supporter of aggressive, preemptive counterterrorism, and he played a significant role in starting the CIA off on a path that would eventually lead to its prosecution of the drone campaign. Casey was appointed with the aim of boosting the CIA's status, morale, and covert-operation capability through what is referred to as the "Reagan revival."[49] His importance and influence are indicated by the fact that the president granted him parity with the secretaries of state and defense, giving the DCI full cabinet status.[50] Casey has been described as having had an exceptionally detailed intellectual grasp of the terrorist threat, drawn from an understanding that though the Soviet Union may have been the CIA's first intelligence priority, it certainly was not the only threat to the United States.[51] His views on the matter evolved considerably during his first few years at the agency.[52] Initially, Casey agreed with the position put forward by Haig that international terrorism was a Cold War issue linked intrinsically to the USSR. "The Soviet Union has provided funding and support for terrorist operations via Eastern Europe and its client nations like Libya and Cuba," Casey reported in a speech in May, 1982.[53]

Haig and Casey had formed a close political alliance, as well as being friends and confidants. They saw themselves as policy heavyweights, standing out from the rest of the nonconfrontational, image-obsessed Reagan crowd. More important, they saw one another as serious foreign-policy thinkers who had a responsibility to shape the international agenda.[54] When, however, the intelligence failed to support the theory of the Soviets masterminding international terrorism, Casey refused to manipulate it to support a position he and his friend had agreed on, and he dutifully delivered the report distancing the Soviets from international terrorism to the president.[55] Still, the DCI never entirely abandoned the view that the Soviet Union had a hand in international terrorism, accusing the Soviets, in a speech he gave in April, 1985, of indoctrinating those who trained in terrorist training camps with a Marxist-Leninist ideology.[56]

The departure of Haig was a blow to Casey, but he quickly established common ground with Haig's successor, George Shultz, with whom he had served in the Nixon administration. Casey saw Shultz as a man of keen intellect, like himself, and someone with whom he could continue to shape U.S. foreign policy. Shultz had even agreed to attend a testimonial dinner to support Casey when the DCI was facing calls for his resignation early in his tenure, over inconsistencies in his reporting of his business dealings.[57] The bombings of the U.S. embassy and Marine barracks in Beirut galvanized the two men into a pursuit of a new counterterrorism policy. Though it had been the State Department's embassy that was bombed, the CIA contingent within the embassy had been the terrorists' target. The bombers had been tipped off that the agency staff were due to meet there by a female accomplice who had managed to gain employment in the embassy's kitchen.[58] Seven of the American dead were CIA staffers, including the station chief, Kenneth Hass, and the agency's foremost Middle East expert, Robert Ames, eulogized in the memorial service by Casey as "the closest thing to an irreplaceable man."[59]

Compounding Casey's anger, William Buckley, the agency's foremost terrorism expert, who was handpicked by the DCI to replace Ames, was kidnapped shortly after arriving in Beirut on March 16, 1984. He died at the hands of Islamic Holy Jihad (Hezbollah) sometime in mid-1985.[60] It is doubtful Buckley should ever have been sent back into the Middle East—his cover had been blown in at least two countries, one of which was Syria; therefore, it could be assumed his kidnappers had known exactly who he was.[61] His blood was indirectly on Casey's hands. The DCI described the losses as being like "a personal wound," with "nothing quite so devastating" ever having happened in an organization he previously headed.[62] Casey's response was to adopt a very similar hardline mindset to that of Shultz. The synergy in the aggressive policies both men went on to advocate serves to demonstrate the raw emotional impact the terrorist attacks had upon both leaders whose respective departments and staff had been targeted. A desire for forceful retribution and need to stop future attacks drove them. The fact that these two men were so instrumental in the development of the Reagan administration's counterterrorism policy also illustrates how limited the input on the issue was within Reagan's government.

In the wake of the Beirut bombings Casey was under no illusions about what fighting terrorism meant: it was a dirty job, but it had to be done.[63] Consistent with Shultz's views, the director briefed a closed session of the

Senate Intelligence Committee on June 19, 1985, arguing that "the United States is at war" with terrorism.[64] He took this argument public with a speech at Tufts University on April 17, 1985, in which he described international terrorism as having become "a perpetual war without borders."[65] Matching the frustrations of the secretary of state, Casey publicly hinted at the lack of support for proactive counterterrorism measures within the Reagan administration and Congress, equating it to "the absence of a national will to fight terrorism at its roots." The DCI warned that the consequence of this lack of will was that the United States would have to be content to cope only with terrorism's effects—not its cause. For Casey, that was not enough. Echoing Shultz's criticism of the Hamlet-like paralysis in U.S. counterterrorism decision making, Casey acknowledged that the decision of when and how to respond to terrorism posed difficult and sensitive problems, but he argued that this difficulty should not be allowed to "freeze us into paralysis. . . . That is exactly what the terrorists now expect and would like us to do."[66]

The DCI also supported the most controversial element of Shultz's proposals, that effective counterterrorism called for the United States to be prepared to act preemptively in the face of a possible terrorist threat. If the government's response was too "bogged down in interminable consultations or debates," Casey reflected, then the United States would in fact not have a deterrent.[67] Effective counterterrorism was not just about the will to act but also about the speed with which the government could act. Though the DCI went to some lengths to make clear that the United States did not use lethal action indiscriminately, he also presented the use of force against terrorists as a necessary act of self-defense: the United States, he said, "cannot and will not abstain from forcible action to prevent, pre-empt, or respond to terrorist acts where the conditions justify—the knowledge justifies—the use of force."[68] On the matter of deploying this force, the DCI backed Shultz's criticism of Reagan's secretary of defense. Casey too voiced his concern that Weinberger's Soviet Union–focused military buildup left the United States without flexible options in the face of an asymmetric foe and the sorts of unconventional threats which terrorism posed.[69] The criticism from Casey on this matter is particularly relevant, as he and Weinberger were otherwise very close political allies within the Reagan cabinet.

Where Casey's language did deviate from that of Shultz, fittingly, given his role as director of central intelligence, was in his focus upon the necessity of possessing the right knowledge to effectively deploy preemptive force. Knowing where and when to hit terrorist groups would require the United States to know and understand the various terrorist groups, as well as to acquire a knowledge of operating methods, support structures, and the location of training camps around the world.[70] When charged with this mission, Casey's analysts warned the director that penetrating small, paranoid terrorist cells was extremely difficult. Casey refused to accept this could not be done, however. If agents from his Office of Strategic Services (OSS) days had managed to infiltrate Nazi-controlled France, Italy, and Japanese-controlled Burma to gather intelligence during the Second World War, his agency could surely find ways of gathering more intelligence on terrorist groups operating in the Middle East and Eastern Europe.[71]

Lt. Col. Oliver L. North, National Security Council Staffer, 1983–1986

It serves as a good indicator of what a low priority terrorism was accorded in the Reagan administration that an obscure lieutenant colonel was given responsibility for "coordinating the policy and plans of the U.S. Government for low-intensity conflict and counterterrorism."[72] Despite his rise to public notoriety through the Iran-Contra scandal, Oliver North actually spent more hours working on counterterrorism than on anything else. He was, in his own words, "the de facto counterterrorism coordinator" of the United States and played a significant role in the introduction of an approach that would set the precedent for the CIA's drone campaign.[73] His significance in the development of this policy has been lost amid the clamor and noise over his involvement in the Iran-Contra affair, but North played a remarkable role in crafting a hardline counterterrorism approach that reflected the views of Shultz and Casey—one that well outstripped both his rank and the unglamorous surrounds of his cramped Old Executive Office Building (EOB) room. "No lieutenant colonel ever had been given as much power—to rewrite U.S. counterterrorist policy—and to have had such a huge impact on our foreign policy," claimed one senior State Departmental official close to North. Notably, the same official regarded this as a "colossal mistake."[74]

Upon his arrival in Washington, D.C., North quickly established a reputation as someone who could get things done. Major General Uri Simhoni, the former Israeli military attaché to Washington, described him as "a fulcrum in the White House, a man of action in a system often paralyzed by checks and balances."[75] North's profile was quickly raised through his development of a close relationship with DCI Casey. By chance, North was a close neighbor of Casey's in the EOB. The DCI had a corner office to which he would retreat at times when he wanted fewer calls and visitors.[76] Casey approved of North's energy, aggression, and apparent willingness to take risks. North on the other hand saw Casey as something of a mentor—"a man of immense proportions and a man whose advice I admired greatly." "He wasn't a boss," North later reflected, "so much as he was a personal friend and an advisor, and a person with whom I could consult and get good advice."[77] This close relationship saw North brought in to find a way of continuing U.S. support to the Contras in Nicaragua following the December 1982 Boland Amendment, Congressional legislation that outlawed American assistance to the rebels for the purpose of overthrowing the Nicaraguan government and ended the CIA's official role.[78] North played a significant part in helping the Reagan administration circumvent this amendment and continue supplying arms to the rebels—an action that would eventually come to light in the Iran-Contra scandal.[79] The lieutenant colonel's success in finding a solution to the problem of the Contras encouraged Casey to recommend the relatively junior NSC staffer take up the task of drafting a Decision Directive focused upon combating international terrorism.[80] It is North's role as the chief architect of this document that makes him so significant to both the counterterrorism efforts of the time and the CIA's drone program today. In North's own words, "It was time to kill 'cocksucker' terrorists."[81]

By the time North was tasked with producing the directive that would encapsulate the views of the counterterrorism hardliners, there was already growing support among government officials for a more active strategy against terrorism. The Long Commission—an investigation into the mistakes made leading up to the Beirut barracks bombing—was established under the principal guidance of retired admiral Robert Long and published its hard-hitting findings on December 20, 1983. The position taken by the commission was entirely in line with that which the hardliners within the

Reagan administration had been pushing. The report argued that far from being a criminal act, "state sponsored terrorism is an important part of the spectrum of warfare." It went on to recommend that an "adequate response to this increasing threat requires an active national policy which seeks to deter attack or reduce its effectiveness."[82] The commission also endorsed the possibility of preemptive force as a legitimate approach, arguing, "It makes little sense to learn that a state or its surrogate is conducting a terrorist campaign or planning a terrorist attack and not confront that government with political or military consequences if it continues forward."[83] The members of the commission even expressed the same criticism of Weinberger's DoD as Shultz and Casey, arguing that the Pentagon needed to diversify its forces in order to be able to undertake warfare at the "low ends of the conflict spectrum" that terrorism occupied. Finally, as with the position shared by the hardliners, the report concluded that "the most effective defense is an aggressive anti-terrorism program supported by good intelligence, strong information awareness programs and good defensive measures."[84]

Building upon this growing support for action against terrorism, North drafted what has been described as the most ambitious U.S. counterterrorism policy to have been introduced during the Reagan administration. NSDD 138, signed by Reagan on April 3, 1984, amounted in effect to a declaration of war against international terrorism.[85] Noel Koch, who worked with North as director of special planning with responsibility for antiterrorism and counterterrorism at the Pentagon, described the new policy as "a quantum leap in countering terrorism."[86] Or at least it would have been had the policies it proposed ever been fully executed. Koch went on to lament that NSDD 138 "was simply ignored. No part of it was ever implemented."[87] Reagan's government never fully embraced the proactive approach set out in the pages of the directive.[88]

While much of what was proposed in NSDD 138 was never implemented, for three key reasons this does not reduce its significance. First, it perfectly summarizes the views of the counterterrorism hardliners within the Reagan administration in 1984. The directive declared that international terrorism was a "threat to [American] national security."[89] In line with the desire for a more proactive U.S. approach against terrorism, it aimed to "shift policy focus from passive to active defense measures."[90] "Whenever we have evidence that a state is mounting or intends to conduct an act of terrorism against us," the document stated, "we have a responsibility to take measures to protect our citizens, property and interests."[91] Though "take measures" was vague, the willingness to act preemptively, evidenced by the reference to the intentions of terrorists and their sponsors rather than the actions they have undertaken, was unmistakable. The document took this further by requiring the secretary of defense, in consultation with the secretary of state, the DCI, and the attorney general to "develop a military strategy that is supportive of an active, preventive program to combat state-sponsored terrorism before the terrorists can initiate hostile acts." The perceived lack of military flexibility was addressed with the call for "improvements to the U.S. capability to conduct military operations to counter terrorism" and the development of "a full range of military options to combat terrorism throughout the entire terrorist threat spectrum."[92]

Second, the administration's failure to implement the recommendations serves as useful evidence of just how divided and paralyzed the various branches of government were on the issue and demonstrates the weakness of the NSC in being unable to enforce its own recommendations. Robert Gates, at the time the deputy director of intelligence at the CIA, has attempted to provide some insight into why the NSC was so ineffective, placing much of the blame on the man serving as Reagan's assistant to the president for national security affairs (APNSA) at the time, Robert McFarlane. According to Gates, McFarlane lacked the open access to the president that previous advisers had enjoyed. This was partly a deliberate structural change intended to avoid what Gates describes as the "internecine warfare" that had beset the Nixon, Ford, and Carter administrations as a result of unduly influential national security advisers. Beyond the deliberate efforts to reduce the influence of the chair of the NSC, McFarlane was weak and isolated in a cabinet dominated by overbearing personalities with their own agendas. This, Gates concluded, led to some of the serious problems of the Reagan presidency in foreign affairs, including its inability to implement its own counterterrorism policy.[93]

Noel Koch's account supports Gates's analysis, also placing the blame for the failure to implement NSDD 138 on a lack of leadership from Reagan's weak NSC. The Pentagon official describes McFarlane as "just kind of a space cadet" and argues that his equally incompetent deputy,

John Poindexter, was "covering up that he was totally at sea."[94] As Koch observed, the contentious nature of how best to respond to terrorism, coupled with the lack of firm leadership from the NSC, meant that introducing a counterterrorism policy was always going to "devolve on one or two guys that were willing to do what the President wanted done."[95] As things turned out, Oliver North was that guy. With Casey's support, the relatively low-ranked and inexperienced North was able to take advantage of the power vacuum at the heart of the NSC and dictate the counterterrorism policy of the Reagan administration.[96] However, the same lack of leadership that enabled North to wield so much influence in the first place was also the reason so little of NSDD 138 was implemented, as the weak NSC proved incapable of forcing the other branches of government to accept the new counterterrorism policies.

Whereas the counterterrorism hardliners may have been unsuccessful in their first efforts to formalize the policy of preemptive military force in U.S. counterterrorism, the creation of NSDD 138 would eventually have a noteworthy impact upon the CIA's drone program. North, supported by some elements at the Pentagon, had argued that surgical strikes against known terrorists who had been identified as culprits of attacks against the United States, or who were planning future assaults, were the most efficient and effective means of dealing with the terrorist threat. As North saw it, this enabled the United States to go to the root of the problem, while eliminating unnecessary civilian loss of life.[97] North included a section in the draft NSDD 138 which declared that the DCI, in consultation with the secretaries of state and defense and the attorney general, should "develop . . . capabilities for pre-emptive neutralization of anti-American terrorist groups which plan, support, or conduct hostile terrorist acts against U.S. citizens, interests, and property overseas." Should the United States prove unable to strike terrorist targets *before* they attacked, the draft also called for the development of "a clandestine service capability, using all lawful means, for effective responses overseas against terrorist acts committed against U.S. citizens, facilities, or interests."[98]

North's use of the term "neutralize" was controversial. The strongest reaction came from Casey's deputy, John McMahon, who, upon reading the draft at home one evening, reportedly called North to express his anger.[99] McMahon's concerns were not misplaced. The term seemed to put the CIA at risk of breaching Reagan's Executive Order 12333, which stated: "No person employed by or acting on behalf of the United States Government shall engage in, or conspire to engage in, assassination."[100] Reagan's order, signed on December 4, 1981, was a reiteration of President Ford's EO 11905, signed six years earlier in the wake of three separate Congressional investigations into suspected abuses by the intelligence community. The first was the Rockefeller Commission, established following an article by Seymour Hersh, published on the front page of the *New York Times* on December 22, 1974, linking the CIA to domestic spying operations against antiwar protestors—an action that breached the agency's charter.[101] The second was the Church Committee, established on January 27, 1975, and chaired by Idaho Senator Frank Church to investigate allegations of CIA involvement in assassination attempts against foreign leaders.[102] The third, established the following month on February 19, 1975, was the Pike Committee, headed by Democratic representative Otis Pike of New York, with the aim of investigating the CIA's effectiveness and cost to the taxpayer.[103] CIA historian Gerald Haines has argued that this heavy-handed oversight was as much about the wider political events of the time, with Congress battling what it regarded as the imperial presidency, as about the CIA's actions. As Haines apologetically explained, the agency was caught up in the greater power struggle that was taking place between the legislative and the executive branches at the time, as Congress tried to regain greater control over U.S. foreign policy.[104]

While it was undoubtedly the case that the CIA was caught up in the wider political struggle, the various committee's findings did identify serial abuse on the part of the agency, as well as a history of complacent oversight from the relevant congressional intelligence committees.[105] Regarding its role in covert killings, the final report of the Church Committee "condemn[ed] the use of assassination as a tool of foreign policy" and concluded that the CIA had "violate[d] moral precepts fundamental to [the American] way of life."[106] Despite the various committees' investigations being focused upon specific abuses, the combined findings highlighted a consistent direction behind the questionable activities of the intelligence community, and that direction came from the top.[107] Acknowledging the role of the executive in perpetuating the abuses, Ford's executive order was designed to curtail the CIA's involvement in any sorts of future activities that could be construed as assassination in order to protect both the agency and future presidents from the seduction of relying upon lethal covert operations to try to resolve foreign policy problems. In order to maintain better oversight of the activities of the intelligence community in the future, the Permanent Select Committee on Intelligence was established, becoming active on July 14, 1977.[108] As the deputy director saw it, NSDD 138's call for covert teams tasked with the preemptive neutralization of targets risked dragging the CIA back into the riskiest elements of pre–Church Committee covert operations. McMahon's response was one of a man who cared about the reputation of the agency to which he had dedicated thirty-three years of his life, and who had lived through its previous castigation over the issue of assassination.[109]

There were others who shared McMahon's concerns over embracing lethal force against terrorists, but for reasons very different from the vulnerability of the CIA. Rand's resident terrorism expert, Brian Jenkins, argued strongly against the targeting of specific terrorists. In criticism that would be echoed two decades later with regard to the drone campaign, Jenkins contested that while retaliatory actions against terrorists could take several forms, the U.S. should "rule out terrorist-type actions such as campaigns of assassination aimed at known or suspected leaders or members of terrorist groups." His reasoning for this was that, in addition to the moral and political difficulties such actions raised, "these are not appropriate actions to be carried out by armed forces." Furthermore, Jenkins argued that preemptive force should be used only in extraordinary circumstances.[110]

Casey saw things differently. Perhaps because he had not been at the CIA during the days of the Rockefeller, Church, and Pike Commissions, or as a result of his OSS background, the director regarded his deputy as a timid, overcautious pessimist who failed to grasp the emerging action-orientated mentality he sought to restore to the agency. With regard to Jenkins's warning that the targeted killing of known terrorists was not something the United States' armed forces should be involved in, one could argue that NSDD 138's placement of the CIA at the forefront of developing the necessary measures would have ensured the armed forces were not directly involved. It is extremely unlikely, however, that this was a concern the DCI was acting upon. It is more likely that the selection of the CIA as the lead agency for counterterrorism reflected the willingness of its director to get involved in the task. Casey had the CIA's chief counsel, Stanley Sporkin, one of the first men he had handpicked to work alongside him and someone whom he referred to as his "right hand man," investigate the issue.[111] "Don't tell me it can't be done legally," was Casey's frequent refrain to Sporkin; "find me a legal way to do it."[112] Sporkin came to the conclusion that targeting terrorists did not count as assassination, which technically only referred to foreign political leaders.[113] As long as the CIA got the right targets, did all it could to minimize civilian casualties, and kept the proper congressional committees notified, there was no legal problem. With regard to preemptive strikes, Sporkin concluded that if the CIA had hard intelligence proving that the terrorists were planning to attack, preemptive action could be undertaken under Article 51 of the U.N. Charter—the right to self-defense.[114]

Despite this legal reassurance and the growing support from other parts of the administration, the wording of the final version of NSDD 138 was altered to avoid further opposition. The aim to "neutralize" terrorists was replaced with a less controversial reference to the adoption of "proactive efforts" to counter them instead—the same phrase Shultz had used in his speech at the Jonathan Institute previously. Regardless of this change, what NSDD 138 represented was the first attempt to transform the rhetoric of the counterterrorism hardliners into actionable policy. The directive sought to establish the practice of retaliating against terrorist attacks with a flexible military force and preemptively targeting known terrorists as the primary counterterrorism approach of the United States. Though it may not have been successfully implemented upon its introduction, NSDD 138 set many important precedents, which have since contributed to the CIA's emergence as the United States' lead counterterrorism agency and the preemptive targeted killing of terrorists as an acceptable element of U.S. counterterrorism and cornerstone of current drone policy.

Caspar W. Weinberger, Secretary of Defense, 1981–1987

Weinberger's significance for the development of U.S. counterterrorism policy, in particular the CIA's drone program, lies not in what he did as secretary of defense but rather in what he did not do. The title of his autobiography is *Fighting for Peace*, a designation that gives a significant insight into what his view was on the purpose of the U.S. military and his own role at the Pentagon. In the wake of American defeat in Vietnam and subsequent loss of confidence in U.S. military dominance, Weinberger aimed to, in his own words, "regain, as quickly as possible, sufficient military strength to convince our friends to stay closely allied with us and to convince the Soviets they could not win any war they might start against us or our allies."[115] To the secretary of defense, overwhelming American military strength was not a tool to start wars but the best hope for peace. This was not just a strategic position for the Pentagon but a popular political one for the White House. By 1980, in no small part due to speeches delivered by Reagan during his campaign, 56 percent of Americans polled supported significantly increased defense spending.[116] The narrative the Reagan campaign had created, with varying degrees of truth, was that U.S. defense spending had declined by more than 20 percent between 1970 and 1980, causing America's military to wither.[117] It was Weinberger's first priority to address this decline. After what he described as "strenuous arguments" with Congress, the secretary of defense, with the full support of the president, oversaw the largest peacetime military buildup in American history.[118]

During his first press conference as secretary of defense, Weinberger described his second priority as "strengthen[ing] . . . the support of the American people for all the men and women of our armed forces—both uniformed and civilian."[119] Post-Vietnam, a defeatist attitude had entrenched itself both in the services and among civilians. What was more, a chasm of mistrust, and in some cases hatred, had opened up between the American people and the military.[120] To rectify this, Reagan employed dramatic rhetoric to create a world that American citizens could easily understand, in which the United States was the "good," fighting for freedom against the "evil empire" of the Soviet Union. Within this simplistic, Manichean, good-versus-evil version of the world, Weinberger's military investment was vital to keep America "strong and free."[121]

The challenge for Weinberger was that international terrorism, with its messy state sponsorship, unclear rules of engagement, and high potential for collateral damage, did not fit in with this simplified narrative. The secretary of defense did not want to get the U.S. military involved in anything that could interfere with his primary and secondary goals by turning Congress or public goodwill against the Pentagon. The political and public backlash against Carter's failed attempt to use the military's elite counterterrorist force in Iran, which the media had labeled a "fiasco," calling Carter and the Pentagon "incompetent, devious and incoherent," served as a lesson for Weinberger.[122] As a result, he stubbornly refused to allow the military to get drawn into counterterrorism. Instead, he labeled international terrorism a criminal activity rather than a threat to national security, and suggested that it ought to be dealt with through the United Nations.[123] Refusing to accept the ongoing conflict with terrorism as a war and a very significant national security, as Weinberger argued: "[We] should only engage our troops if we must do so as a matter of our own vital national interest."[124] In a speech delivered just a few months after Shultz set out his "Hamlet of nations" argument, Weinberger questioned whether military force should be used as a tool for diplomacy or problem solving. His comments also served to reinforce his secondary goal of healing the split between the U.S. public and the military: "Employing our forces almost indiscriminately and as a regular and customary part of our diplomatic efforts—would surely plunge us head-long into the sort of domestic turmoil we experienced during the Vietnam war, without accomplishing the goal for which we committed our forces. . . . Such policies might well tear at the fabric of our society, endangering the single most critical element of a successful democracy: a strong consensus of support and agreement for our basic purposes."[125]

In order to define what the appropriate conditions for the use of military force were, Weinberger used the same speech to set out the specific requirements that would need to be met before the United States would deploy its forces. These strict conditions, known as the Weinberger Doctrine, established that the United States should commit its forces to combat overseas only "as a last resort," if the engagement or occasion was "deemed vital to our national interest or that of our allies." As Weinberger had already rejected the notion that terrorism was a threat to America's national security, the doctrine made clear the DoD would not be engaging in counterterrorism. In addition, the doctrine dealt with the nature and scale of America's military response. In a rejection of Lyndon Johnson's gradual escalation of American forces in Vietnam, as well as small-scale operations such as the Eagle Claw mission authorized by Carter, Weinberger stated that if and when the United States engaged in military action, it should do so "wholeheartedly, and with the clear intention of winning." Weinberger maintained that such an approach would ensure the United

States could always bring its overwhelming strength to bear and avoid protracted conflicts. Shultz and Casey criticized this approach for creating an all-or-nothing mentality within the DoD—exactly what Weinberger intended. By limiting the U.S. military to large-scale interventions, he ensured the Pentagon would not get tasked with any other Eagle Claw–type missions. Finally, in a rejection of Shultz's argument for the preemptive use of force, possibly before public support has been gained, Weinberger maintained there must be reasonable assurances that the executive would have the support of the American people and their elected representatives in Congress before force is deployed.[126]

Just over a year later Weinberger returned to the issue of force and counterterrorism at a conference on low-intensity warfare hosted in Washington, D.C. Reflecting the increased tempo of terrorist attacks, the secretary of defense's language marked an acknowledgment that the threat was growing, referring to the conflict with terrorism as "this war, as it now reveals itself to us." Nevertheless, despite this acknowledgment of the growing threat posed by terrorism and the determination of its protagonists, Weinberger maintained that the use of military force for counterterrorism purposes would be an error. Striking a chord similar to George Ball's criticism of Shultz's emotional response to the anti-American attacks, Weinberger argued that terrorism was "prosecuted in such a way as to erode and destroy the values of civilization," putting "a special obligation to act as to uphold these values" upon the United States. America must not, for the sake of expediency in the pursuit of terrorists, get dragged into committing "blind acts of revenge that may kill innocent people who had nothing to do with terrorism." To do so, Weinberger warned, was exactly what those responsible for the attacks wanted, and the resulting loss of American moral standing would play into their hands.[127]

Prior to the conference, Weinberger had already publicly raised his concerns about the likelihood that any sort of retaliation would have the necessary quality of evidence to be certain beyond reasonable doubt, that the strike was hitting the guilty party.[128] In a particularly damning appraisal of the approach, he had argued that "retaliation would be analogous to firing a gun in a crowded theater in the slim hope of hitting the guilty party."[129] He built upon this theme during his conference address with another simile, comparing the difficulty of preemptively neutralizing a terrorist threat to that of a "medical profession to be able to diagnose an illness in its earliest stages, and then to act to cure it before it becomes dangerous."[130] Successfully identifying and then countering the political and geostrategic ills that fueled terrorism at its incipient stages was, in Weinberger's view, the greatest challenge of low-intensity warfare.

The Joint Chiefs of Staff (JCS) generally shared Weinberger's skepticism regarding the utility of force in counterterrorism. At the Pentagon terrorism was regarded as a low priority in terms of the national security threat it presented but as potentially a high risk in public relations and political costs. Like Weinberger, General John Vessey, the chairman of the Joint Chiefs, believed that the primary purpose of the U.S. military was to counter the most significant threat to America's national security—the Soviet Union.[131] Furthermore, in line with Weinberger's concerns, Pentagon planners feared that, should the U.S. military be dragged into asymmetric warfare in places such as Lebanon, there was a strong likelihood that American servicemen would die and innocent civilians would be killed or injured. Such an event could easily undermine the public's support for the military and perhaps jeopardize backing for the ongoing buildup.[132] Despite this, not all members of the senior military leadership were in agreement with Weinberger's position. Admiral James D. Watkins, the chief of the navy, described terrorism as "an already declared war," and Admiral William J. Crowe, who took over as chairman in 1985, testified during his confirmation hearing that if faced with the choice of using force in a terrorist situation, his view would be to "project ahead to determine whether force would produce the results desired."[133] If the analysis suggested America's counterterrorism objectives could be achieved, Crowe stated that he believed the use of force would be justified. As General Vessey made clear, however, the chairman was "a servant . . . to the National Command Authority: the president and the secretary of defense." The chain of command ran clearly from the secretary of defense, and the Joint Chiefs fell in line.[134]

How much of what Weinberger said was actually his heartfelt position, and how much was about not jeopardizing his military project is unclear. It has also been suggested that Weinberger's attitudes were informed by his Arabist inclinations; he worried constantly about the possible disruption of U.S. relations with Arab countries should the United States take any sort of significant action against terrorist groups and their sponsors.[135] Whatever the reasoning behind it, Weinberger was the gatekeeper of America's armed forces, and he was entirely unwilling to see those forces deployed in low-intensity conflicts or assume a counterterrorism role, be that preemptive or retaliatory. So, despite parity in thinking among the secretary of state, the DCI, and members of the NSC, Weinberger was able to block the use of military force against terrorist groups or their sponsors. Stalin had once mockingly asked of the pope, "How many divisions has he got?" when the pontiff had attempted to pressure him to stop his suppression of Catholics within the USSR.[136] For all Shultz's talk of preemption, Casey's desire for retribution and North's plans for neutralization, without the support of the secretary of defense and the Joint Chiefs, the counterterrorism hardliners had no armed divisions, and without them, no means to execute the policies they were promoting so vehemently. Only the president of the United States had the authority to overcome the paralysis that the deep split in opinions had created. But he refused to exercise it.

President Ronald W. Reagan, 1981–1989

There has been much debate over what kind of leader Ronald Reagan really was. Supporters and critics alike have described him as a hands-off president, with Jane Mayer and Doyle McManus going as far as to suggest his two terms in office bordered on a "no-hands presidency." Reagan adopted a remarkably passive approach to running the most powerful office in the Western world, delegating many of the responsibilities of the presidency to his aides and advisers.[137] As the president's unabashed disconnection from many of the responsibilities associated with the office became increasingly common knowledge, he sought to justify his managerial approach in a *Fortune* magazine interview on September 15, 1986. He defined the "cabinet model" approach he adopted as: "Surround yourself with the best people you can find, delegate authority and don't interfere as long as the overall policy that you've decided upon is being carried out."[138] Reagan expanded upon what he saw as his leadership style in his autobiography, published a year after he left office: "The chief executive should set broad policy and general ground rules, tell people what he or she wants them to do, then let them do it," he wrote. "I don't believe a chief executive should supervise every detail of what goes on in his organization." Should a problem arise in the pursuit of these broad goals, Reagan explained, he would make himself available so that members of his team could come to him and, if necessary, fine-tune the policies. If things went wrong and "somebody dropped the ball" in the pursuit of these goals, Reagan noted that he would "intervene and make a change."[139]

In theory Reagan's description of his management style is a sound way to run an executive cabinet, but there is considerable disagreement among observers and former participants over the extent to which this description reflected the reality of the president's leadership. A novice in international affairs, Reagan had foreign policy goals that were, in the words of Sean Wilentz, "long on style and symbolism," setting broad policy aims that he describes in his memoirs but that were also notoriously lacking in executable instructions.[140] Some critics have attributed the lack of a coherent approach in Reagan's foreign policy to the president's general disinterest in details and notoriously short attention span.[141] In a more damning indictment, Hayes Johnson has argued that the focus upon rhetoric over action resulted from the fact that Reagan was ultimately more interested in being popular than in being recognized as dealing with hard problems of governance; "he preferred to reign rather than rule."[142] Garry Wills has presented a slightly more nuanced reading of this behavior, still ascribing the reliance upon lofty rhetoric to a focus upon popularity and good headlines but connecting it to a showman's efforts to keep his audience— the American public—engaged. Being "the great communicator," explained Wills, was "a matter of remaining understood, step by step, never breaking the sequence of easy exposition."[143]

The White House's pursuit of popularity, observed James A. Nathan, served to create an environment of ad-hocism, where foreign policies generally seemed to be judged according to just three criteria: Had the president made some vague statement of support; would it sound good on the evening news; and would it make the president look good?[144] "The investment in long-term results," noted Wills perceptively, "is not congenial to a performer attuned to a constant interplay with his audience."[145] Rather than implementing any sort of grand strategy or vision, Reagan, as Francis Fitzgerald puts it, "liv[ed] in a world of rhetoric, performance and perceptions."[146] Foreign policy was spawned from the populist speeches delivered by the president, not the other way around. As newsworthy events occurred around the world, Reagan would deliver a speech, drawing upon simplistic but lofty rhetoric of American exceptionalism and military strength and how the United States' willingness to confront evil would prevail. It was then left to his advisers to make policies that would be generally consistent with the speech; Reagan's job, as he saw it, was then to sell the policies to the public.[147]

One consequence of basing policy making upon presidential rhetoric was that officials acted randomly on what they believed to be the operant assumptions of "Reaganism," calculating what Reagan really intended.[148] "He trusted his lieutenants to act on his intentions, rather than on his spoken instructions," wrote Donald Regan, who had the challenging task of serving as Reagan's chief of staff from 1985 to 1987. Reagan's detachment from the policy-making process gave his staff enormous power to shape the administration's agenda, but it also placed a tremendous burden upon his closest advisers.[149] "My God," exclaimed Frank Carlucci, Reagan's NSC adviser in 1987, following a meeting between himself, his deputy, General Colin Powell, and a passive Reagan, "we didn't sign up to run this country!"[150]

Though Reagan's reliance upon his advisers was well known to the press and public, it was generally believed at the time that he ran a subtle hidden-hand presidency, similar to that of Dwight Eisenhower, whom Reagan admired.[151] Despite the members Reagan's inner staff being privately astounded by how the president really operated, they rarely discussed their concerns with one another, and never with outsiders—not even their departmental subordinates.[152] This culture of denial emerged partly out of loyalty toward Reagan, who, despite his complacency, was still widely regarded as what Lou Cannon dubbed a "wholesome citizen hero"; partly out of a concern of losing their positions of influence; and partly as an act of self-protection.[153] The administration's senior staffers did not want to admit the extent to which they personally were making national policy, instead adapting to the president's lax management style by forging ahead with their own personal agendas.[154]

On the occasion that one vague goal set by the president seemed to clash with or contradict another vague goal, dysfunction emerged within the administration. For example, Shultz believed he was in line with the president's goals when he spoke out forcefully about the need to counter terrorist acts. The president had, after all, delivered a number of tough-talking speeches on the matter. Weinberger also believed, however, that he was doing the president's bidding when he did all he could to avoid endangering support for the military and its buildup by engaging in risky counterterrorist actions. Assertively combating terrorism and restoring the military were both key agenda items that Reagan had consistently campaigned on, and both Shultz and Weinberger were correct to see their respective causes as core policy issues for the administration. While it seems that Reagan's description of his hands-off management style is accurate, closer inspection reveals that what was missing were the executive interventions and policy fine-tuning. "The catch," notes Malcolm Byrne, "was Reagan rarely followed through in monitoring whether his aides were properly implementing his orders."[155] Supporting this criticism, Reagan's former secretary of state Haig compared the policy-making culture within the administration to life aboard a "mysterious ghost ship; you heard the creak of the rigging and the groan of the timbers and even glimpsed the crew on deck. But which one of the crew was at the helm? It was impossible to know for sure."[156]

Beyond the structural problems with policy and decision making in the Reagan administration, Haig confided in DCI Casey that he did not believe the president himself to be qualified for the job. He felt his attention span and interests were too narrow, going as far as to describe him as "an amorphous mass that had no substance, no opinion." "Ronald Reagan," Haig reflected, "knew little about foreign policy and cared less." Haig claimed Casey shared these frustrations, though the DCI, loyal to Reagan, never stated as much on the record.[157] Lee Hamilton, a veteran congressman who had served with five presidents, agreed with Haig's assessment when describing working with the president. Reagan was "the most inarticulate" president he had worked with, "unable to formulate any problem precisely." Hamilton observed that Reagan had "an unsubtle, incurious mind" and that "he clearly didn't worry much about ends and means. So long as he believed the goal a worthy one, he didn't seem troubled by the way it was achieved." Hamilton, too, observed the hands-off management approach but believed that, even with this approach in mind, the president delegated his authority to an extraordinary degree.[158]

Weinberger countered such criticism during an August 1990 interview in which he maintained that Reagan was one of the most underestimated men in the world. Putting this down partially to elitism and snobbery, the former Harvard Law School scholar railed against "intellectuals or academics or experts generally," who had what he described as "preconceived attitude[s] towards him [Reagan] of either contempt or amusement, along with a feeling that a person with his background [movie actor] couldn't do anything else."[159] Rather than being a sign of weakness or a product of a lack of intellectual curiosity, Weinberger believed, the president's style was a product of his skill in putting people at ease so as to create a congenial and productive atmosphere. In Weinberger's view, this not only helped get the best out of the president's own staff but also enabled him to secure vital agreements from congressmen, senators, and even difficult heads of state, "which neither logic, nor table pounding, nor cajoling could bring about."[160]

Reagan's handling of the policy split between Shultz and Weinberger on how best to respond to terrorism does not, however, suggest the approach the secretary of defense describes. Perhaps because Weinberger himself was so instrumental in this division, he is at lengths to play it down in his account of the administration; but other witnesses tell a very different story. Robert McFarlane, the former national security adviser, describes a meeting with the president in which he tried to discuss the Shultz/Weinberger split. "I must tell you, Mr. President, we do not have a team in national security affairs," said the beleaguered adviser. "Your Secretaries of State and Defense agree on next to nothing, whether it's East-West relations, the Middle East, economic matters, trade, or dealing with terrorism. . . . When a decision goes against one of them, he does all he can to obstruct its implementation." The national security adviser also informed Reagan of the acrimonious nature of the split, confessing that there was personal hostility between the two secretaries, which resulted in a constant air of confrontation. McFarlane concluded by informing Reagan this was resulting in the paralysis of the NSC.[161] He presented the president with two recommendations—he should either sack one of the secretaries or get more involved in the policy process. "So what I want you to do," the president told the exasperated McFarlane in response, "is just make it work."[162]

Reagan's abrogation of a meaningful leadership role reflected neither the fine-tuning intervention the president wrote about nor the harmonious smoothing of relations Weinberger presented. Instead it revealed something both Shultz and Weinberger understood and exploited about the president:

Reagan disliked coming down firmly on behalf of one of his principal subordinates at the expense of another.[163] Reagan's vaguely defined and contradictory foreign policy goals, combined with his unwillingness to wade in when senior advisers squabbled, meant that the administration's counterterrorism policy was defined not by an overarching strategic approach but by whoever could mobilize or block the use of resources.[164]

The contested nature of counterterrorism policy during the Reagan administration is perhaps best evidenced by the failed attempt to launch a retaliatory strike against the Iranian Revolutionary Guard stationed in the Sheikh Abdullah Barracks in the Bekaa Valley of Syria after intercepted intelligence proved Iranian complicity in the bombings of the U.S. embassy and Marines in Beirut.[165] NSDD 109, signed on October 23, 1983, the same day as the barracks bombing, had a military component entitled "Responding to the Attacks on the USMNF Contingents."[166] Despite the fact that this document is still classified, its very existence reveals that the administration was considering military action. Furthermore, other sources reveal that military options were being debated—and according to one account, were agreed upon. McFarlane provided an account of a National Security Planning Group (NSPG) meeting he chaired on November 14 during which the president gave his approval for a retaliatory strike to be conducted on 16 November. The National Security Adviser describes it as a "direct, unambiguous decision."[167] The strike was to be conducted in coordination with the French. According to Weinberger's account, he received a call from Charles Hernu, the French minister of defense, on the morning of 16 November, telling him that French planes were going to attack Syrian positions "in about two or three hours." Contrary to McFarlane's account, Weinberger claimed that he had received no orders or notifications from the president prior to the French call, so wished his counterpart luck and informed him, "Unfortunately it is a bit too late for us to join you in this one."[168]

McFarlane's account of the aborted retaliation makes the serious accusation that the secretary of defense directly violated a presidential order due to his personal disagreement with the policy. According to Reagan's national security adviser, Weinberger launched into a long series of obfuscations about misunderstandings with the French and all the things that could have gone wrong with the attack. McFarlane quotes the secretary of defense as saying, "I just don't think it's the right thing to do." In the security adviser's view, it was Weinberger's knowledge of his old friend Ronald Reagan that gave him the confidence to behave in this way, whereas Weinberger puts the incongruity in accounts down to "McFarlane's well known flexible 'recollections.'" The president's reaction was minimal. "Gosh, that's really disappointing," McFarlane recalls the commander in chief saying. "That's terrible. We should have blown the daylights out of them. I just don't understand."[169] Reagan's own account only briefly mentions the canceled strikes. In his memoir, he states that *he* canceled the strikes after his experts informed him that they were not absolutely sure they had the right targets.[170] It is likely this was a face-saving exercise on the part of the former president, looking to cover the fact that one of his orders as commander in chief was ignored by his secretary of defense. The record shows that the United States had convincing intelligence about the culpability of the Iranians and Syrians—strong enough for the French to decide to act. Furthermore, still angry over what they regarded as America's cancellation of a planned joint strike, the French later denied the United States overfly rights when they launched a bombing raid on Libya in 1986, suggesting there is truth in the frustrated national security adviser's account that confirmation had initially been given.[171]

Richard Armitage, at the time assistant secretary of defense for international security affairs, has also attested to the role Weinberger was able to play in blocking any military action on counterterrorism. Armitage told the 9/11 Commission that in the mid-1980s there had been those at the Pentagon who "wanted to put a cruise missile into the window of the Iranian ambassador in Damascus."[172] Not only does Armitage's testimony show that there were DoD staff who supported the use of military force against those responsible for terrorist attacks, it also shows that the proposed weapon was the same as would later be used by President Clinton in his efforts to kill Osama bin Laden in Afghanistan during Operation Infinite Reach. Armitage reveals too that there was concern about whether or not such an attack would have been classed as assassination, thus violating EO 12333, something with which Clinton would also later wrestle.[173] Ultimately, Weinberger decided against the use of Tomahawk cruise missiles, arguing that if one failed to detonate the Syrians would give it to the Soviets for reverse engineering.[174] This reasoning further supports the view that although Weinberger was committed to building up the U.S. military's capabilities, his primary focus was on deterrence of the USSR, not on how these capabilities could be used to provide responses to terrorism and other low-intensity challenges.

One other factor worthy of consideration when assessing Reagan's indecision on America's counterterrorism actions is the frailty of the president's state of mind. Reagan, sixty-nine years old when elected to the Oval Office, suffered frequent mental lapses throughout his presidency, which, following the former president's diagnosis of suffering from Alzheimer's disease in 1994, have come to be seen by many scholars as early signs of his deteriorating mental condition.[175] Anecdotally, Ron Reagan Jr. recollected his concern about his father's fumbling performance during the first presidential debate with Walter Mondale on October 7, 1984, and recalled another occasion when Reagan was upset at his inability to remember the names of what should have been familiar landmarks near Los Angeles in August 1986.[176] John Poindexter, who served as the president's deputy national security adviser from 1983 to 1985 before taking the primary post from 1985 to 1986, has since confessed that "by the end of 1986 . . . Alzheimer's had really begun to be a factor."[177] More recently, an analysis of the president's use of language by researchers at Arizona State University revealed subtle changes in Reagan's speaking patterns linked to the earlier onset of dementia.[178] Despite being heavily masked by Reagan's experience as a professional actor, and blurring with his general indifference and indecisiveness, the possible impact of dementia is nonetheless relevant to an assessment of Reagan's performance on counterterrorism issues.

Regardless of the actual reason behind Reagan's hesitation, the administration's failure to launch any sort of response against Iran, Syria, and Hezbollah for their collective roles in the killing of hundreds of Americans chimes with Shultz's argument that the United States was becoming the "Hamlet of nations." The key members of Reagan's administration were not, however, paralyzed by indecision. Strong decisions had been made on each side of the debate. The problem for the administration was that the decisions were diametrically opposed. James Nathan summed this up insightfully when he wrote: "F. Scott Fitzgerald once observed that the mark of intelligence is the capacity to hold two contradictory ideas and still retain the ability to act effectively. The Reagan administration, to the degree it was marked by any ideas at all, was caught in its own antinomies."[179]

The administration's counterterrorism policy was paralyzed by a lack of leadership. Whether it was the result of his intellectual limitations, a deliberate leadership strategy to bring the best out of his officials, or the result of his declining mental state, Reagan handed an extraordinary amount of autonomy to each appointed department or agency manager, then failed to provide them with even the vaguest of instructions to bring their fiefdoms in line with his philosophical thinking. Casey was to revive the CIA as an aggressive, risk-taking anti-Communist tool. Shultz was to use the State Department to promote U.S. power and to undermine and alienate the Soviet Union, and Weinberger was to build the U.S. military into an unchallengeable martial force. For such an approach to work, three things were required. First, due to the extent of autonomy each manager was granted, headstrong and confident individuals were needed to lead the branches of government. Casey, Shultz, and Weinberger all fitted this bill. Second, it required an element of collegiality, a common goal between these leaders. The problem here was that the individual mandates of the leaders were frequently contradictory and antagonistic to one another. Shultz believed that military force was a necessary backup to any diplomatic actions the United States undertook, whereas Weinberger thought any use of military force that was not directly related to the Cold War was a waste of resources and will. Finally, such a system needed an authoritative leader at the top who was able to make decisions and force action when the managers' views were diametrically opposed, as was the case with Shultz and Weinberger. How to best counter terrorism was and remains an extremely complex moral and strategic issue.

When it came to making definitive decisions, Reagan failed. His ambiguity left key debates relating to counterterrorism policies unresolved. As the criticism rightly leveled against him by the Tower Commission identified, "the NSC system will not work unless the President makes it work." Established specifically to investigate the NSC's role in the Iran-Contra crisis, the commission's findings nonetheless provided a damning indictment of the Reagan administration's general approach to national security policies. Created to serve the commander in chief, the quality of America's national security decision making is determined, the report's conclusion noted, by his leadership. Reagan's negligent management ensured policy debates did not undergo the degree of critical review the NSC participants and the process were capable of. Without a president who was willing to force his will on the deliberative process, the most powerful features of the NSC system—comprehensive analysis, policy alternatives, and analytical follow-up—were not utilized.[180] Because of his unwillingness or inability to bring his advisers behind a single conclusion, any members of Reagan's national security team could stubbornly refuse to abandon their position and allow paralysis to ensue, ensuring they did not have to follow a policy they personally did not agree with or felt undermined their own mandated responsibility. This resulted in a standstill in the formal policy-making process.

Shultz, Casey, and North had crafted the fundamentals of a counterterrorism strategy they believed matched the strong rhetorical and philosophical position Reagan had established. As set out in NSDD 138, the United States was to use military force to retaliate against terrorists and their sponsors as a deterrent. Furthermore, the United States was to actively gather intelligence to inform covert and clandestine forces to neutralize terrorist leaders and their followers preemptively before threats fully materialized. By 1985, the policy was in place. What its creators lacked was the military muscle to make it happen, but the president was unwilling to force his secretary of defense to provide it. It is this deadlock and failed government policy that would lead to the CIA filling the counterterrorism void, with far-reaching consequences.

Conclusion

Though the Reagan administration may have failed to implement an effective counterterrorism policy, it did succeed in establishing the core principles that would eventually become the backbone of George W. Bush's War on Terror and the CIA's drone campaign. NSDD 138 was the first U.S. government document to introduce the concept of an act of terrorism being equivalent to an act of war and a legitimate threat to U.S. national security. Likewise, it was this document that called for the use of lethal force as both a retaliatory deterrent and a preemptive tool to neutralize anti-American terrorist groups. It is a misrepresentation, however, to describe the adoption of these principles as a "neo-Reaganite foreign policy," as Robert Kagan and William Kristol, the influential founders of the neoconservative think tank Project for the New American Century, have done.[181] It is more accurate to attribute this approach to the counter-terrorism hardliners within the Reagan administration, namely Shultz, Casey, and North. What were the distinctive views of the three most hard-line advocates of aggressive counterterrorism policies have since become the mainstream, if still controversial, policies of U.S. counterterrorism.

Following the 9/11 attacks, Congress passed Joint Resolution 23, the Authorization for Use of Military Force against Terrorists (AUMF), which granted the president the authority "to use all necessary and appropriate force" against those determined to be responsible for 9/11.[182] This war footing continued into the Obama administration, which asserted in its 2011 National Strategy for Counterterrorism: "[The] United States deliberately uses the word 'war' to describe our relentless campaign against al-Qa'ida."[183] On June 1, 2002, George W. Bush used a speech delivered at the U.S. Military Academy at West Point to endorse a policy of using preemptive force against emergent threats. Later encapsulated within his 2002 National Security Strategy, Bush's view was: "The greater the threat, the greater is the risk of inaction—and the more compelling the case for taking anticipatory action to defend ourselves." Echoing the words of Shultz's Park Avenue speech almost two decades earlier, Bush warned that at times "even if uncertainty remains as to the time and place of the enemy's attack . . . the United States would if necessary, act pre-emptively."[184]

Post-9/11, the Department of Justice (DoJ) has argued for a "broader concept of imminence" in terms of the threat that enables the United States to act in self-defense. The reasoning behind this, set out in a DoJ white paper leaked to NBC News in February 2013, revolves around the concept of the "limited window of opportunity" to successfully hit al-Qaeda and associated targets, while limiting civilian casualties.[185] The fact that al-Qaeda members have been instructed by the writings of bin Laden that it is their duty to "kill the Americans and their allies" whenever they can, and that Ayman al-Zawahiri has instructed that all members should seek to "inflict maximum casualties against the opponent, . . . no matter how much time and effort such operations take," is taken as evidence that an al-Qaeda member is always plotting.[186] Furthermore, Khalid Sheikh

Mohammed, the mastermind of the 9/11 plot, revealed that the operation took five years to plan, adding further credence to the argument that America's terrorist foes are engaged in continual and long-term planning.[187] Essentially, the DoJ argued that the United States can use force against al-Qaeda members or affiliates at any point it knows where they are and can hit them without disproportionate collateral damage, because those targets will try to hide themselves among civilians at a later stage.[188] "Delaying action against individuals continually planning to kill Americans until some theoretical end stage of the planning for a particular plot," stated the DoJ's white paper, "would create an unacceptably high risk that the action would fail and that American casualties would result."[189] The belated capture of Khalid Sheikh Mohammed on March 1, 2003, and the killing of bin Laden on May 2, 2011, serve as examples of the DoJ's argument.

The Obama administration's preferred method of undertaking these preemptive attacks has been through the CIA's drone campaign. It is here that the policies first set out in NSDD 138 can most clearly be seen. John Brennan, at the time serving as Obama's assistant for homeland security and counterterrorism, echoed North's ideas of targeting terrorist leaders in a rare speech on the topic delivered in June 2011. The purpose of the drone campaign, he explained, has been to "dismantle the core of al-Qaeda—its leadership in the tribal regions of Pakistan."[190] Evidence of this strategy can be seen in the numbers of significant al-Qaeda and Taliban leaders who have been reported killed in drone strikes by the end of the Obama administration's first term.[191] Again echoing North's proposals, Brennan has vigorously defended these strikes on the ground that they are surgically precise: "With the unprecedented ability of remotely piloted aircraft to precisely target a military objective while minimizing collateral damage, one could argue that never before has there been a weapon that allows us to distinguish more effectively between an al-Qaida terrorist and innocent civilians. . . . By targeting an individual terrorist or small numbers of terrorists with ordnance that can be adapted to avoid harming others in the immediate vicinity, it is hard to imagine a tool that can better minimize the risk to civilians than remotely piloted aircraft."[192]

Regarding the legality of these strikes, the justification that the Obama administration has used is identical to that set out by Stanley Sporkin for

Casey and NSDD 138—that the targeted killing of a terrorist is a lawful act of national self-defense, which does not violate the assassination ban in EO 12333. The DoJ also maintains the United States' right to act unilaterally in the territory of another nation, should the host nation's government have given consent or have been determined to be "unable or unwilling to suppress the threat posed by the individual targeted."[193]

Hence, the CIA's drone campaign follows the exact proposals and legal arguments set out by the Reagan administration in NSDD 138. What accounts for the significant change from the policy and attitude of the Reagan administration's failed implementing of its own guidance to George W. Bush and Barack Obama's wholesale adoption of targeted killings as a means of waging war against terrorists? The events of 9/11 stand out as the most obvious cause of this change. With NSDD 138, Shultz, Casey, and North were focused upon trying to persuade doubters that counterterrorism was a national security issue. The preamble to NSDD 138 declared: "The U.S. government considers the practice of terrorism by any person or group in any cause a threat to our national security."[194] At the time, however, this language reflected the emotional and psychological impact of the threat of terrorism to the United States rather than its actuality. The counterterrorism hardliners had built terrorism into something it was not, and they were therefore unable to sell NSDD 138 to Weinberger and the Joint Chiefs on national security grounds. As horrific and costly in American lives as the Beirut bombings were, they did not threaten the national security of the United States. When America withdrew its forces shortly afterward, there were not significant consequences to the geopolitical balance of power, demonstrating that Lebanon was in fact not a core security issue for the United States. Terrorism was a threat to America's image and to that of the president who was in office when overseas attacks occurred. It was also a threat to American citizens around the world. It was not, however, something that could significantly undermine the national security of the superpower.

The killing of more than three thousand people on American soil on September 11, 2001, served to escalate the threat terrorism posed. Following the multiple attacks that day, including one upon the Pentagon itself, terrorism shifted in peoples' perceptions to a demonstrable challenge to the security of the United States, as well as to the lives of its citizens and reputation of its president. George W. Bush's vice president, Dick Cheney, summed up this new thinking in an interview two years after the event: "9/11 changed everything. It changed the way we think about threats to the United States. It changed [*sic*] about our recognition of our vulnerabilities. It changed in terms of the kind of national security strategy we need to pursue, in terms of guaranteeing the safety and security of the American people."[195]

It is tempting to regard 9/11 as the turning point where the counterterrorism policies first pushed by the Reagan-era hardliners were suddenly and wholeheartedly adopted either as a kneejerk response motivated by fear and a desire for vengeance or as some sort of delayed realization that Shultz and company had been right all along. The reality is much more complicated, however; this chapter indicates that America's adoption of such aggressive counterterrorism methods did not occur overnight. In the two decades between the introduction of NSDD 138 and the start of the CIA's lethal drone campaign, the counterterrorism policies of the United States had been gradually evolving, moving ever closer to the policy approach advocated by Shultz, Casey, and North as the terrorist foe proved increasingly dangerous and difficult to counter; indeed, it can be plausibly argued that the much strengthened 9/11 foe was the predictable outcome of this very policy. It was no coincidence that by the time the hijacked planes slammed into the twin towers of the World Trade Center, the headquarters of the U.S. military, and a field in Pennsylvania, the CIA was ready to take the lead in George W. Bush's newly declared War on Terror. One way or another the agency had been fighting a war against terrorism since the signing of NSDD 138. The legal architecture that authorized the hundreds of drone strikes that the Obama administration launched, the CIA's Counterterrorism Center, which has overseen and coordinated the strikes, the network of informants on the ground who have guided them, and the technological innovations behind the drones that have executed the strikes are all the products of two decades of persistent experimentation and evolution in America's ongoing effort to neutralize the threat posed by burgeoning anti-American terrorist groups.

2

"LET'S FIND A WAY TO GO AFTER THEM": THE FADLALLAH AFFAIR AND THE *ACHILLE LAURO* HIJACKING, 1985

Those sons of bitches, let's find a way to go after them.
—*President Ronald Reagan*

In the wake of the NSC's failure to implement the recommendations of NSDD 138, terrorist attacks against the United States continued, as did the Reagan administration's ad hoc, conflicted, and disorganized response. Without a clear consensus America's approach to counterterrorism during the mid-1980s is better characterized as a series of disjointed operations than as a strategy. As the Reagan administration's hardliners continued to try to implement their agenda, they were forced to bypass those who opposed it, in particular Secretary of Defense Weinberger. The lack of access to the resources of the DoD forced them to explore other avenues through which they could undertake their proactive counterterrorism policy. Though ultimately this failed to significantly reduce the threat posed to U.S. citizens by international terrorism, it did serve as a useful opportunity for those engaged in counterterrorism operations to explore different approaches and methods. Two operations in particular stand out as having been extremely influential in terms of the lessons they taught the counterterrorism community. The first is the Fadlallah affair—a failed targeted killing that shaped the way the CIA has worked with foreign agents in counterterrorism operations ever since. The second is the response to the hijacking of the *Achille Lauro* cruise liner, which revealed that the American public and wider political establishment is less concerned about the international fallout caused by a counterterrorism operation if the guilty parties are seen to have been brought to justice. The influence of both of these operations and the lessons they taught can be clearly seen in the CTC's conduct of its drone campaign under the Obama administration.

The Failed Targeted Killing Attempt on Ayatollah Mohammed Hussayn Fadlallah

In his text on civilian casualties in America's wars, John Tirman argues that Reagan's preferred method of military intervention was to use "other people to do the fighting."[1] If one were to look at the wording of NSDD 75, the document that established the policy approach that would eventually become known as the Reagan Doctrine, the administration's intention to rely upon proxies to confront the Soviet Union and far-left forces is clear. Referring to the newly emergent Cold War battle lines in nations such as Afghanistan, the directive states that the Reagan administration will "support effectively those Third World states that are willing to resist Soviet pressures or oppose Soviet initiatives hostile to the United States, or are special targets of Soviet policy." Rather than committing U.S. forces to such struggles, the Reagan Doctrine goes on to highlight that this support will come in the form of "security assistance and foreign military sales" to the indigenous forces resisting the USSR.[2] Evidence of the administration's commitment to fighting through surrogates is provided by the huge financial and military aid that was funneled by the United States to the mujahideen fighters in Afghanistan and the Contra rebels in Nicaragua during Reagan's time in office.[3] In his 1985 State of the Union address, Reagan identified the role these surrogates were playing in America's national defense, stressing to Congress: "[The United States] must not break faith with those who are risking their lives, on every continent—from Afghanistan to Nicaragua—to defy Soviet aggression and secure rights which have been ours from birth. . . . Support for freedom fighters," he reasoned, "is self-defense."[4]

The extent to which this willingness to use surrogates was actually a policy preference, as Tirman describes it, or rather a forced necessity due to wider policy goals and their consequences, is debatable. As discussed in the previous chapter, Secretary of Defense Weinberger's primary objective during Reagan's tenure was to oversee the largest peacetime buildup of the U.S. military in American history, with the aim of deterring any Soviet aggression. Achieving this goal required not only significant financial support but also the regaining of public support in the wake of the damage caused by the Vietnam War. Committing U.S. forces to an engagement in a country such as Afghanistan would both have risked a dangerous escalation of Cold War tensions and have jeopardized the support for the buildup through another potentially bloody conflict. The counterterrorism hardliners within the Reagan administration faced the same choice when looking to undertake the preemptive neutralization of terrorist leaders, which NSDD 138 advocated. Denied the use of the Pentagon's resources by Weinberger's parsimonious attitude toward the deployment of the U.S. military, Shultz, Casey, and North turned to other people to do their fighting. Tirman's assessment that the Reagan Doctrine, a "policy mishmash" born of necessity not strategy, "led to blunders and dashed expectations" is borne out by the administration's first effort to use surrogates to implement its policy of preemptive neutralization of terrorist leaders, namely, against Ayatollah Mohammed Hussayn Fadlallah, Hezbollah's so-called spiritual guide in Lebanon.[5]

On March 8, 1985, a car bomb was detonated in the Bir al-'Abd quarter of Beirut, Lebanon, close to the apartment building where Fadlallah, a hugely influential Shiite cleric, lived. Fadlallah's writings and preaching had led to him being identified, mistakenly some have argued, as the spiritual guide of Hezbollah, and the CIA believed he was responsible for numerous attacks against Western targets, including the 1983 Beirut barracks bombing.[6] The car bomb killed more than eighty people and injured another two hundred. Fadlallah, however, was not among the casualties, escaping the explosion unharmed, if he was present at all. His followers hung a huge banner in front of a blown-out building emblazoned with the words "MADE IN THE USA." Though the bombing itself was carried out by local operatives recruited by the Lebanese intelligence agency, G-2, sources have since argued that Fadlallah's supporters were correct to link the bombing to the United States.[7]

Bob Woodward, whose account of the bombing in his book *Veil* places DCI Casey at the center of the plan to kill Fadlallah, explains that in the absence of a consensus on how to deal with the anti-U.S. terrorism emerging from Lebanon, Casey, supported by Shultz, proposed a plan to the president whereby local operatives would be recruited by the CIA to gather intelligence on terrorist targets in Lebanon and then undertake lethal actions to neutralize them, under the CIA's chief counsel Stanley Sporkin's preemptive self-defense rationale.[8] That the U.S. government would adopt such a plan is perfectly likely, as the proposal is in line with the position set out in the Reagan Doctrine and NSDD 138, signed a year prior to the bombing. According to Woodward, Casey's deputy, John McMahon, expressed unequivocal disapproval of the plan, voicing the same doubts he had raised when NSDD 138 was first drafted by Oliver North—could the United States trust foreign nationals, particularly the Lebanese? Could the CIA control the G-2 agents? Even if it could guarantee the Lebanese would follow their orders, wouldn't this put the agency in breach of EO 12333, which banned it not only from participating in assassinations but from facilitating them too?[9] If the CIA were not in control, how could it ensure that the foreign agents were using adequate intelligence to justify a preemptive attack, and how could it ensure that this would be proportionate, taking all care to avoid unnecessary civilian casualties?[10]

Woodward reports that despite these concerns Reagan signed the necessary Finding and accompanying NSDD, and he has argued that Casey himself was heavily involved in planning the attack. He provides a detailed account of how Casey, frustrated with the CIA's unwillingness to take the necessary risks to make the plan work, made a deal with the Saudi government to use a car bomb to target Fadlallah.[11] This account is contested by Joseph Persico, whose alternative version of the lead-up to the car bombing makes no mention of a deal with Saudi Arabia, instead identifying the NSC's Oliver North as the primary architect of the plan to train Lebanese agents to target terrorists.[12] The findings of the official investigation into the bombing found no evidence of Casey's or the wider CIA's complicity in the attack, instead attributing the killings to the surrogate forces acting as "rogue operatives" without sanction from their CIA handlers. The CIA did not escape the affair unscathed, however. Bernie McMahon, a thirty-year intelligence veteran who ran the Senate inquiry into the affair, concluded that while the CIA had not participated in the bombing, it had created a mechanism that ultimately got out of control and led to the bombing.[13]

Whatever Casey's personal level of involvement and the culpability of those within the NSC and CIA, the Fadlallah affair represented the Reagan administration's first attempt to put into practice the NSDD 138 policy of preemptively neutralizing terrorists. The operation's failure identified some important lessons for the CIA in relation to counterterrorism, which can be identified in the way their drone campaign has been conducted. First, it illustrated the risks of building too close a relationship with the press, driven by the American government's desire to present its counterterrorist operations as just and legal. Second, it highlighted the importance of the CIA staying in control of any potentially lethal counterterrorist operation. Third, it highlighted the advantages, but also the risks, of using foreign agents. Fourth, it showed the importance of interagency cooperation and analysis, and the drawbacks of a lack of consensus. Finally, it showed that any effort to use lethal force against terrorists came with significant risks, particularly the potential for blowback.

Building a Domestic and International Legal Case

The link with the G-2 agents represented the first attempt to turn the NSDD 138 theory of anticipatory self-defense into a legal, workable policy. It was justified by the CIA's Sporkin on the grounds that preemptive attack was enshrined within the jurisprudential standard of *jus ad bellum,* or right to (wage) war. The agency's lead lawyer also argued that the policy followed Article 51 of the U.N. Charter, which granted "the inherent right of individual or collective self-defense."[14] Yet, the policy flaunted elements of jus ad bellum.[15] By using foreign agents to conduct the preemptive attack, while having little to no control over their actual methods, the United States failed to meet the requirement of proportionality by ensuring all due care was taken to minimize civilian casualties and damage to infrastructure. Regardless of whether the car bombing was executed without CIA sanction, it was nonetheless carried out within the context of the U.S. counterterrorism program, itself under U.S. oversight.[16] Robert Oakley, director of the State Department's Office of Combating Terrorism at the time of the bombing, later recounted that two assessments were made of the reliability of the foreign counterterrorism units—one by U.S. Special Forces advisers and another by the CIA itself. In both instances, Oakley attests, discipline standards were found to be unacceptable, thereby precluding direct U.S. provision of money and armaments to those units. Oakley recalls the general feeling around the plan being that "we can't count upon G-2 and who they recruit to have the sort of discipline which we think is essential to conduct a really targeted operation."[17]

By using international law to make the case for preemptive counterterrorism actions, while simultaneously flouting the same law in the conduct of these actions, the CIA was acting in accordance with a notion that has surfaced again in America's conduct of its drone campaign—that for the United States international law is subordinate to geopolitical considerations and American objectives, and is something to be bent and shaped accordingly to add a veneer of legal authority to actions of dubious legality. While George W. Bush flagrantly disregarded international law in favor of unilateral action, issuing a legal memo that stated "I accept the legal conclusion of the Department of Justice that none of the provisions of Geneva apply to our conflict with al-Qaeda," the Obama administration sought to follow the approach established by Casey and Sporkin, creating a legal argument that aimed to place the drone campaign within the moral constraints of "just war" and the legal constraints of international law, while putting American geopolitical considerations first.[18] Obama used the acceptance speech for his Nobel Peace Prize in Oslo on December 10, 2009, to clearly define America's war against al-Qaeda as a just war, firmly within the bounds of international law and meeting the requirements of jus ad bellum. "We have a moral and strategic interest in binding ourselves to certain rules of conduct," explained the president. "The United States of America must remain a standard bearer in the conduct of war, that is why . . . I have reaffirmed America's commitment to abide by the Geneva Conventions."[19] In spite of these claims, the CIA's conduct of its drone campaign, as with the Fadlallah affair before, has frequently been criticized as operating outside international law.

Maintaining Direct Control of Counterterrorism Operations

Despite the Obama administration's conduct of the drone campaign in Pakistan leaving it vulnerable to allegations of hypocrisy—preaching international law on the one hand while flouting it on the other—America's insistence upon conducting the drone strikes itself, as opposed to allowing Pakistan to take control, is drawn from the second hard lesson learned from the fallout of the Fadlallah bombing. In the immediate wake of the failed targeted killing, the CIA dismantled the entire foreign agent counterterror program.[20] From this point onward, if and when lethal counterterrorist operations were conducted they occurred with a much greater level of American oversight. With regard to the drone program in Pakistan, this is illustrated by the United States' refusal to provide the technology to the Pakistani government to conduct the strikes itself on its own territory. In a confidential cable sent from the Islamabad embassy on November 13, 2008, the Pakistani prime minister Yousaf Gilani was reported to have asked the United States to help Pakistan hit targets within its own borders, reiterating that the struggle against extremism was "Pakistan's war." Gilani argued that U.S.-conducted drone attacks were "counterproductive in winning the public's support." The cable went on to make it clear that the prime minister was not pushing for the suspension of the drone campaign but rather pushing for the Pakistani military to be given the technology to conduct the strikes itself. "We have the will," said Gilani, "but not the capacity."[21] In spite of such requests, the CIA has maintained tight control over the program, reflecting a shift from the Reagan administration's approach of using surrogates to do the fighting.

That is not to say foreign agents are no longer used in American counterterrorism, in particular in the drone campaign. It is a frequent misconception of the campaign that the decision to strike is being made solely by drone pilots thousands of miles away from the target—drone strikes need spotters. In line with the original recommendations set out in NSDD 138, the drone campaign uses foreign agents to provide intelligence and targeting information for preemptive attacks against terrorists and militants. The difference now is that, after the debacle of the Fadlallah assassination attempt, the foreign agents do not undertake the neutralization themselves. This network of foreign agents has taken time for the CIA to rebuild. Following Casey's shutting down of the program in Lebanon and the CIA's disengagement from Afghanistan following the Soviet withdrawal, the agency did not have foreign agents it could use to infiltrate the tribal areas where al-Qaeda and the Taliban were sheltering. When facing the 9/11 Commission on April 14, 2004, then CIA director George Tenet was asked how long it would take for the CIA to be in a position to counterattack al-Qaeda. "It's going to take another five years," Tenet confessed, "to build the clandestine service the way the human intelligence capability of this country needs to be run."[22]

Since Tenet's testimony, the CIA has cultivated a large web of Afghan proxy forces, Pakistan-focused informants, and foreign agents to help wage its drone campaign, just as NSDD 138 initially envisioned, but with new procedures and safeguards put in place to guarantee greater U.S. control.[23] In response, al-Qaeda produced a document called "Guide to the Laws Regarding Muslim Spies," which provided guidance on how to identify the agents and the GPS locators that the CIA equipped them with to guide the drone strikes.[24] The existence of this guide, coupled with the fact that it was written by Abu Yahya al-Libi, at the time al-Qaeda's third in command (since killed in a drone strike), with a foreword by Ayman al-Zawahiri, offers a clue as to how effective these foreign agents have been and reveals the seriousness with which al-Qaeda's senior leadership regarded the penetration.[25]

America's refusal to provide the weapons to Pakistan also illustrates a lesson learned from the blowback of the CIA's role in the Afghan jihad throughout the 1980s, which saw vast amounts of U.S.-supplied military hardware end up in the hands of anti-American jihadists and the Taliban via Pakistan's Inter-Services Intelligence (ISI).[26] In *Knights under the Prophet's Banner* Ayman al-Zawahiri mocked the role played by the CIA in equipping its future enemy, stating that the United States' backing of Pakistan and the mujahideen with money and equipment made the Afghan jihad "a training course of the utmost importance" in preparing the Muslim mujahidin to "wage their awaited battle" against the United States.[27] While America's primary objective in the region was to drive the Soviets out of Afghanistan, Pakistan's overriding interest in its western neighbor's fate related to its strategic relationship with its intense rival, India. Pakistan's traumatic history since its foundation on August 15, 1947, had led to the loss of East Pakistan as Bangladesh in 1971 and to Kashmir becoming disputed territory. Further fears of Pashtun calls for independence together with the possibility of Indian influence on Pakistan's western border led General Mohammed Zia-al-Haq, Pakistani's dictator during the time of the anti-Soviet jihad, to seek an Afghanistan dominated by an allied Sunni Islamic power, regardless of whether that power was anti-American or otherwise.[28]

As the Soviets withdrew, staunch anti-American jihadists such as Gulbuddin Hekmatyar, backed by Pakistan's ISI and thousands of Arab volunteers from a dozen countries, went about systematically wiping out their more moderate rivals within the Afghan resistance. "For God's sake," warned one secular Afghan tribal leader to the CIA agents coordinating American military aid, "You're funding your own assassins!"[29] As the country transitioned from an anti-Soviet jihad to a bloody civil war, the ISI eventually shifted its support to the Taliban, equipping the Kandahar-based militants with millions of dollars' worth of weapons left over from the jihad on the orders of President Benazir Bhutto, who mistakenly believed the religious scholars would make pliant clients. The long-term costs to the United States of this error have been well documented.[30] As well as the blowback from the CIA's policy of strengthening al-Qaeda and its Taliban hosts, evidence has also emerged that has linked training camps established by the ISI with CIA funds for terror attacks upon India. For example, following the Mumbai massacre of 2008 in which 164 were killed and more than six hundred injured, Ajmal Kasab, the only attacker who was captured alive confessed under interrogation that the attacks were conducted with the support of the ISI, who allowed the militants to train in Afghan-jihad-era camps.[31] The CIA's failure to identify the future implications of flooding both Afghanistan and Pakistan with millions of dollars' worth of weapons is not an error the United States wants to repeat. As a result, the drone campaign has been kept under strict control, even when passing the technology on would likely limit anti-Americanism and legal criticism in the short term.

Building a Cross-Governmental Consensus

The Fadlallah affair also marked a vital moment of policy shift for the Reagan administration, when its paralyzing lack of consensus on counterterrorism policy turned into a fumbling foreign policy. In what Graham Allison describes as the "pulling and hauling" dynamics of bureaucratic politics, the fierce competition between the different high-level decision makers to protect their own departmental and agency interests while pushing their own policy preferences meant there was too little cooperation within the federal government on the issue of counterterrorism.[32] There were profound differences of opinion between the CIA, the DoD, and the DoS. Even within these institutions there were full-blown disagreements about the utility of force and the targeted killing of terrorist leaders as a counterterrorism strategy. The failure of the administration's leadership to either form a consensus or nominate a lead agency to take charge meant the different branches of government failed to work together toward a common goal. This created a situation in which the personal preferences of senior government officials took precedence over informed analysis.

The effort to implement NSDD 138 through the use of foreign agents was not guided by a carefully reasoned set of counterterrorism measures consistent with wider American foreign policy interests in Lebanon and the wider Middle East. The primary architects of the directive—namely, North, Casey, and Shultz—were so embroiled in the bureaucratic dispute with opponents of the plan, such as Weinberger and Casey's deputy, John McMahon, that all criticism of the operation, no matter how valid, was ignored. When two separate assessments, one from the DoD and another from within the CIA itself, warned of the risks associated with using ill-disciplined G-2 agents coupled with the potential blowback of a targeted killing in Beirut, they were rejected on the grounds that they represented the views of those who opposed proactive counterterrorism. Robert Kupperman, Jeff Kamen, Paul Pillar, Martha Crenshaw, and many other terrorism experts all stress that without synergy between counterterrorism practices and broader foreign policy objectives based on solid analysis, severe political setbacks are likely.[33] The affair taught a valuable lesson that any future counterterrorism action would need to involve a much more effective connection between policy recommendations, analysis, and operational planning. The creation of the CIA's CTC less than a year after the failed bombing was a direct response to this and marked an effort to provide an analytical, centralized, and coordinated approach to counterterrorism operations.[34] The center, sometimes described as "an agency within an agency," has since become the hub for the War on Terror in the Afghanistan/Pakistan (AfPak) region, and has responsibility for overseeing the drone campaign in Pakistan's territory, a mission the CTC has undertaken with a near total cross-branch and cross-party consensus.[35]

Consideration of the Potential Blowback

The fallout from the car bombing also demonstrated that undertaking lethal counterterrorist operations ran the risk of escalating America's conflict with terrorism, especially when, as with the Fadlallah bombing, the United States was seen to have been responsible for civilian casualties. The impact of this was heightened by the attack taking place in a country where anti-Americanism was already prevalent. The linking of the CIA to the attempted killing of an influential religious figure and the deaths of so many civilians, regardless of how culpable the Agency actually was, further diminished an already lowly American reputation in Lebanon and the wider Middle East. The contradiction between American pronouncements about democracy and international law and the exposed reality of the Reagan administration's plotting and partnership with the brutal Lebanese government was exposed.[36] Three months after the bombing Hezbollah and Islamic Jihad members hijacked TWA 847, an American airliner, using the hostages to demand the release of seven hundred Shiite Muslims from Israeli custody. During a standoff at Beirut airport the hijackers beat and killed Robert Stethem, a U.S. Navy diver. When challenged by a control-tower operator for killing an innocent passenger, the lead hijacker replied angrily, "Did you forget the Bir al Abed massacre?"[37] This sense of injustice and the view that the United States had deliberately targeted Shiite civilians aided Hezbollah recruitment throughout Lebanon and the wider region.

There is evidence to suggest that America's subsequent use of drones in counterterrorism operations generates a similar sense of anger and resentment. When Faisal Shahzad, a thirty-one-year-old U.S. citizen who lived in the Connecticut suburbs, was arrested for parking a car full of explosives in New York City's Times Square in May 2010, he identified the CIA's drone campaign as the key factor that motivated him, telling the judge: "The drone hits in Afghanistan and Iraq, . . . they don't see children, they don't see anybody." In blaming the drone strikes for the deaths of Muslim women and children, Shahzad added: "I am part of the answer to the US terrorising the Muslim nations . . . I'm avenging the attacks because the Americans only care about their people, but they don't care about the people elsewhere in the world when they die."[38] Suggesting the negative image of drone warfare is radicalizing more Muslims to take up arms against the United States, a Lahore-based terrorist recruiter quoted in the *Observer* in September 2010, stated, "We don't have to do anything to attract them—the Americans and the Pakistani government do our work for us." According to the recruiter, the perception that the drone attacks targeting "the innocents who live in Waziristan" meant that many were already sympathetic to their cause. Recruiting these sympathizers was "simply a case of converting these sentiments into action."[39]

There is contradictory evidence, however, that the drone campaign has had the opposite effect and has played a significant role in drying up the flow of recruits to the al-Qaeda training camps in the Federally Administered Tribal Areas (FATA) of Pakistan. Four months after the attempted Times Square bombing, *Newsweek* interviewed Hafiz Hanif (an alias), a young Afghan who attended an al-Qaeda training camp in Waziristan. He described his initial time at the camp in a positive light, testifying to the wide appeal the terrorist group held with his cohort of "30 students of various nationalities—Chechens, Tajiks, Saudis, Syrians, and Turks, two Frenchmen of Algerian extraction, and three Germans, one of whom was European and the other two of Arabic or Turkish extraction. . . . New recruits are always arriving."[40] Yet, when the same reporter retuned to interview Hanif eighteen months later, after the Obama administration had dramatically escalated the CIA's drone campaign, his account was very different. "Only four of the cell's 15 fighters were left, huddled in a two-room mud-brick house, with little or no money or food." The cell's commander, a Kuwaiti named Sheik Attiya Ayatullah, had gone into hiding, and the other members had either run off or been killed in drone strikes.[41] "Why should we call you back just to get killed in a drone attack?" Hanif's jihadi brothers asked. Due to the secrecy surrounding the CIA's campaign, it is extremely difficult to discern which account is more accurate, whether hatred, resentment, and negative propaganda relating to the drones drive terrorist recruitment or whether fear of attack has deterred recruitment to al-Qaeda's training camps. What is clear is that achieving the correct balance of force to deter rather than recruit has been a challenge facing U.S. counterterrorism policy from the Reagan administration into the present day.

The *Achille Lauro* Hijacking

On October 7, 1985, the Italian cruise liner *Achille Lauro* was hijacked by four men representing the Palestinian Liberation Front (PLF), a radical faction of the Palestinian Liberation Organization (PLO), as it sailed from Alexandria to Port Said along the Egyptian coast. During the hijacking Leon Klinghoffer, a sixty-nine-year-old wheelchair-bound Jewish American, was murdered by the hijackers and his body thrown overboard.[42] The hijacking was in retaliation for a bombing raid by the Israeli Air Force upon the PLO headquarters in Tunisia six days earlier, an attack strongly condemned by the United Nations; it was reported to have killed between sixty and seventy-five Palestinians and injured a similar number.[43] The attack upon the PLO headquarters was itself in retaliation for the killing of three Israelis by members of the PLO's commando force, Unit 17, on a yacht off the coast of Cyprus on September 25, 1985.[44] The incident serves as an example of the spiral of violence that retaliation can cause, a key argument put forward by many who oppose America's use of lethal force with drones against al-Qaeda and its affiliates.[45]

Supporters of America's use of force in counterterrorism take a different lesson from the *Achille Lauro* affair. America's response has been seen by some to have been one of the high points of American counterterrorism under the Reagan administration. "We Bag the Bums," declared the *New York Daily News,* and a simple "Got 'Em" took the front page of the *New York Post* when news broke that the hijackers had been captured through American intervention.[46] Reagan himself summed up the jubilant feeling in the nation when he wrote, "Americans as well as our friends abroad are standing six inches taller. We're flooded with wires and calls."[47] Despite the eventual success, the Reagan administration's intervention in the affair had very nearly gone the way of its previous interventions. On the advice of his NSC, Reagan had ordered Joint Special Operations Command (JSOC) forces, namely, Delta Force and the U.S. Navy's SEAL Team Six, be dispatched to undertake a rescue operation should the cruise liner reenter international waters.[48] When, however, the ship returned to Egypt before an operation could be launched, it looked as though the chance to mount an effective intervention had passed. Egypt's President Hosni Mubarak negotiated a peace deal that allowed the hijackers to escape by air in return for the safe release of the remaining hostages. Mubarak deliberately misled the White House, keeping reports of Klinghoffer's murder from the administration in order to ensure the deal went ahead. Upon learning of Klinghoffer's killing from the U.S. ambassador to Egypt, however, the White House demanded the hijackers be detained as murderers. Once more Mubarak misled the Reagan administration, informing it that the EgyptAir plane with the hijackers aboard had already left Cairo. When Israeli intelligence informed the United States otherwise, members of the NSC and the Joint Chiefs liaised to come up with a plan to deploy military aircraft that would divert the airliner once it was in the air to a NATO base at Sigonella, Sicily, where the hijackers could be detained.[49]

According to an account provided by the then assistant secretary of state for intelligence, Morton Abramowitz, who witnessed the deliberations, NSC staffer James Stark was the first to come up with the idea of intercepting the hijackers' escape aircraft. Stark briefly discussed the idea with fellow NSC member Howard Teicher before the two men took the idea to Oliver North. North, unsurprisingly, liked the idea of aggressively pursuing the terrorists, and he took it to the APNSA, Robert McFarlane. McFarlane supported the idea but, having to leave for Illinois with the president, left the planning of the operation to his deputy, Admiral John Poindexter. Poindexter contacted Vice Admiral Art Moreau, the assistant to Admiral William Crowe, the newly appointed chairman of the Joint Chiefs, and asked him to begin constructing a plan to take action against the hijackers' plane. Once the plan was formed and the logistical requirements arranged, they contacted McFarlane, who then reported the information to Reagan, who agreed planning should proceed. Upon returning to the White House, McFarlane convened a meeting of the NSC's National Security Planning Group, which decided upon the rules of engagement, discussed the diplomatic implications, and then ordered the intercept plan to go ahead.[50] The most vital part in the execution of this operation was that the planning had been very swift, taking a matter of hours from conception to implementation, bypassing the normal interagency and interdepartmental process, thus enabling the NSC to overcome the institutional stalemate discussed in the previous chapter. The speed outmaneuvered Weinberger, who, when he heard of the plan from the chairman of the Joint Chiefs, was livid. "What the hell's going on?" he demanded. "That's a terrible idea. I'm dead set against it, interfering with civilian aircraft. We'll be castigated all over the world."[51]

In spite of his protests, Weinberger was too late to stop the operation. McFarlane, with the support of Shultz, had been able to convince Reagan of the viability of the plan. Without the dissenting voice of his secretary of defense, the president had agreed that the operation should go ahead. An intercept package of six F-14s, two A-6 tankers, and two E-2C aircraft was dispatched from the USS *Saratoga* and, using sophisticated radars, located the EgyptAir 737, forcing it to land at the NATO base in Sicily.[52] On the landing strip U.S. Delta Force troops attempted to seize the hijackers for extradition to the United States, but the Italian authorities intervened and, after a tense standoff with the American forces, took the PLF members into custody.[53] Though the United States subsequently asked the Italian government to extradite the hijackers, Italy asserted jurisdiction over them and put them on trial. Much to the Reagan administration's frustration the Italians chose to release Abdul Abbas, as he had not actually been directly involved in the hijacking, despite American intelligence identifying him as the mastermind of the terrorist attack.[54] He remained at large until his eventual capture by JSOC in Baghdad in 2003, following Operation Iraqi Freedom, and died in U.S. custody in March 2004.[55]

Be Prepared to Bend the Law

There are a number of reasons why the *Achille Lauro* incident is significant for the CIA drone campaign. First, it demonstrates that the United States had already set a precedent for breaking international laws in the pursuit of terrorists. The diversion of the Egyptian airliner and planned abduction of the hijackers technically violated international sovereignty, territorial integrity, and internationally protected due-process rights, and in doing so set off a legal debate as to what the United States could and could not do in its efforts to counter international terrorism.[56] In his legal analysis of the event at the time, attorney Andrew Liput predated the criticism by contemporary legal scholars who rail against America's unilateral conduct of the drone campaign outside the boundaries of international law. He stressed that the primary purpose of modern international law was to enable states to live together despite their varying and frequently opposing political views, and that action which was based upon universal norms that do not depend on any particular, narrowly focused political ideology was key to maintaining stability while being able to eliminate the threat of international terrorism.[57] Both the drone campaign and the forced landing that predated it illustrate America's long-standing willingness to operate outside these universal norms when it comes to carrying out counterterrorism operations and seizing the opportunity to neutralize anti-American terror threats.

Putting a Strain on Alliances

Egypt's response to the hijacking illustrated that, as with Pakistan and the drone campaign today, an American ally's national interests may prevent it from supporting America's counterterrorism cause, even when the United States presents a legal case for doing so, thus placing America in a position of either abandoning its efforts or undermining its ally in the pursuit of its goals. In the case of the *Achille Lauro* hijackers, the United States could claim the legal right for Egypt either to prosecute the hostage takers or to extradite them to the United States under the terms of an extradition treaty signed between Egypt and the United States in 1874 and also of the International Convention against the Taking of Hostages, ratified by Egypt in 1981 and the United States in 1984.[58] Egypt neither prosecuted nor extradited the hijackers but instead provided them with their means of escape. Mubarak had a twofold reason to try to bring the hostage crisis to a peaceful end. First, as the president of an Arab nation he was aware that there was considerable sympathy among his people for the Palestinian cause, and any role on his part that might look too much like pandering to the Americans or Israelis could severely damage his credibility. Moreover, Egypt had a long history of radical Islamic leaders and groups operating within its territory, and any action seen to further align his government with the United States at the expense of the Palestinian people could cause a violent backlash.[59] Second, Mubarak was involved in the ongoing Arab-Israeli peace process. It was in his interests and those of the peace process itself to promote the Palestinian leader Yasser Arafat and his PLO as a legitimate partner in the negotiations. The hijacking and subsequent murder of an American citizen, though not instigated by the PLO, served to illustrate Arafat's continued links to international terrorism. If Egypt were to have any chance of mediating the peace process, the tension around the situation had to be relieved in a way that maintained Egyptian influence over the Palestinian cause.[60]

Pakistan faces a similar clash between U.S. counterterrorism efforts, supported by international law, and its own national interests. Legally, the United States claims international authority to pursue al-Qaeda and the Taliban through United Nations Security Council resolution 1368. Passed the day after the 9/11 attacks, the resolution identifies those responsible [al-Qaeda] as "a threat to international peace and security" and "calls on all States to work together . . . to bring to justice the perpetrators." The resolution also goes on to stress that "those responsible for aiding, supporting or harboring the perpetrators [the Taliban] will be held accountable."[61] Like Egypt, however, Pakistan is home to a significant number of Islamic radicals who have directed attacks against the Pakistani government and its citizens as punishment for collaboration with the United States. A report published by Brown University's Watson Institute for International Studies has estimated that between 14,780 and 43,149 Pakistanis were killed and a further forty thousand injured by Taliban, al-Qaeda, and other militant insurgents using suicide attacks, assassinations, and ambushes within their own country between 2004 and 2010.[62] Among these casualties was former Pakistani prime minister and leader of the main opposition party at the time, Benazir Bhutto. Upon her assassination on December 27, 2007, Mustafa Abu al-Yazid, an al-Qaeda commander, claimed responsibility, telling several news outlets: "We terminated the most precious American asset which vowed to defeat [the] mujahideen." The commander went on to link the killing to Pakistani cooperation with the United States on counterterrorism: "This is our first major victory against those who have been siding with infidels in the fight against al-Qaeda and declared war against mujahideen."[63]

Furthermore, as with Mubarak's calculations regarding Egypt's position in the region, Pakistan has always had a strategic interest in having a Sunni Islamic power in control of Afghanistan. This also serves to put its national interests at odds with U.S. counterterrorism aims. As already discussed above, it has been a long-term strategic goal of successive Pakistani governments to ensure that Afghanistan is dominated by an allied Sunni Islamic power, regardless of whether that power is the Afghan Taliban or another anti-American force. This clash of interests has left the United States in a position where, as with Egypt in the 1980s, it claims a legal right to pursue operations against its terrorist enemies but finds this to be at odds with the national interests of its ally. On a number of occasions, Pakistani officials have publicly condemned the drone strikes and demanded they end.[64] Following three days of consultation with senior figures in the Pakistani government between March 11 and 13, 2013, Ben Emmerson, serving as the United Nations' special rapporteur on human rights and counterterrorism, issued a formal statement on behalf of the Pakistani government, reporting that Pakistan did not consent to the use of drones by the United States on its territory, and that it considered the actions to be a violation of the country's sovereignty and territorial integrity. The lawyer's report noted that Pakistan officially called upon the United States to cease its campaign immediately.[65]

Contrary to the government of Pakistan's public position, there is evidence that suggests covert Pakistani complicity in the CIA drone campaign. In 2009, Google Earth images revealed CIA Predator drones at a Pakistani military airbase in Baluchistan. In addition, a classified cable sent from the U.S. embassy in Islamabad on February 11, 2008, revealed that the Pakistani chief of army staff, General Ashfaq Parvez Kayani, had asked America's Admiral William J. Fallon for "continuous Predator coverage of the conflict area [Waziristan]."[66] This presents a number of possibilities. It could suggest the public criticism of the drone campaign is posturing on the part of the Pakistani government in an effort to limit the fallout that could result from overt collaboration with the United States, a country that is deeply unpopular with the majority of the Pakistani public. Alternatively, it could be argued that any Pakistani cooperation is actually the result of fear of its ally. Former Prime Minister Perez Musharraf revealed in an interview on CBS's *60 Minutes* that immediately following the 9/11 attacks, his director of intelligence was warned by Deputy Secretary of Defense Richard Armitage that Pakistan would be "bombed back to the stone age" should it fail to cooperate with U.S. efforts.[67] Finally, it could demonstrate America's willingness to undermine the democratically elected Pakistani government by subverting its will and dealing directly with the Pakistani military instead.

Whatever the truth behind the Pakistani government's relationship with the CIA's drone campaign, the United States claims lawful authority to undertake its counterterrorism actions. Just as the Reagan administration drew upon the existing Bilateral Extradition Treaty with Egypt and the United Nations' International Convention against the Taking of Hostages to justify its pursuit of the PLF terrorists, so too has the Obama administration drawn upon existing legislation to provide legal authority for the CIA's use of drones in an ally's territory, regardless of whether that ally is complicit or not. Obama administration officials have made reference to

U.N. Security Council Resolution 748 passed on March 31, 1992, in the wake of Libya's failure to cooperate with the investigation into the bombing of Pan Am 103. The resolution states, in accordance with Article 2 of the U.N. Charter, that "every State has the duty to refrain from organizing, instigating, assisting or participating in terrorist acts in another State or acquiescing in organized activities within its territory directed towards the commission of such acts, when such acts involve a threat or use of force."[68]

The acquiescence portion of Resolution 748 is the most relevant to the American/Pakistan case. Pakistan's double game between the United States and elements of the Taliban was acknowledged by the influential Obama adviser Bruce Riedel in 2009, and again by former chairman of the Joint Chiefs of Staff Admiral Mike Mullen in one of his final sessions before Congress. Mullen informed Congress that the United States possessed clear evidence that "the Haqqani network [a branch of the Taliban] acts as a veritable arm of Pakistan's Inter-Services Intelligence agency" and that it "operates from Pakistan with impunity."[69] This legal position was further explained in a report published on May 28, 2010, by Philip Alston, the U.N.'s former special rapporteur on extrajudicial, summary, or arbitrary executions. Alston, whose report was generally critical of the CIA's drone campaign, conceded that under the law of interstate force: "A targeted killing conducted by one State in the territory of a second State does not violate the second State's sovereignty if either (a) the second State consents, or (b) the first, targeting State has the right under international law to use force in self-defense under Article 51 of the UN Charter, because (1) the second State is responsible for an armed attack against the first State, or (ii) the second State is unwilling or unable to stop armed attacks against the first State launched from its territory."[70]

The Obama administration has argued that Pakistan is both unwilling to act against the Taliban, either through fear of reprisals or due to their links to the Haqqani network, and unable to act against al-Qaeda, as evidenced by the CIA's discovery of bin Laden in a compound in Abbottabad, home of the Pakistani military training academy.[71] Whether or not the threat the United States faces from the Taliban and al-Qaeda is enough to meet the self-defense terms of Article 51 of the U.N. Charter is a separate debate from whether or not the CIA's drone strikes represent a proportionate enough response to meet the terms of international humanitarian law. What is clear from this example is that the United States will put the act of neutralizing terrorists above the national interests of its allies, even when it may cause considerable social and diplomatic unrest. This strategy is not unique to the drone campaign but based upon a precedent set under the Reagan administration.

When Weinberger warned the president that intercepting the Egyptian airliner would damage American relations with Egypt, other moderate Arab states and their NATO ally Italy, Reagan told him the operation was going ahead regardless: "Cap, it is pretty cut and dried. This is a guilty party; we cannot let them go."[72] As the secretary of defense had warned, the fallout from the American intervention was significant. Both radical and more moderate elements within Egypt marched through the capital demanding Mubarak cancel upcoming U.S.-Egyptian military exercises and deny America access to Egyptian military facilities, putting intense pressure on their president to undertake some sort of retaliatory measure against America. As America's close ally, Mubarak was condemned by stone-throwing protestors in Cairo as a "coward" and "U.S. agent." He responded by denouncing the American intercept as "an act of piracy" and declaring that relations between the two countries would be strained "for a long time to come."[73] The CIA's assessment concluded this was more than just a bluff, predicting: "Egyptian relations with the United States and the PLO are likely to remain strained for some time."[74] Wider relations in the region were also harmed by the action. Mubarak was furious with Syria's Hafaz al-Assad after the Syrian leader ordered Klinghoffer's body, which had washed up on Syria's coastline, be returned to the United States, and the CIA reported that numerous leaders in the region who were previously pro-American such as Tunisia's President Habib Bourguiba were likely to "be obliged by internal political considerations to seek to lower U.S. profile" in their countries.[75]

The fallout did not stop in the Middle East. The Reagan administration had wanted Abul Abbas even more than the four hijackers. Reflecting the language in NSDD 138, and the later deliberate targeting of al-Qaeda leaders in drone strikes, the Reagan administration saw Abbas as the leader and mastermind of the attack, and therefore the most important part of the terror network to neutralize. The others were merely militants following his orders.[76] When the Italian government released Abbas, the Reagan administration issued a stinging censure: "The U.S. government finds it incomprehensible that the Italian authorities permitted Abul Abbas to leave Italy despite a U.S. Government request to the Italian Government for his arrest and detention." Larry Speakes, the presidential spokesman, went on to describe Abbas as "one of the most notorious Palestinian terrorists and has been involved in savage attacks on civilians. . . . The U.S. government is astonished at this breach of any reasonable standard of due process and is deeply disappointed."[77] For the ruling coalition of the Italian government, already strained by differences over economic policy, the argument over the release of Abbas proved the final straw. Giovanni Spadolini, the leader of the Republican Party and minister of defense, accused the prime minister, the Socialist Party's Bettino Craxi, of releasing Abbas before the judge could review the U.S. extradition request and thus failing to meet Italy's international responsibilities. In the ensuing political row, Spadolini withdrew his party from the coalition, and the government collapsed on October 17, 1985.[78]

The American Public Supports Proactive Counterterrorism Actions

Despite the chaos left in its wake with America's allies, the interception of EgyptAir flight 2843 is seen domestically as one of the most successful counterterrorist actions undertaken by the Reagan administration. It represented America's first clear-cut victory against terrorism. As Mubarak faced stone-throwing anti-American protesters, Reagan was showered with praise from the American public, as well as from fellow Republicans and Democrats alike.[79] This demonstrates that a counterterrorism policy may play out badly overseas, but if it is popular with the American people, then it will be regarded as successful. Such a perspective helps put the CIA's drone campaign in Pakistan in a different and better-informed light. Rather than reflecting a new level of disregard for America's allies, the calculations behind the drone campaign build upon the populist approach that was successfully adopted by the Reagan administration. Despite the drone campaign's many critics among law school academics, human rights campaigners, antiwar protestors, and some sections of the media, a *Washington Post*–ABC News poll conducted in February 2012 showed that an overwhelming 83 percent of the American public approved of Barack Obama's use of unmanned drone aircraft against terrorist suspects overseas; 59 percent strongly approved, and only 4 percent strongly disapproved.[80] A poll on international attitudes conducted by Pew the same year also put American approval for the drone campaign at 62 percent.[81] Although increased debate around the controversies of drone use appears to have reduced support, with a July 2014 *Washington Post* poll revealing that support had dropped to 52 percent, the majority of the American public accepts the positive narrative of the drone campaign—that it is legal, precise, and the most effective method of combating the terrorist threat posed by al-Qaeda and its affiliates.[82]

The Pakistani public's attitude toward the CIA's drone campaign has been, unsurprisingly, markedly different. A poll from the University of Maryland's World Opinion project released in July 2009 following the first sixth months of increased drone activity under the Obama administration showed that 82 percent of Pakistanis questioned believed that drone strikes were unjustified, and that 69 percent had an unfavorable view of the U.S. government.[83] These negative attitudes toward the drone campaign can be seen to have impacted upon the domestic political situation in Pakistan. President Zardari, who held office during the Obama administration's expansion of the drone campaign, saw his personal approval ratings drop from a respectable 64 percent in 2008 to a mere 14 percent in a June 2012 poll. Though this rapid decline can be attributed to a number of factors, such as the faltering economy and accusations of corruption, America's actions almost certainly played a part in undermining Zardari. Pakistanis were forced to choose between regarding their president as too weak to stop the United States doing what it wanted in Pakistani territory or being complicit in its attacks upon fellow Muslims.[84] The role of the CIA's drone campaign in undermining Zardari can also be evidenced by the significant rise in popularity of Imran Khan during the same period. Khan built much of the support base of his Pakistan Tehrik-e-Insaaf party through anti-drone rallies and promises to end the drone campaign.[85] In the same poll, Khan achieved a 70 percent favorability rating in June 2012, up from 52 percent in 2010. Furthermore, in parallel with the anger stirred in Egypt and the wider Middle East by Reagan's interception of the *Achille Lauro* hijackers' plane, further polling by Pew reveals widespread opposition globally to America's drone campaign. Of twenty countries polled, seventeen registered 50 percent or higher disapproval of America's use of drones to target extremist leaders, with some predominantly Muslim countries, such as Turkey, Egypt, and Jordan, scoring more than 80 percent disapproval.[86]

As with Reagan's risky *Achille Lauro* venture, the CIA's drone campaign, despite major international repercussions, is regarded at home as a success. For Reagan, the United States' relationship with Egypt recovered shortly afterward, with President George H. W. Bush formally recognizing Mubarak's state as the second ever major non-NATO ally in 1989. Anti-Americanism in the region remained, but it had been present before the intervention, and the protestors soon returned to their homes. For Obama, the same polls that showed significant international disapproval for the American drone campaign also showed widespread support for his presidency, and the United States in general.[87] For example, in 2012 though 63 percent of the French disapproved of drone strikes, 69 percent had favorable views of the United States. This was up from 42 percent in George W. Bush's last year as president. In Germany, there was similar pattern, with 59 percent against drone strikes, but 52 percent then holding a positive view of the United States. This again represents an increase from Bush's final year as president, when German approval of the United States was as low as 31 percent. Although approval ratings in predominantly Muslim countries—in particular Pakistan—continue to fall, the United States is following a proven pattern of maintaining domestic support while temporarily sacrificing relations and stability of certain allies in order to neutralize its terrorist enemies. As Reagan said, "It is pretty cut and dried"—if America has a clear opportunity to strike guilty parties in one way or another, it "cannot let them go." For the Obama administration, anti-Americanism already existed in Pakistan, Egypt, and other Middle Eastern states for a wide range of reasons beyond the drone campaign.[88] The use of the drones may have worsened these relations in the short term, but the administration applied the *Achille Lauro* approach of following an America-first policy in counterterrorism. It is neither unprecedented nor necessarily the cause of a terminal decline in America's relations with its Muslim allies, as the fallout of the EgyptAir interception proved.

Conclusion

During the reelection campaign in 2012, some media reports began to compare Obama's presidency to that of Ronald Reagan, looking at similarities in unemployment figures, voting patterns, rhetoric, and the way in which both presidencies sought to change the direction of the United

States.[89] None of these reports drew any connections between the counterterrorism policies of the two administrations, yet this is one area in which some major similarities can be found. The two case studies discussed above go some way toward demonstrating that the counterterrorism policy adopted during Obama's "drone presidency" has its roots in the policies advocated and pursued by the counterterrorism hardliners of the Reagan administration. That is not to say that the drone campaign simply reproduces these policies: important lessons have been learned since the patchy and haphazard execution of counterterrorism in the mid-1980s, showing the evolutionary nature of U.S. counterterrorism over the past three decades. While the drone campaign is still focused upon the preemptive neutralization of terrorists, this has become a much more centralized effort, run through the CIA's CTC with lethal operations undertaken with much tighter control. In the case of high-profile targets the final authorization for strikes is reserved for the president himself in an effort to ensure the campaign abides by the requirements of a just war. The Obama administration placed a much greater emphasis on consensus building, ensuring that clear lines of cooperation and responsibility exist between the Departments of Justice, Defense, and State and the CIA. Foreign agents still play a role, providing vital intelligence and targeting details, but are no longer trained or equipped to undertake the lethal operations themselves. Obama did follow Reagan, however, in being willing to sacrifice relations with allies in the short-term in the pursuit of terrorists.

In 2008, then senator Obama offered his own comparison between his hoped-for presidential legacy and that of Reagan.[90] "He put us on a fundamentally different path because the country was ready for it," reflected Obama. "I think we are in one of those times right now, where people feel like things as they are going, aren't working, that we're bogged down in the same arguments that we've been having and they're not useful."[91] Though Obama's comments were intended to relate to domestic concerns, they could just as easily have referred to America's War on Terror at the time. Following George W. Bush's neglect of Afghanistan in favor of regime change in Iraq, the mission against al-Qaeda had stalled. The terror group's core leaders and their Taliban hosts were resurgent in the AfPak border region, using the momentum gained from the conflict in Iraq to recruit and train the next wave of jihadist fighters. By embracing drone warfare on a huge scale Obama was in many ways setting U.S. counterterrorism on a different path. Never before had the targeted killing of terrorists been undertaken on such a large and well-coordinated scale. And yet at its core the drone campaign was based upon the principles that had governed the Reagan administration's first efforts at preemptive neutralization of terrorists. Rather than a revolution in U.S. counterterrorism, the Obama administration was following an evolutionary path, drawing upon the lessons that the CIA and their colleagues in the national security fraternity had learned from their predecessors.

Perhaps the most important lesson learned from the Reagan administration was to do with the American public. Ultimately, the United States is a democracy. While a president may go against the will of the American public at times, spending political capital to pursue a particular path, in the end approval ratings, midterm elections, and reelection campaigns ensure the president follows policies of which the majority of Americans approve. When Reagan authorized U.S. warplanes to intercept a civilian airliner carrying terrorists who had murdered a U.S. citizen, the American public supported the action in spite of the international political fallout. When George W. Bush announced American forces would be entering Afghanistan to find bin Laden "dead or alive," his approval ratings soared higher than those of any president before him. Obama's use of drones had the highest approval rating of any policy from his two terms in office. When justifying the operation against the Egyptian airliner, Reagan told his close friend Weinberger, "This is a guilty party; we cannot let them go."[92] In using the word "cannot," Reagan was reflecting not just his personal view but also the will of the majority of Americans. The preemptive neutralization of terrorists, or at least the judicious application of force after an attack, was not just the desire of a small group of hardliners within the Reagan administration. It is an expectation of the American public, Republican and Democrat alike.

3

"WE HAVE TO FIND A BETTER WAY TO SEND A MESSAGE": THE CIA'S COUNTERTERRORISM CENTER, FROM THE EAGLE PROGRAM TO THE PREDATOR DRONE, 1986–2001

We have to find a better way to send a message to outlaw nations that we don't like their behavior—short of sending in squadrons of F-111Bs.

—*Duane "Dewey" Clarridge, first director of the CIA's Counterterrorist Center*

The CIA's Counterterrorism Center was spearheading America's War on Terror in the AfPak region before George W. Bush even announced its commencement to the American people in his joint address to Congress on September 21, 2001. Members of the CTC covertly entered Afghanistan days after the 9/11 attacks to prepare the way for the American assault, striking a deal with the Afghan Northern Alliance to oust the Taliban and track down al-Qaeda's fighters. Since then, it has been the CTC that has been responsible for executing the CIA's drone campaign in the region, launching hundreds of strikes and killing thousands of suspected anti-American militants. The center's aggressive pursuit of al-Qaeda and its affiliates has been seen by many to serve as an example of how America's belligerent response to 9/11 has militarized the CIA. Yet, for the CTC, this aggressive stance marked a return to its founding principles rather than a post-9/11 departure.

Established in 1986 by Duane Clarridge—either one of the agency's most capable, aggressive, and innovative case officers or a belligerent risk taker steeped in Cold War myths, depending upon your source and perspective—the CTC was a unique fusion of intelligence, technology, and operations that was, from the outset, intended to put the CIA on an aggressive footing against international terrorism. Supported by the likes of Casey and North, the CTC introduced many of the practices that enable the drone campaign to function today. Even more significant, through its Eagle Program the CTC introduced a prototype unmanned aircraft that set in motion the technological developments that culminated in the creation of the drones used by the CIA in its pursuit of al-Qaeda today. Though the fallout of the Iran-Contra scandal coupled with Casey's death shortly after the center's founding meant it never launched its proactive counterterrorism campaign during the Reagan administration, the demand for action following the September 11 attacks saw the center eventually take up the role it was initially intended to fulfill.

The Founding of the Counterterrorism Center

Despite the split between his advisers and his own periodic indecisiveness regarding retaliation, Reagan seems to have genuinely wanted to punish terrorists who killed Americans.[1] Reagan's APNSA Robert McFarlane recalled that after the bombings of the Marine barracks in Beirut in 1983, the president had entered the Situation Room with "an expression of hatred and wish for revenge" that McFarlane had never seen in him before. "Those sons of bitches," Reagan had said, "let's find a way to go after them."[2] But three years on from the bombings, his administration was still fumbling around to find an effective way of going after terrorists. The *Achille Lauro* affair had demonstrated what could be done when all branches of the government worked together—the Fadlallah affair had shown what happened when splits and divisions ruled.

In January 1986, Casey confided in "Dewey" Clarridge, his favorite CIA officer, that Reagan was putting intense pressure on him and the CIA to show more leadership and initiative in the fight against terrorism.[3] This uncharacteristic directness from the usually passive president revealed Reagan's growing obsession with the plight of Americans held hostage by terrorist groups, as well as the White House's concerns about the political implications of the administration's inaction.[4] Casey had a reason for discussing the issue with Clarridge, who had run Casey's covert Contra war against Nicaragua.[5] The DCI had handed Clarridge the job personally after meeting him during a tour of CIA stations. "Dewey's the only really impressive guy I met out there," Casey had stated to his deputy, Robert Inman. "He's a doer, a take-charge guy." Casey's instincts were correct. Clarridge was already a legend at Langley. When asked to describe him, colleagues would use terms like "a roughrider," "a cowboy," "a hip-shooter," "our General Patton." Robert Gates summed him up by saying, "If you have a tough, dangerous job, critical to national security, Dewey's your man. He's talented. One of our best operations officers." Adding a touch of caution, Clarridge's CIA colleague warned, "Just make sure you have a good lawyer at his elbow—Dewey's not easy to control."[6]

Gates's point was proven in 1984 when Clarridge had to be "promoted" to chief of the CIA's European operations after a storm over the Contra manual *Psychological Operations in Guerrilla Warfare* broke during Reagan's reelection campaign.[7] The book, written by the CIA, was intended to provide guidance to the Contras on how to conduct a guerrilla campaign, but it caused an investigation by the Intelligence Oversight Board (IOB) due to its use of the term "neutralization," implying assassination.[8] The section in the book entitled "Selective Use of Violence for Propagandistic Effects" suggested, "It is possible to neutralize carefully selected and planned targets, such as court judges, mesta judges, police and State Security officials, CDS chiefs, etc." The book recommended: "For psychological purposes it is necessary to gather together the population affected so that they will be present, take part in the act, and formulate accusations against the oppressor."[9]

Far from putting Casey off, Clarridge's aggressive risk-taking approach and willingness to work on the fringes of EO 12333 were exactly what the DCI was looking for in response to the growing pressure from the president. The very same use of the term "neutralization" had caused controversy with the drafting of NSDD 138, so Casey could be confident Clarridge shared a similar outlook to him and his fellow counterterrorism hardliners.[10] The DCI believed that the agency as a whole had been shackled by the Church Commission and heavy-handed congressional oversight, which had brought aggressive covert action "to a screeching halt."[11] He felt his Directorate of Operations (DO) had become too timid and bureaucratic. This was perhaps best illustrated for Casey by the 130-page book of guidelines governing covert operations. "This kind of crap is smothering us." Casey had ranted to his DO officers, waving the book in the air. "You practically have to take a lawyer with you on a mission. I'm throwing this thing out." To Casey, who had cut his teeth in the OSS, covert operations were the equivalent of his gambles on the stock market—you had to take risks to get big results. He replaced the guidelines with a single memorandum, which essentially told agents: "In conducting a covert operation, use your common sense."[12] In the view of the esteemed intelligence scholar Loch K. Johnson, this rejection of oversight measures marked a slide backward from the reforms that had been achieved to curb agency excess, but Casey was unrepentant.[13] "This is a rough business," he had argued during an NSC meeting. "If we're afraid to hit terrorists because somebody's going to yell 'assassination,' it'll never stop. The terrorists will own the world. They'll know nobody's going to raise a finger against them."[14] Clarridge had already proven he was not afraid of somebody yelling assassination.

When the director told Clarridge he wanted to put the CIA on the offensive in a global campaign against terrorist groups, Clarridge told Casey what the director already believed: to succeed, the CIA had to attack the terrorist cells preemptively. If not, Clarridge said, "the incidents would become bolder, bloodier, and more numerous." This is exactly what Casey wanted to hear, and Clarridge was given the task of writing up a proposal for a new covert CIA counterterrorist strategy.[15] By late January, Clarridge had his first draft for the creation of what would become the CTC. The eight-page blueprint was accepted by Casey without a single edit, and the CTC, which would go on to become the command center for the modern drone campaign, was born on February 1, 1986, with Clarridge appointed as its first director.[16]

Reviewing Clarridge's line of argument for the creation of the CTC helps to give a clear picture of how the center functions today, and reveals just how influential Clarridge was in shaping the development of the agency's drone campaign. The new director argued that the CIA had four main problems in confronting the international terrorist threat. The most significant of these was what he referred to as the agency's "defensive mentality." Terrorists were able to operate from safe havens around the world "knowing there was little chance of retribution or of their being brought to justice."[17] Terrorism, he explained in his blueprint, was effective even when the targeted nation just tried to contain it, as doing so provided time and space for would-be attackers to recruit and plan. Instead the CTC would need to be established on an offensive footing, building upon the preemptive self-defense argument made by the hardliners in NSDD 138. "We thought it was time to go on the offensive against terrorism," said Frederick Turco, Clarridge's second in the European division, who would join him as deputy at the CTC.[18] Clarridge called for new legal parameters for the CIA that would enable it to undertake offensive strikes against terrorist groups worldwide, depriving them of safe havens. With the support of Casey and Robert Gates—then head of the Directorate of Intelligence (DI)—Reagan signed a highly classified Finding granting the CIA the authority to do so.[19] This Finding was accompanied by a new NSDD, based partly upon Clarridge's recommendations, and also a review of counterterrorism practices undertaken by Vice President Bush. NSDD 207—entitled "The National Program for Combatting Terrorism"—was signed on January 20, 1986.[20]

NSDD 207 did not introduce any particularly new measures to what was already set out in NSDD 138, but it did serve to reinforce some of the policy positions and tighten up procedures. It recognized terrorism as "a potential threat to our national security" and acknowledged the growing support for the preemptive self-defense argument, stating: "Whenever we [the United States] have evidence that a state is mounting or intends to conduct an act of terrorism against us, we have a responsibility to take measures to protect our citizens, property and interests."[21] To lead this new offensive, Clarridge proposed the formation of two secret action teams that would track terrorists globally. The action teams would have the authority to kill terrorists, if doing so would preempt a terrorist event, or arrest them to bring them to justice if possible. The plan called for one team to be made up of foreign nationals so that they could blend more easily into hostile environments overseas, and the other to be made up of Americans.[22] These policies on preemptive strikes can be seen at the heart of the CTC's drone campaign two decades later, as can the aim to deny a safe haven within which terrorists can operate.

Second, Clarridge argued that the CIA in its current form was organizationally incapable of dealing with international terrorism. Since its creation the agency had primarily focused upon the Soviet Union, and it was set up to deal with geopolitical Cold War issues and structured along regional geographic lines. Terrorism, argued Clarridge, was so effective precisely because it did not follow geographic lines. "An Arab terrorist group may be based in Libya or Syria, but its operations are likely to take place in Rome or London or Athens. Within the Clandestine Services, which division has jurisdiction—the Near East or Europe?" he asked rhetorically.[23] This was especially true with regard to a stateless terrorist group, in Clarridge's case Palestinian militants, and in the modern example, al-Qaeda.[24] The CIA was not structured in a way that enabled it to pool knowledge and resources to deal with a transnational problem like international terrorism. Clarridge's solution was for the CTC to serve as a "fusion center," which combined the resources of the different agency directorates and broke down the traditional divides.[25] It would be located within the DO to reflect its action-orientated purpose but would also include analysts from the DI and engineers from the Directorate of Science and Technology (DST). Clarridge defined this as "nothing less than a revolution" within the CIA, and as being "without precedent."[26] The plan met with considerable resistance, especially from some members of the DO who reportedly saw counterterrorism as police work best left to the FBI, but Casey's unflinching support for Clarridge was enough to ensure the changes were made.[27]

The fusion concept went beyond the different directorates of the CIA and reached to the DoD. Clarridge requested that the entire antiterrorist unit of the CIA's Special Operations Group (SOG) be moved to the CTC to enable better coordination of covert and military action. The SOG comprised the CIA's paramilitary specialists, who regularly worked with JSOC: Delta Force and the Navy SEALs.[28] Casey agreed, and the formal link between CIA counterterrorism operations and JSOC was established. Even though cooperation between the agency and JSOC was patchy at times, the formal connection between the two reveals that the concept of a militarized CIA had existed well before September 11, 2001.[29]

It serves as evidence of just how influential Clarridge's fusion center concept was that when the United States sought a more proactive strategy against al-Qaeda the CTC grew sevenfold, from three hundred staffers in

2001 to more than two thousand in 2011. The center now accounts for more than 10 percent of the agency's workforce and has been described by the *Washington Post*'s Greg Miller as "an agency within an agency."[30] Most of these additional staff members are focused upon the CIA's drone campaign, which fully embodies the global focus Clarridge called for the CTC to have. At Langley, the CTC staffers are divided into subunits focused upon specific areas, for example, the Pakistan/Afghanistan Department, known as PAD, and equivalent units for Yemen, Somalia, and other potential terrorist hotspots. Nick Turse and Tom Engelhardt claim their research has uncovered at least sixty CTC drone bases around the world, mostly concentrated around North and East Africa, the Middle East, and Central Asia.[31] In addition, the *New York Times* broke an informal agreement with the CIA in February 2013 and revealed that the agency even had a drone base in the most sensitive of locations—Saudi Arabia. This global network ensures that the CTC can, as Clarridge had originally demanded, counter the transnational nature of nonstate terrorism.[32]

NSDD 207 sought to create a similar level of fusion and cooperation on a governmental level. While identifying the CIA as the lead agency in planning and executing counterterrorism operations, the directive sought to ensure the other departments would work together too. "The entire range of diplomatic, economic, legal, military, paramilitary, covert action and informational assets at our disposal must be brought to bear against terrorism," the document stated.[33] In making this declaration, the Reagan administration was providing an answer of sorts to a question that had been undermining its counterterrorism efforts since it had come to power. Was terrorism a law enforcement problem or a national security issue? Should the CIA try to capture terrorists alive in order to try them on criminal charges in open courts, or should the goal be to kill them? NSDD 207 came down on both sides—in some cases terrorism was a legal matter, in others it was an act of war. Terrorists should be captured for trial if possible, but that would not always be a requirement. Though NSDD 207 did not provide clarity, it at least acknowledged that the situation was complex and that agencies and departments would need to work together on a case-by-case basis.

The third issue identified by Clarridge was what he described as the "inadequate analysis of data to support counterterrorist operations—and a corollary failure to centralize the available data."[34] The new CTC director believed that there was excellent analysis being undertaken in parts of the DI, and within the individual sections of the DO, but that there was no mechanism for bringing this all together. Counterterrorism, Clarridge explained in his blueprint, was "a business of minutiae—collecting bits and pieces of data on people, events, places."[35] The CTC was to act as this central point of collection and analysis, where the puzzle pieces that identified terrorist plots, leaders, and members could be assembled. Echoing Clarridge's observation on the importance of intelligence analysis twenty-five years on, an anonymous CTC official interviewed in relation to the drone campaign told the *Washington Post:* "The kinetic piece of any counterterror strike is the last 20 seconds of an enormously long chain of collection and analysis. Traditional elements of espionage and analysis have not been lost at the agency. On the contrary. The CT effort is largely an intelligence game. It's about finding a target . . . the finish piece is the easy part."[36]

In order to provide the analysis that informs the drone strikes, approximately 20 percent of the staff in the agency's analytical branch now work as "targeters," scanning data for individuals to recruit, arrest, or, most important, target for drone strikes.[37] While the final missile strike from a drone delivers the preemptive neutralization of terrorists that Casey and the other counterterror hardliners pushed for in NSDD 138, it is the existence of the CTC and its careful attention to intelligence gathering that Clarridge instilled which forms the keystone of the campaign.

The final area Clarridge's blueprint identified as needing reform was the technological support from the Directorate of Science and Technology. The DST usually worked on lengthy development cycles. High-tech projects such as the U-2, SR-71 Blackbird, and Corona spy satellite system were each in development for five years or more.[38] Though this was still an impressive pace given the experimental nature of so much of the DST's work, fighting terrorism, Clarridge argued, required a much more straightforward and rapid response. The CTC director called for what he termed the "Radio Shack approach to research and development—taking pieces of existing things off the shelf and putting them together to deal with a particular situation in no more than a year." Clarridge wanted "simple, cheap, low-tech stuff that worked even in parts of the world where electricity and flush toilets were luxuries."[39] Jeffrey T. Richelson has argued that the history of the DST is a key element in the history of both the CIA and the entire intelligence community, and it has too often been overlooked. This is undoubtedly true, though ironically it is arguably these low-tech solutions to the CTC's problems that signify the DST's most significant contribution to CIA history in the past decade, in that they led to the CIA's development and deployment of the drones that serve as the primary tool of its counterterrorism efforts today.[40]

The CTC's First Challenges: Libya and Gaddafi

Examination of the CIA's electronic archive reveals that by far the most persistent topic discussed by the terrorism experts transferred to the newly formed CTC within its first months of operation was Libya's state-sponsored terrorism. A CIA review of Libya's terrorist capabilities described the state's leader Mu'ammar Gaddafi as being "committed to terrorism as an instrument of policy." Furthermore, the review estimated that his influence and reach were growing as he expanded his network of terrorist agents through sponsorship.[41] This was demonstrated by growing evidence of Libya's hand in a number of international terrorist incidents throughout the 1980s. In a speech delivered on July 8, 1985, Reagan told the assembled audience that the United States had "evidence which links Libyan agents or surrogates to at least 25 incidents last year."[42] Despite this, the American response to this growing terrorist threat had been as stuttering and limited as the rest of the administration's counterterrorism policy, for the reasons already discussed. In June 1985, McFarlane, Poindexter, and North at the NSC had tried to force the issue through the creation of Operation Flower, split into two components: Tulip and Rose. Tulip was a covert operation that was to be run by the CIA aimed at overthrowing Gaddafi with the aid of Libyan exiles. Pre-CTC, however, there was little support within the agency for the plan, which was a product of the NSC, not the DO. The second element, Rose, was a proposed preemptive military strike against Libya, planned to be conducted in unison with the Egyptian military. Weinberger, his deputy, Richard Armitage, and the Joint Chiefs believed the plan was ridiculous, and the secretary of defense ensured its preparation was so complex and resource intensive that it was never a viable option. What is more, the Egyptian president Hosni Mubarak had no interest in attacking Libya on America's command.[43]

Weinberger's hand was eventually forced in the early hours of April 5, 1986, when Libyan agents bombed *La Belle* nightclub in West Berlin, a venue popular with American service personnel. Two Americans and a Turkish woman were killed, and a further 229 people were injured, including seventy-eight U.S. military personnel.[44] Libyan cables intercepted by the NSA and Britain's Government Communications Headquarters (GCHQ) provided "the smoking gun" that proved Gaddafi's responsibility for the attack.[45] Reagan, who despite his combative rhetoric had to that point not authorized a retaliation against a terrorist attack, decided it was a matter of self-defense to respond. He later wrote, "I felt we must show Qaddafi that there was a price he would have to pay for that kind of behavior, that we wouldn't let him get away with it."[46] With such a blatant attack upon the U.S. military and clear evidence identifying the culprit, the secretary of defense had little choice but to back the use of military force in retaliation against a terrorist attack.

The reprisal came in the form of a large bombing raid code-named Operation El Dorado Canyon, launched on April 14, 1986. While the counterterrorism hardliners celebrated Reagan's decision to use military force, others did not: the NSC's Howard Teicher referred to the raid as a "disproportionate response."[47] There is still debate regarding the utility of the attack. Politically, the raid achieved its objective. Gaddafi and the governments of Iran and Syria were put on notice that they could no longer attack the United States via state-sponsored terrorism with impunity. Reagan reinforced this point with an address from the Oval Office on the night of the raid, addressed as much to Gaddafi and other state sponsors of anti-American terrorism as it was the American people. "Despite our repeated warnings" Reagan explained, "Qadhafi continued his reckless policy of intimidation, his relentless pursuit of terror." The Libyan leader had counted upon America remaining passive, explained the president, but he had miscalculated the country's resolve to act. "I warned that there should be no place on Earth where terrorists can rest and train and practice their deadly skills," stressed the president, "I said that we would act with others, if possible, and alone if necessary to ensure that terrorists have no sanctuary anywhere . . . I meant it."[48] Reagan sought to make it clear to

America's enemies that while his administration may have been slow to act upon his strong antiterrorist rhetoric, this was a sign of restraint, not weakness, and that his administration was perfectly willing to use force if it deemed it necessary.

Though the bombing raid did not end the threat of state terrorism, a detailed case study of its impact more than ten years later suggests that it certainly reduced the frequency of terror acts. Furthermore, the study argues that Gaddafi's reputation was severely damaged. The weaknesses of his regime were exposed, and his international isolation was highlighted.[49] The leaders of Syria and Iran had to think twice before risking the same happening to them. In addition, the escalation of eye-for-an-eye violence, which critics of military retaliation had warned would result, did not materialize. In fact, the opposite occurred: Brian Davis has argued that rather than provoking escalation, the U.S. attack may have helped break the cycle of accelerating Middle Eastern terrorism dating from 1983.[50] Domestically, as with the EgyptAir incident, the perception of Reagan as being willing to stand up to terrorism played very well with the American public. Ratings from a *New York Times*/CBS poll taken the day after the raid showed that 77 percent of Americans questioned approved of the action.[51]

Militarily, however, the raid can be seen in a very different light. America's retaliation may have forced Gaddafi to be more covert and less publicly belligerent with his sponsorship of terrorism, but his employment of it as a tool of foreign policy was not stopped entirely. The actual damage to Libya's terrorism infrastructure was relatively modest.[52] The CTC concluded that Gaddafi would consistently return to terrorism as a primary tool for achieving his goals.[53] The 1988 bombing of Pan Am flight 103 over Lockerbie, Scotland, which resulted in the death of all 259 passengers and eleven people on the ground served as confirmation of the point. Gaddafi remained a threat because the raid had failed to hit what has since been reported to have been its primary target—Gaddafi himself. After three months of careful investigation and interviews with more than seventy serving and former officials from the White House, DoS, CIA, NSA, and Pentagon, Seymour Hersh published an article in the *New York Times* that clearly set out evidence that "a small group of military and civilian officials in the NSC," including Poindexter (then serving as Reagan's APNSA),

North, and "several senior CIA officials" had sought to ensure the raid would specifically target Gaddafi's personal quarters, without this information being known to the public, the military, or even those flying the mission.[54] It is unclear whether Clarridge was one of the senior CIA officials, but as the newly appointed director of the CTC and a close confidant of Casey it is highly likely he would have been involved in the discussions regarding retaliation. If so, his views regarding the need to take the offensive against terrorists and their sponsors, together with his willingness to bend the law to achieve his goals, would suggest his involvement.

Gaddafi may have been a proven sponsor of terrorism, but he was still a head of state, which therefore put him outside the legal criteria for CIA counsel Sporkin's preemptive self-defense argument. This made the deliberate targeting of Gaddafi's home an illegal assassination attempt, in breach of both international law and America's own EO 12333. There is no suggestion that Reagan himself was involved in the plotting, though he did help select the target sites. Gaddafi's residence, built in the grounds of the Al Azziziyah barracks, was described in the targeting information as "the command, control and communications center for Libya's terrorist-related activities," as opposed to the dictator's home.[55] A U.S. Air Force officer described Gaddafi's survival to Hersh as "a fluke." Post-strike photography showed that the bombs dropped by the F-111s left a line of craters right past both his two-story house and his Bedouin tent. Another air force officer quoted by Hersh blamed the failure to kill Gaddafi on equipment failure. The high-tech laser-guidance system on four of the nine F-111s attacking Gaddafi's quarters malfunctioned. The Rules of Engagement, written to try to minimize the chances of ordnance going off target and causing civilian casualties, stated that if any aircraft were not 100 percent functional then they were to withdraw from the mission, forcing the pilots to abort the attack before they could drop their ordnance. With four two-thousand-pound bombs per F-111, this eliminated sixteen more bombs that should have been dropped on the quarters. "The very high-technology system that was meant to insure Qadhafi's death," mused Hersh, "may have spared his life."[56]

Though the bombs may have missed Gaddafi, unfortunately the same could not be said for a number of civilians around the target sites. Reportedly, all eight of the dictator's children as well as his wife Safiya were hospitalized, suffering from shock and various injuries. In addition, it was claimed that his fifteen-month-old adopted daughter, Hana, died several hours after the raid, though Weinberger later contended that the claim was completely unverified and based upon "highly suspect Libyan reports." Later reports supported the secretary of defense's cynicism, suggesting Hana was alive and well, working as a doctor within Libya's health ministry.[57] Regardless of the fate of Gaddafi's daughter, overall the raid was less surgical than Reagan had hoped. Nearly five tons of explosives had landed in residential neighborhoods by mistake and thirty-seven civilians were reported killed, with approximately a hundred more injured. The French embassy in Tripoli was also damaged by stray munitions, and two U.S. aviators were killed when their F-111 was shot down.[58]

Despite the general approval from the American public for the operation, the Reagan administration did face a backlash from its European allies and a minority of Americans over the unintended human costs, which, in the words of Mark Kosnik, "become a major part of later criticisms of Reagan's decision to use armed force against the Libyan regime."[59] Former attorney general Ramsey Clark heavily criticized the Reagan administration in an article in the *Nation*, pointing out that Reagan's "surgical strike" had "killed at least twice as many Libyans in one night as all Americans killed by terrorists world-wide in 1985." "Unless it is lawful for the President to use military bombers in an attempt to assassinate a foreign leader and to kill and mutilate scores of human beings sleeping innocently in their homes," Clark concluded, "Ronald Reagan must be impeached and tried for high crimes and misdemeanors."[60] Like Reagan, Obama too faced similar criticism for the collateral damage caused by CIA drone strikes against suspected terrorists during his administration. Echoing a similar sentiment to Clark, Medea Benjamin, cofounder of the anti-war NGO Code Pink, publicly stated that she regarded the authorizing of the drone campaign to be an impeachable offense and believed Obama should have had to answer to the International Criminal Court (ICC) for "war crimes committed via his authorization" against "blameless people in foreign countries, against which the U.S. is not at war."[61]

As both examples show, ensuring any strikes against terrorists are proportional and precise and do not entail extensive collateral damage is vital, both to meet the requirements of jus ad bellum, and to maintain public support for such action. The counterterrorism hardliners did show an awareness of this. Decades before the CTC's decision to use local agents on the ground to guide drone strikes against al-Qaeda and the Taliban, North, Poindexter, and a small group of special-operations planners at the Pentagon proposed the insertion of Navy SEALs into Tripoli to place homing devices on Gaddafi's headquarters. These devices would then guide smart bombs precisely to their target, hitting Gaddafi with minimal risk of collateral damage. The plan was rejected, however, by Admiral William Crowe, chairman of the Joint Chiefs, as too risky for the SEAL team.[62] North also told colleagues that he proposed a number of other alternatives to Crowe, including the use of Tomahawk cruise missiles launched from submarines off the coast of Libya, but was told that there were too few conventionally armed variants, and that they lacked the necessary accuracy to be used in a built-up environment at that stage, having been configured to deliver nuclear warheads—which required considerably less precision.[63] Once more North's rejected proposal anticipated what eventually became official counterterrorism policy; following later reconfiguration to deliver conventional warheads with greater precision, cruise missiles became the primary tool adopted by Bill Clinton administration in its efforts to target bin Laden and al-Qaeda members in Afghanistan.[64]

Clarridge was acutely aware that if the CIA was to take an offensive footing against terrorists, it would need to do so with more precise tools than bombers and a carrier battle group. "We have to find a better way to send a message to outlaw nations," he told Casey. The counterterrorism chief criticized what he regarded as the hypocrisy of the government's ban on assassination, asking the DCI why "an expensive military raid with heavy collateral damage to our allies and to innocent children" was "more morally acceptable than a bullet to the head."[65] Clarridge believed the DST staff attached to the CTC had the answer to the problem. Rather than the high-tech, high-risk strategies North and the NSC had pitched, Clarridge's proposal reflected the Radio Shack approach he had encouraged the DST staff to adopt. Due to the classified nature of the material, Clarridge is unusually reserved in his memoir about what exactly this solution was, but he does mention that "one of the technical geniuses in the Center thought he could develop a system that would send a clear message, with poignant effect, but with minimal loss of life for the recipients, and none for our delivery personnel."[66] Clarridge provides no more details about what this system actually was, other than to mention that Casey was fascinated with the idea, and that, after securing his approval, the CTC had a working model within a year. In an interview with the former journalist Steve Coll shortly after 9/11, however, Clarridge did elaborate on something dubbed the "Eagle Program."

The CIA will neither confirm nor deny the existence of the Eagle Program, but the dates Clarridge provides in his account of this program match those of the system he mentions in his memoirs, indicating they are one and the same. The Eagle Program, Clarridge told Coll, was a highly classified pilotless drone equipped with intercept equipment, an infrared camera, and low-noise wooden propellers. Clarridge's plan was to equip the drones with small rockets that could be fired at predesignated targets to take them out in a surgically precise way, avoiding the potential collateral damage of a full-scale bombing raid while keeping American assets safe from harm. The CTC was even experimenting with loading the drones with two hundred pounds of C-4 plastic explosives and a hundred ball bearings with the aim of flying them into sensitive areas such as Tripoli's airport at night and blowing them up to sabotage aircraft, though the lethal implications of such a device would be significant, too.[67] The development was to have far-reaching implications for the CIA, as there is little doubt from his description that what Clarridge and the CTC had developed was the precursor to the Predator drones and their Reaper successors the center now uses to combat al-Qaeda.

The CTC's First Challenges: The Lebanon Hostage Crisis

The experimental drones of the Eagle Program did not have only a lethal purpose. Just like the CIA's current fleet of drones, Clarridge planned to use them for reconnaissance purposes and intelligence gathering, in this instance to help locate the Western hostages in Lebanon. Since the summer of 1982, Hezbollah had been systematically kidnapping Westerners, among them numerous American citizens, including William Buckley, the local CIA station chief. Retrieving these hostages had become something of an obsession for the White House. On December 20, 1985, Reagan's new

APNSA, Poindexter, had ordered the creation of a Hostage Locating Task Force, bringing together staff from the CIA, the Pentagon's DIA, NSA, and JSOC.[68] The CIA was the lead agency in this task force, which was to report its progress on a weekly basis. Clarridge's plan was to attempt to penetrate Hezbollah and to gather enough intelligence to mount a hostage rescue mission undertaken by U.S. Special Forces. The Pentagon's generals, however, were still scarred from the catastrophic hostage rescue attempt of Operation Eagle Claw during the Carter administration, and they did not like the plan.[69] Citing what they saw as weak intelligence regarding the actual location of the hostages, the military commanders refused to deploy their troops based upon information supplied by local agents. They insisted that there would need to be American eyes on the target confirming the location of the hostages twenty-four hours before any rescue operation would be launched.[70]

Getting accurate, up-to-date intelligence on Hezbollah from an American source was exceptionally difficult. Beirut was a dangerous place for U.S. citizens to operate in, and a terrorist group like Hezbollah is notoriously paranoid about spies and foreign agents. As pressure to do something about the hostages grew after months without any results, Casey used a speech before the Jewish American Committee to address the issue of gaining human intelligence from a terrorist group somewhere like Beirut: "In order to prevent terrorist plans or disrupt their activities, we need information about them. But the very nature of terrorist groups and their activities makes this task extremely complicated. Terrorist groups are very small, making penetration a very difficult task for police or intelligence agents. Moreover, the operating life of any single group of terrorists is often no more than a few years. Likewise, typical terrorist leaders have a relatively short business life."[71] Clarridge hoped that the CTC's new drones might be able to fly twenty-five hundred feet over Beirut and locate the American hostages through careful monitoring of the movement and habits of Hezbollah militants. If a CTC agent operating the drone from a safe distance could get the American hostages on camera, the generals would have their American eyes on the target without the risk of exposing more CIA agents to kidnapping.[72]

This use of unmanned drones for gathering intelligence on terrorists' locations would be the exact role the RQ-1A Predator, the initial reconnaissance version of the later armed MQ-1 Predator, would come to play in Afghanistan in a joint CIA-DoD effort to locate Osama bin Laden. As with Clarridge in Beirut, the problem facing the CTC in its efforts to hunt down bin Laden following his declaration of jihad and subsequent bombing of U.S. embassies in Kenya and Tanzania in 1998 was locating the al-Qaeda leadership in Afghanistan long enough to enable the United States to act.[73] The operation, dubbed "Afghan Eyes," was championed by Richard Clarke, Clinton's national coordinator for counterterrorism, and involved a sixty-day trial of Predators flying over Afghanistan. The first flight took place on September 7, 2000. Upon seeing the video captured by the drone, Clarke described the imagery to Sandy Berger, Clinton's national security adviser, as "truly astonishing."[74] Cofer Black, the head of the CTC, and Charles Allen, assistant director of the CIA's intelligence-collection operations, were also enthusiastic about what they saw. A total of fifteen missions were flown over Afghanistan, ten of which were judged successful. On the Predator's first flight, the drone filmed a tall man in white robes matching bin Laden's description surrounded by a security detail on his Tarnak Farm compound. Footage from a second sighting of the man in white, taken on September 28, 2000, was later judged by analysts from the intelligence community probably to have been of bin Laden.[75] Had the drone been armed, consistent with Clarridge's initial vision for unmanned aircraft, bin Laden could have been neutralized that day. As it was, it would not be until October 2001 that the CTC flew an unmanned drone over Afghanistan equipped with missiles.

Origins of the Predator Drone

Abraham Karem's Amber and GNAT-750

Ironically, Clarke claims to have initially struggled to get the CTC to adopt the use of the Predator drone for locating bin Laden. The center was set on the use of human assets on the ground in Afghanistan, but these were proving unreliable. Drones, Clarke says, were seen as: "Too risky. Too costly. Too not-invented-here."[76] According to Clarridge's account, this could not have been further from the truth, with the very drones used to undertake the CTC's strikes being directly evolved from the Eagle

Program and invented within the CTC. In his memoirs, Clarridge describes how the system (the Eagle Program) saw five operation prototypes built at a cost of $8 million. Part of the funding for this project, he explains, came from Charles Hawkins, then assistant to secretary of defense for intelligence oversight, who saw the device's applicability to military problems.[77] According to Clarridge, the army looked into developing a "gold plated" version of the device, spending $900 million but eventually canceling the project some years later due to technical difficulties. Though Clarridge never names the army's project, the *Directory of U.S. Rockets and Missiles* lists the Amber UAV (unmanned aerial vehicle) as a system developed by the U.S. Army, with work starting in 1986—the same year Hawkins picked up on the Eagle Program—and being canceled in 1990. Like the Eagle, the Amber was a propeller-driven, remote-controlled aircraft equipped with a daylight television camera and a Forward-Looking Infrared system. In addition, as with the CTC's prototype, the Amber was designed to be lethal if necessary, with the nose section containing a warhead. When over its target, the Amber could jettison its wings and fall as a form of cheap, highly accurate cruise missile.[78]

The aviation historian Richard Whittle's exhaustive account of the origins of the Predator drone offers an alternative narrative to that put forward by Clarridge, interestingly eschewing any mention of the still classified Eagle Program, or Charles Hawkins's intervention. According to Whittle's findings, which are supported by the earlier investigations of the Smithsonian Institution's aerospace historian Curtis Peebles, initial development of the Amber was enabled by a $5 million grant from the Defense Advanced Research Project Agency (DARPA), awarded to the drone's creator, Abraham Karem, a gifted Israeli American engineer and owner of the California-based aeronautics company Leading Systems Inc. (LSI), in December 1984.[79]

DARPA was established following the Soviet Union's successful launching of the Sputnik satellite in 1957, a traumatic experience of technological surprise for both American policy makers and the public. The new DoD-linked department was charged with the mission of ensuring that "from that time forward, [the United States] would be the inheritor and not the victim of strategic technological surprises."[80] DARPA's employees were trained to identify revolutionary concepts that could have strategic benefits for the United States, and help develop them into practical capabilities. The agency's Bob Williams saw such potential in Karem's designs for high-endurance drones, believing the model could help meet the specific needs of the U.S. Navy, Marine Corps, and Army. Marine commanders, still reeling from the devastating attack upon their barracks in Lebanon the previous year, were determined to find a better way to conduct tactical reconnaissance to enable them to detect enemies before they were able to attack fixed U.S. positions; the navy sought a way to facilitate the gunners on battleships to identify targets for over-the-horizon fire; while the army had been working on its own drone, named Aquila (interestingly, Latin for "eagle"), to help direct laser-guided artillery shells—a decade-long effort that had cost a significant sum but failed to deliver acceptable results.[81]

Ascertaining how direct a role, if any, the Eagle Program played in the initial development of the Amber is rendered near impossible due to the still classified nature of the project. The CIA has invoked the National Security Act to decline Freedom of Information (FOI) requests related to the program, and the DoD responded to inquiries into documents relating to Charles Hawkins and the Eagle Program with a statement that the request had been "indefinitely delayed" due to "unusual circumstances" which included "the need for consultation with one or more other agencies or DoD components having a substantial interest in either the determination or the subject matter of the records."[82] Though this initial stage of the CIA's involvement in drone development is obscured by classification and secrecy, the next step of the drone's evolution is more transparent.

In October 1987, a report published by the United States General Accounting Office into the army's Aquila drone revealed that the UAV only fulfilled its mission parameters in seven of 105 test flights in a period from November 1986 to March 1987, leading the evaluators to question the viability of the drone.[83] The damming assessment caused a public scandal, with a withering critique of DoD procurement in the *New York Times* using the grossly over budget and underperforming project to illustrate what it described as "one of the persistent flaws that plague the military research and development process."[84] With the costs reportedly having tripled since the project's commencement in 1974, and delays putting development seven years behind schedule, Congress intervened: Aquila was scrapped; the following year's drone research budget was slashed from $103 million to $52.2 million; and the Pentagon was directed to consolidate all drone research and development under a single body, the new multi-service Unmanned Air Vehicle Joint Program Office (JPO), to avoid wasteful duplication.[85]

The austere new drone budget was devastating for LSI. The company responded by seeking new financial opportunities, beginning development on a smaller, less capable drone—the GNAT-750—which could get past the State Department's export licenses to generate orders from foreign militaries. To fund the project, Karem signed a credit agreement with the larger and wealthier defense contractor Hughes Aircraft Company in August 1989. But the modest international interest in the GNAT was not enough to sustain LSI, and when the JPO contract was eventually canceled on September 14, 1990, Karem was forced to file for bankruptcy, owing Hughes approximately $5 million. LSI's creditor foreclosed in late 1990, taking legal possession of all the company's physical infrastructure and, more important, Karem's intellectual property and patents.[86]

The Blue Brothers, General Atomics, and the Balkans

Having witnessed the collapse of the UAV market, and anxious about the impact of the looming $64 billion defense cuts the Democrat-controlled Congress had demanded George H. W. Bush authorize as part of the post–Cold War "peace dividend," Hughes sought to sell LSI's holdings as quickly as possible. For expediency's sake, the company settled on a cut-down price of $1.8 million with General Atomics (GA) in March 1991—approximately 10 percent the value of the physical property alone. The deal included six GNAT-750s, already in built-up stage, and new contracts for Karem and ten other LSI engineers and pilots, creating General Atomics Aeronautical Systems Inc. (GA-ASI).[87] At first glance, the intervention of GA, a company known more for its development of nuclear power reactors than unmanned aircraft, seemed nonsensical. But GA's owners, Neal and Linden Blue, already had a history with unmanned aircraft.

Following their graduation from Yale in the mid-1950s, the Colorado-born Blue brothers had traveled in Latin America, confident the region was ripe with entrepreneurial opportunities. They eventually established a ranch for cultivating cocoa and bananas on Nicaragua's Caribbean coast, but they were forced to relinquish it after being called up to serve in the U.S. Air Force in 1961. The brothers opted to return to their home town of Denver once their service was complete, where they followed in their parents' footsteps in establishing a successful real estate business, eventually diversifying their commercial portfolio to include construction, gas, and in 1986 the purchase of GA from Chevron for $50 million.[88] Ardent anti-Communists, the brother's maintained a keen interest in the political events of the Cold War and were incensed by the 1979 coup that saw the Soviet-backed Sandinistas seize control of the brothers' former place of business, Nicaragua. When the Reagan administration authorized the CIA to begin aiding the Contra rebels in their efforts to overthrow the Sandinista government, the Blues were, in Neal's words, "enthusiastic supporters," replete with what the elder Blue described as "top-secret clearance with the United States government."[89]

Whether their support for the Contra operation brought the Blues into contact with Casey, Clarridge, and North—the main proponents of the CIA's intervention—is unclear, but perfectly plausible. When later pressed by the *New York Times*'s Charles Duhigg to disclose if the brothers had worked for the CIA, Neal declined to discuss the matter.[90] Acting either under the auspices of the agency or on their own motivation, the Blues sought to use their recent acquisition of GA to intervene against the Communist Sandinista government, briefing their puzzled new employees that they wanted the company to develop GPS-equipped unmanned planes that could be launched from behind the line of sight on kamikaze missions to blow up the country's gasoline storage tanks, damaging the Nicaraguan economy while giving the Americans total deniability.[91] A prototype of the drone, named "Predator," was successfully built, but once the breaking of the Iran-Contra scandal forced the end of America's covert support for the Contras, its deployment proved a dead end for the Blues.[92]

Despite the fact that unmanned aircraft were out of favor in the wake of the Aquila calamity and end of the Cold War, GA's shrewd entrepreneurial owners were confident that the technological possibilities offered by drones were only just beginning to emerge; all they needed was a high-quality drone and a scenario in which a remotely piloted, high-endurance aircraft would be valuable. The purchase of LSI's holdings and subsequent assimilation of Karem's team delivered the first, while the tragic events that followed the post–Cold War breakup of the former Yugoslavia in 1992 provided the later. As decades of repressed hostility between the multiethnic peoples of the Balkans erupted into the worst violence seen in Europe since World War II, the leaders of the once dominant Yugoslav province Serbia, in league with ethnic Serb insurgents, sought to carve their own republic from Bosnian territory. To aid in their strategic objective of ending Bosnia as an independent entity, the Serb forces laid siege to the capital Sarajevo on April 5, 1992, blockading the city and shelling the citizenry with heavy artillery.[93] The chaos in the region, and the challenges it presented to U.S. military power, provided an ideal scenario for the Blues' newly purchased drone to make its debut.

George H. W. Bush, a realist with a strong sense of America's limitations, had maintained a pragmatic policy of indifference to the Balkan region's strife. Widely quoted, his secretary of state, James Baker, had summed up the administration's position with a blunt statement of inconsequence— "We don't have a dog in this fight."[94] But American attitudes, and with it perceptions of the United States' global responsibility, were changing. As Bill Clinton assumed office in January 1993, Washington's emerging post–Cold War policy consensus saw neoliberals and neoconservatives coalesce around a sense of the United States' moral obligation and strategic responsibility to use American military power to enforce international peace. The creation of a Pax Americana, mixing altruism and self-interest with a desire to help the global community, became regarded as a legitimate and achievable aim.[95] Thus, when the United Nations, unable to stop the Balkan carnage, formally requested NATO's military support, the recently elected Clinton, riding a wave of post–Cold War triumphalism, felt inclined to intervene.[96]

In order to demonstrate the United States' resolve as global peacekeeper, Clinton sought to use NATO's predominantly American airpower to break the siege of Sarajevo and force the Bosnian Serb Army to engage in peace negotiations.[97] But the president was shocked by the scarcity of information the Pentagon and his intelligence agencies could provide regarding events on the ground around Sarajevo. Weather conditions over Bosnia, a territory as large as Clinton's home state of Arkansas and as mountainous as Colorado, hampered American reconnaissance efforts, with the cameras mounted on America's high-tech network of spy satellites and older U-2 aircraft unable to penetrate the heavy cloud cover.[98] The problem was compounded by the Russian-manufactured 2K12 Kub (DoD designation SA-6) mobile surface-to-air launchers the Serbs possessed, which rendered low-level aerial scouting too dangerous (a USAF F-16 was shot down by the weapon system over hostile Serb territory on June 2, 1995, though the pilot was able to eject and evade Serb troops for almost a week before being rescued by U.S. Marines).[99] The Serbs also employed simple but effective countersurveillance measures, frequently changing their firing positions, hiding their weapons in barns, woodland, and dense foliage during the short periods each day they knew America's satellites would be orbiting overhead, and only moving their artillery during the cover of night.[100]

A 1999 study by the U.S. Naval War College's Department of Joint Military Operations identified that American forces had encountered very similar difficulties during Operation Desert Storm, two years before the Bosnian intervention. Despite waging arguably the most dominant military campaign of the twentieth century against Saddam Hussein's forces, the USAF and its coalition partners had struggled to locate and destroy Iraq's mobile Scud ballistic missile launchers. Saddam's military went to great lengths to protect the launchers, employing guillies, wadis, culverts, and highway underpasses to thwart aerial reconnaissance.[101] As a result of these simple deceptions, the Iraqi military, despite being soundly and swiftly defeated, was able to launch eighty-eight Scuds into the territory of America's regional allies—Israel, Bahrain, and Saudi Arabia. The strikes resulted in the deaths of twenty-eight American service personnel when their Saudi-based barracks were hit, the killing of two Israeli civilians, and the injuring of more than three hundred, along with extensive property damage.[102] Combined with Clarridge's unsuccessful pursuit of hostages in Lebanon, this marked three times in the preceding decade that militarily superior U.S. forces had been prevented from acting due to an inability to get American eyes on the target, exposing an attenuating gap in America's intelligence-collection capabilities.

As attention shifted to finding a solution to this intelligence gap, the Blues' shrewd investment in LSI, or more accurately in Karem, paid off.

Charged by the president with providing the military with the necessary intelligence to enable the destruction of the Serb heavy weapons, James Woolsey, Clinton's recently appointed DCI, made contact with the Israeli American engineer. Woolsey had first become aware of Karem's work with drones in 1981 when serving on the Townes Commission, a high-level group exploring how best to deploy America's new MX nuclear missiles (later named the LGM-118 Peacekeeper) in order to maintain the balance of mutually assured destruction in the event of a Soviet first strike.[103] One unconventional idea the commission had investigated, which won approval from both Reagan and Secretary of Defense Weinberger, was to use a large high-endurance drone, dubbed "Big Bird" by the media, to carry a ninety-megaton MX. The unmanned aircraft would stay aloft for nearly seven days at a time, and cycle to provide constant coverage.[104] Sensing a challenge to its bomber fleets, the air force rejected the concept, but Woolsey, impressed with the engineer's unorthodox problem solving, became an advocate for Karem. As a result, when a problem emerged that required a persistent aerial presence and real-time surveillance, the new DCI went to the innovator behind Big Bird, and GA-ASI's recently acquired GNAT-750 was selected for the mission.[105]

To meet the Bosnian mission requirements, the GNAT had to be enhanced with additional sensors. Reflecting the strict Pentagon budget (capped at $2.5 million per drone following the Aquila scandal), the pressing deadline, and legitimate concerns about the loss of the experimental aircraft through mechanical failure or to antiaircraft fire, the decision was made to appropriate a commercial, off-the-shelf sensor package, thus avoiding the risk of classified technology being lost.[106] The Pentagon planned to purchase two of the modified variants—known as GNAT-750-45s—but cumbersome DoD procurement procedures and the strict controls of the JPO meant that the drones could not be bought and deployed by the air force within the necessary timescale. Instead, the two GNAT-750-45s were transferred to the CIA due to the agency's greater budgetary discretion. In addition to its faster procurement schedule, the CIA brought its own expertise to the customization of the GNAT. According to former CIA technician Frank Strickland, the CIA's DST had been conducting its own research on remote piloting since the 1980s, quite possibly a reference to the Eagle Program. The agency's engineers promptly

merged their expertise with that of Karem's staff.[107] Attesting to the significant contribution of the agency's engineers, Karem later noted: "We
certainly would not have a Predator today without a great team from the
CIA." Revealing the Radio Shack mentality Clarridge had championed,
Karem also praised the CIA's willingness to take risks with rapidly modi-
fying and deploying the GNAT 750 system, as well as its readiness to
experiment with a reasonably complex new technology.[108]

Even with their combined knowledge, the drone's engineers were
working on a cutting-edge concept, and mistakes were inevitable. One
GNAT was lost in an accident, and technical delays meant the remaining
drone was not deployed until early 1994. In February of that year, a small
team of CIA and GA-ASI personnel began the first drone operations—
code-named Lofty View—in the skies above Sarajevo from the safety of
the semiderelict Gjader military air base in western Albania, 140 miles away
from the Serbian forces.[109] Initial results were mixed, with the GNAT's
range and endurance undermined by its telephony, which relied upon a
C-band radio frequency. This required a virtual line of sight between the
UAV and its relay station, with a limited range of 150 nautical miles.[110]
Interference from a nearby mountain range made a direct data link impossible, forcing the team to relay its command signals through a manned
aircraft—a small Schweizer RG-8 motor glider—which reduced operational
flights to just two hours due to the glider's six-hour round trip to the
Adriatic Sea to relay the signal. The video feed was also frequently degraded
by inclement weather, electronic interference, and the distance the relatively
weak data signal had to travel.[111] Despite these setbacks, the mission was
regarded as a successful proof of concept. The CIA's GNAT, along with
an additional one leased from GA-ASI, was refitted with improved sensors
and redeployed to the Balkans later in 1994.[112]

Acknowledging the need to upgrade the GNAT before the results of its
first deployment were even in, DARPA awarded GA-ASI a contract in
January 1994 for the production of a drone that would fulfill what it referred
to as "Tier II MAE." MAE stood for medium altitude endurance, classed
by DARPA as operating in an altitude window of ten thousand to thirty
thousand feet, for a duration of between twenty-four to forty-eight hours.[113]
More pressingly, based upon the lessons of Lofty View, the new drone
would need to be able to operate over the horizon, beyond line of sight.[114]

Reviving the name of the Blue brothers' first drone, GA's Predator was completed by June 1994, meeting DARPA's deadline and demanding criteria. Derived from the GNAT, the Predator's larger frame meant that it could loiter at twenty-six thousand feet for between twenty-four and forty hours within a range of four hundred nautical miles. But its most transformative feature was the addition of a Ku-band SATCOM link, housed in a bulbous dome at the top of the aircraft.[115] This link meant the drone could—in theory at least—be operated by a pilot in a different region or continent, and ensured its sensor operator and any linked displays received a dramatically improved real-time feed. Even though these improvements came at an increased cost, GA's new drone still reflected much better value for money than previous UAV programs, with four aircraft, a ground control station, and primary satellite link costing $20 million.[116]

The air force's first four Predators were deployed to the Balkans under Operation Nomad Vigil in 1995, operating from a base in Albania to provide intelligence on Serbian installations. Despite the improvements, the first Predators still had major limitations. They were not equipped with radar systems that enabled them to see through cloud cover. To compensate, operators flew the Predators low beneath the clouds, making them easier targets and resulting in the loss of one to Bosnian Serb antiaircraft fire. There were still frequent technical problems too: a second Predator was deliberately crashed to prevent its capture when it developed engine problems and lost power over Bosnia; the UAVs fragile wings were prone to freezing up, causing a loss of control of the aircraft's ailerons and flaps; and despite the addition of the satellite receiver, difficulties with the data signal meant the first Predators still lacked the facility to be controlled from any distance via satellites, causing the pilots, sensor operators, and intelligence analysts to cram into huts beside the runway, watching a camera feed that only transmitted every third or fourth second.[117] Despite these limitations, the Predator's ability to remain on station for up to twenty-four hours still provided an invaluable intelligence capability, evidenced by the 159 missions and 1,169 flight hours logged in its first year of service.[118]

A year later, in March 1996, Predators were redeployed over Bosnia as part of Operation Deliberate Force, an intense NATO airpower campaign against the Serb forces, this time operating out of the Taszar airfield in

Hungary.[119] So great was the demand for aerial surveillance during the bombing campaign that both the Predator and its predecessor, the GNAT-750-45, were used simultaneously.[120] Following the end of the Predator's second tour, the of head of the Pentagon's Defense Airborne Reconnaissance Office (DARO), Air Force Major General Kenneth Israel, summed up the drone's impact by noting: "For years warfighters have articulated the needs for situational awareness, target identification, dominant battlefield awareness, dominant battlespace knowledge, and information superiority." The introduction of the Predator, the general reflected, had enabled the United States "to move from words to deeds."[121] With the CIA, the air force, and DARPA all satisfied with the overall performance of the Predator, a full contract for production of the drone, designated RQ-1 ("R" designating reconnaissance, "Q" signifying remote piloted), was awarded to GA-ASI in August 1997.[122]

Big Safari and WILD Predators

In spite of the CIA's early involvement with the GNAT and the vital contribution of its engineers in the initial development of the Predator, its role in the drone's evolution was sidelined on June 18, 1997. Impressed by the potential of the aircraft but concerned about the wasteful UAV spending of the previous decade, the House Intelligence Committee ruled that all functions related to the Predator Unmanned Aerial Vehicle be transferred to the air force, which assumed control on October 1 that year.[123] Official responsibility for the project was passed to the air force's 645th Aeronautical Systems Group, otherwise known as Big Safari, a small, elite unit staffed by specialized engineers and technicians whose mission is to exploit existing technology and produce high-tech solutions to defense challenges, typically by working in unison with its contractor partners.[124] It was through this close cooperation with GA-ASI and Raytheon—the manufacturer of the Predator's sensor ball and camera (and interestingly, the company that bought out LSI's original creditor, Hughes Aircraft Company)—that arguably the most important upgrade to the Predator system was introduced, one which brought it back full circle to the CTC's initial Eagle concept—the arming of the drone with air-to-ground missiles.

Contrary to later accounts, the decision to arm the Predator had nothing to do with the CIA's covert operations against Osama bin Laden and al-Qaeda that were taking place at the time, though the Predator's eventual role in locating the terrorist leader was a decisive factor in the decision to accelerate the process.[125] Instead, once more the impetus behind the development was NATO's ongoing campaign in the Balkans. On March 1999 Predators were deployed over Kosovo to fly surveillance missions for Operation Allied Force, a NATO bombing campaign against the still rebellious Serb forces. The drone's satellite feed had been significantly improved, allowing the images from the Predator, which could now be controlled from thousands of miles away, to be beamed live to monitors in NATO's southern headquarters in Naples, U.S. European Command in Germany, offices in the Pentagon, and even the White House Situation Room.[126] In theory the improved imagery meant the Predator's camera could zoom in on objects miles away, making it easier to validate targets and reduce the chance of collateral damage from strikes. But in reality U.S. air commanders struggled to capitalize on this improved reconnaissance capability. The formidable Serbian antiaircraft capabilities forced NATO aircraft to fly at altitudes of at least fifteen thousand feet (even with this precaution, an American F-117 stealth fighter was shot down on March 27, 1999, four days into the campaign, though its pilot was able to eject safely and was rescued).[127] At such altitudes, and hampered by the ever-present fog and cloud cover, NATO pilots struggled to identify their targets, while their controllers on the ground found it nearly impossible to talk pilots onto the targets they could see on their live feeds from the Predator's camera.[128]

Big Safari's technicians were ordered to fit a laser designator to the Predator's camera in order to guide NATO aircraft's ordnance accurately to its target.[129] The idea itself was not new—the army's canceled Acquila drone was meant to have operated a similar system to guide smart artillery shells to their target. Big Safari's engineers were able to make the concept functional, however. On June 2, 1999, in what was more a test than an act of war, a WILD (Wartime Integrated Laser Designator) Predator, as the Big Safari team had dubbed it, used its laser designator to successfully guide a five-hundred-pound bomb from a USAF A-10 Warthog onto a shed in the Kosovo countryside (there is some debate about whether there was a Serb tank present in the shed).[130] The WILD Predator did not play a part in the air campaign itself; Serb forces surrendered shortly after the successful test, agreeing to peace terms the following day. Once more, however, the Balkans had served as an ideal proving ground for the evolving concept of drone warfare.

Impressed with what he had seen the Predator do during his command of the Kosovo air war, the recently appointed chief of the air force's Air Combat Command (ACC), Commander General John P. Jumper, decided to act upon a suggestion first put to him by Major General Michael C. Kostelnikm, head of the Air Armament Center, and his deputy, Brigadier General Kevin Sullivan, on March 15, 2000, that the Predator could serve as a perfect platform for laser-guided smart bombs. On May 1, 2000, Jumper sent an announcement to the air force chief of staff, the secretary of the air force, and other top service leaders, reporting that, having "internalized the Predator lessons from Operation Allied Force," the ACC was "moving out on the next logical step for USAF UAVs using Predator— weaponizing UAVs." "All I wanted to do was be able to cure the problem that we had in Kosovo," Jumper recalled a decade later, "and that is, the Predator is sitting there looking at the target. Why can't you put something on there that allows you to do something about it, instead of just looking at it?"[131] Within a month, Big Safari's engineers had identified a missile they believed suitable for launch from the drone's flimsy wings—the army's AGM-114 Heliborne-Launched Fire-and-Forget Missile—more commonly by its acronym, Hellfire. The laser-guided weapon, which had first been fired in the Gulf War nine years earlier, had been designed to be launched from helicopters against tanks, ensuring it had the penetrative force and explosive power to destroy armored vehicles, while also, vitally for the lightweight Predator, weighing just ninety-eight pounds. Big Safari set about trying to adapt the Predator and missile to meet Jumper's ambitious goal.[132]

Osama bin Laden, Afghan Eyes, and 9/11

Initially unaware of the air force plans to weaponize the Predator, the CIA's interest in the drone was rekindled in early 2000 when Admiral Scott Fry, the director of operations for the military's joint staff, recommended its utility in locating bin Laden to Charles Allen, the assistant director of central intelligence for collection.[133] Fry had become frustrated at what he regarded as the expensive waste of resources of having a U.S. submarine indefinitely stationed in the Indian Ocean in the hope that an opportunity

emerged to strike bin Laden with cruise missiles, and believed the introduction of the drone could be a game changer in producing actionable
intelligence on the al-Qaeda leader. Allen, who already had experience
with drones, was immediately attracted to the idea.

Having served for forty years at the CIA, Allen was regarded by his
advocates as a maverick, workaholic legend, while his detractors saw him
as an egotist lacking basic courtesy and diplomacy.[134] As the national intelligence officer for counterterrorism in the 1980s, Allen had worked closely
with Clarridge during the establishment of the CTC, and he managed to
survive a minor role in the Iran-Contra scandal, having served as the
director of the DCI's Hostage Location Task Force. The task force, established at Langley in January 1986 in accordance with a directive from Vice
Admiral John Poindexter—then serving as Reagan's APNSA—was made
up of full-time representatives from CIA, DIA, NSA, NPIC, and JSOC,
and it was charged with locating the hostages held by Hezbollah in Beirut.[135]
The task force was amalgamated into the CTC when the center was established the following month, and Allen was co-located with Clarridge.[136]
The memos sent from Allen to Poindexter regarding the hostages are
highly redacted, expunging any mention of the still classified program, but
as the director of the task force it is highly probable Allen would have been
briefed on the CTC's plans to locate the hostages through the Eagle
Program drone. A decade later, Allen had also liaised with General Atomics
over the agency's purchase of the GNAT-750s for the Bosnia mission.[137]

Despite support from Allen and Cofer Black—the head of the CTC and
the man most responsible for locating bin Laden in his Afghan safe haven—
the idea of using the Predator met with significant resistance from senior
managers at Langley: "Why would we want to do this?" James Pavitt, the
deputy director of operations, and Allen's immediate superior, had
wondered. "Why do I care whether I have an image of a guy, whether he's
six feet four or six feet?" George Tenet, director of central intelligence,
had asked dismissively.[138] For members of the DO such as Pavitt, the
agency's primary role was to work with human intelligence (HUMINT),
flying aircraft was a job for the air force—an attitude that infuriated
Clinton's counterterrorism adviser Richard Clarke, who rebutted Pavitt's
concerns with his characteristic bluntness: "Your valuable HUMINT program
hasn't worked for years. I want to try something else."[139]

Legally, the consideration of using the Predator's laser designator to guide a cruise missile to bin Laden's location made the CIA's managers uneasy about the extent to which the agency's place in the kill chain would implicate the CIA in an act of assassination, an enterprise banned by executive order. In Clarke's view this was not an issue; the CIA had been granted all the authority it needed to kill bin Laden by a series of Memorandums of Notification (MONs) from Clinton.[140] In financial terms, Langley's senior staff were concerned about the cost of the mission, estimated at approximately $3 million—a paltry sum by DoD standards—but serious money for the CIA's squeezed budgets.[141] Furthermore, the Predator had shown itself to be somewhat accident prone, prompting concerns about the costs of any repairs should a drone malfunction. Sure enough, when a Predator did eventually crash on takeoff, the air force tried to bill the CIA for a replacement, something that Tenet, Pavitt, and Black fiercely protested. Aggravated Pentagon officials battled back with Whit Peters, the air force secretary, accusing the agency of wanting to "run everything and pay for nothing," creating a bureaucratic deadlock.[142]

Frustrated at the lack of decisive action, the White House intervened directly to end the financial dispute over the Predator, submitting amended budget requests in the summer of 2000 for supplementary procurement from the air force and additional counterterrorism funding for the CIA, with the two to split any further costs equally between them.[143] An additional compromise was struck over Tenet's concerns about the agency's involvement in a lethal operation, with Clinton's APNSA, Sandy Berger, agreeing that, in the first instance at least, Clarke's Afghan Eyes mission, as it was referred to in official memos, would purely be a test of the Predator's capabilities, with no attempt to direct a lethal action. The first flight of the sixty-day mission began on September 7, 2000.

The Afghan Eyes flights proved the Predator's worth as an intelligence tool, successfully observing the al-Qaeda leader for a total of four hours and twenty-three minutes over a number of missions.[144] More important, the operation also served as a successful test of the infrastructure that would become the foundation of the CIA's lethal drone campaign. The drones were maintained, fueled, and operated for takeoff and landing by a small maintenance crew of CIA contractors, known as a launch recovery element, supplied by GA-ASI and the air force.[145] The team was secretly based at the remote Khanabad airfield, near the Afghan border in Uzbekistan. The discreet diplomatic negotiations that enabled the secret basing became a key feature of America's expanding drone warfare, as a burgeoning network of small, contractor-staffed bases emerged across Afghanistan, Turkey, Qatar, Saudi Arabia, Yemen, Djibouti, Ethiopia, the Philippines, and even the Seychelles over the next fifteen years.[146] Once the unmanned aircraft was airborne, control of it was transferred, by satellite data link, to operators flying the drones, unbeknownst to the German government at the time, from the U.S. air base at Ramstein. A live feed of the mission was transmitted via fiber-optic cable, from Ramstein, under the Atlantic, to screens in the CTC at Langley, where counterterrorism officials and senior NSC staff observed the missions in real time, and analysts reviewed the footage for signs of the drone's target's movements and patterns of life.[147] When the CIA's armed drones were eventually called into service following 9/11, it was the temporary network that had been conceived for Afghan Eyes that was employed, establishing the initial blueprint for operations that the Obama administration would come to formalize.

Even before the completion of the Predator's first trial as a counterterrorism tool, Clarke, Allen, and the CTC's Black advocated the arming of the drone itself, rather than reliance upon its laser designator to direct a cruise missile or ordnance from another asset.[148] Once Clarke learned that Big Safari was already working on such a project, Clinton's counterterrorism adviser incorporated it into his new "Strategy for Eliminating the Threat from the Jihadisat Networks of al Qida [sic]," written in late 2000. Noting that the armed Predator would permit a previously unavailable "see it/shoot it" capability, the strategy called for flights to recommence in late March 2001, shifting responsibility for authorizing armed drone strikes from Clinton to the incoming Bush administration.[149] In keeping with Clarke's proposed deadline, the Big Safari staffers and their colleagues at GA-ASI and Raytheon successfully conducted the first static ground launch of a Hellfire from a Predator on January 23, 2001, three days after Clinton had left office. The first successful airborne live-fire test was completed a month later, on February 21, 2001, changing the Predator's designation from RQ-1 to the armed variant used to launch the CIA's drone campaign, the MQ-1 ("M" denoting multipurpose).[150]

In spite of Big Safari's engineering achievements and the potential Clarke and others saw in the armed drone, the most significant delays to the deployment of the Hellfire-equipped Predator proved to be legal, political, and even cultural in nature.[151] There was still significant resistance from within the CIA regarding the deployment of the armed Predator under the CTC's authority, not least from DCI Tenet, who, concerned about the potential for the agency to once more be dragged into acts of assassination, claimed it would be a "terrible mistake for the CIA to fire the weapon."[152] The Bush administration was slow to address these concerns and tackle the unfinished legal business of using lethal unmanned aircraft. In the first principal-level meeting regarding al-Qaeda, held on September 4, 2001, eight months after Clarke first requested the gathering, it was agreed to deploy the armed drone. The thorny legal issue of who would be responsible for firing its missiles, however, was deferred to a later meeting, remaining unresolved when the September 11 attacks occurred.[153]

In the aftermath of 9/11, the potency of the resistance of those concerned about the CTC's role in piloting lethal drones could not stand the pressure for retribution, with Cofer Black instructing the CTC to "take the gloves off." The bureaucratic deadlock was broken. On September 17, 2001, Bush signed a Finding that created a secret list of high-value targets (HVTs) that the CIA was authorized to kill without further presidential approval. The Finding was, as former acting general counsel John Rizzo put it, "Unprecedented in my 25 years of experience at CIA. . . . Frankly, the finding was so aggressive and comprehensive that honestly there wasn't much more that could have been added." That included authorizing the CIA's use of the armed Predator for lethal strikes.[154]

The first lethal action conducted by a CIA Predator occurred on October 7, 2001, the opening night of Operation Enduring Freedom and the official beginning of America's War on Terror. In an inglorious debut, the drone's pilot targeted a vehicle outside a building in which the Taliban's spiritual leader, Mullah Omar, was thought to be located, likely in an effort to draw him out and avoid possible civilian casualties by collapsing the building. A number of bodyguards were killed in the strike—the first ever casualties of an unmanned drone strike—but in the chaotic moments that followed, the Taliban leader escaped, triggering a furious row between the air force, CENTCOM, and the CIA over chain of command regarding the unmanned vehicle.[155] Though the United States had missed an important strategic opportunity, the failure had been operational, not technical. After more than a decade of incremental development, the Predator had proven itself capable of loitering, tracking its target, and guiding its lethal ordnance with precision. As America's conflict with the Taliban and al-Qaeda evolved, so did the rules of engagement governing drone use, with the armed UAV's unique abilities becoming increasingly central to America's conduct of the War on Terror.

In a further important milestone that was to illustrate the new nature of America's remote war against al-Qaeda and its affiliates, the first lethal drone strike outside an official war zone occurred on November 3, 2002, in Yemen, where the CIA destroyed a jeep carrying six al-Qaeda militants, including Ali Qaed Senyan al-Harethi, one of the masterminds of the 2000 bombing of the USS *Cole*.[156] The CTC had finally undertaken the sort of global retaliation Clarridge had sought to introduce sixteen years prior.

Conclusion

According to Clarridge's own account, the very weapon the CIA's CTC eventually adopted in order to take the offensive against al-Qaeda began as a concept developed by the center's first director fifteen years prior to 9/11 in the form of the Eagle Program, with the exact same purpose in mind—gathering intelligence on terrorists to enable precision strikes to neutralize them. Even if one were to reject Clarridge's version of events on account of the lack of supporting evidence, it is clear from the evolutionary path of GA-ASI's drones that the CIA played a vital role in the development of the tool it was initially so hesitant to adopt in its effort to hunt down and eliminate Osama bin Laden.

So why did it take the agency so long to adopt a concept that had begun so long ago? First of all, Clarridge was a risk taker. He was appointed for exactly that reason, but his risk taking eventually undermined the very counterterrorism cause he had been selected to pursue. He had interpreted the new Finding and creation of the CTC as authority "to do pretty much anything he wanted against the terrorists," recalled Robert Baer, one of the CTC's first recruits from the DO.[157] At first, that was exactly how things had seemed. The former operations officer recalls how the CTC, based on

Langley's sixth floor, burst into life with "pure frenetic energy." The staff worked in a huge open bay with "all the telephones ringing nonstop, printers clattering, files stacked all over the place, CNN playing on TV monitors bolted to the ceiling, hundreds of people in motion and at their computers." Baer's first weeks in the CTC were like "being in a war room."[158] But this energy was quickly sapped in the wake of the Iran-Contra scandal. North, Casey, and Clarridge, among others, were all implicated in the illegal effort to support Nicaraguan rebels and dispatch arms shipments to Iran in an effort to secure the release of the American hostages in Lebanon.[159]

The irony of this position should not be overlooked. Casey, North, and Clarridge proved themselves to be farsighted in their recognition of the growing threat posed to the United States by ever more militant and well-trained nonstate and state-sponsored terrorist groups, such as the PLF and Hezbollah. Yet their willingness to work—covertly and in breach of both international and domestic law—with the very sponsors of such groups as part of a misbegotten plot to free American hostages served to undermine their cause more dramatically than any terror attack or congressional oversight ever could. In the words of Loch K. Johnson, Iran-Contra represented "a fundamental assault on the U.S. Constitution," which cast "the darkest mark" on the agency's use of covert action.[160] Furthermore, despite their vision of future threats to U.S. national security, the outdated and deeply unpopular thinking on Latin America, drawn from what Senator Frank Church had previously criticized as the persistent myth that "Communism is a single, hydra-headed serpent," led Casey and his coconspirators to believe it remained the CIA's duty to "cut off each ugly head, wherever and however it may appear." This mission caused the hardliners to destroy their credibility by lying to Congress and misleading the public in order to illegally divert funds to overthrow the hapless Sandinistas in the tiny country of Nicaragua.[161] In doing so, the three revealed not only a misguided sense of America's national security priorities but also a complete disregard for the consequences of the Church Committee and its sister investigations, which had sought to curb the excesses of previous agency behavior, including organizing coups in the other Latin American states of Guatemala and Chile.[162]

Resultantly, Casey was looked upon by his congressional overseers as "an adventurer," recalled Vincent Cannistraro, an operations officer who had arrived shortly after the center's founding, "and Dewey as kind of a cowboy."[163] Hours before Casey was due to testify before Congress about his knowledge of the affair, he was hospitalized with what turned out to be a brain tumor. Despite surgery, the DCI never recovered and died at home on May 5, 1987.[164] After Clarridge's testimony the CTC director was formally reprimanded and forced to retire from the CIA on June 1, 1987.[165] North, heavily implicated in the scandal, was dismissed by Reagan in November 1986 and testified the following July.[166] Rhodri Jeffreys-Jones has described the long-term impact of the Iran-Contra scandal on President Reagan and his Republican successor, George H. W. Bush, as little more than a discomfort. Though it may have taken some of the shine off the administration, it ultimately produced no great upheaval.[167] This was not the case for the CTC. With two of the original counterterrorism hardliners gone, and the man selected to lead their offensive against terrorism forced to retire, the appetite for risk taking within the CIA and the congressional oversight committees disappeared. The CTC abandoned the "war room" vision of action teams and its offensive posture that had spawned the Eagle program and returned to the more cautious, analytical, report-writing culture that Casey and Clarridge had disdained. "Casey had envisaged it as something different than what it eventually became," lamented Cannistraro.[168]

Yet somewhat paradoxically, when examining the longer-term emergence of the armed drone it is important not to underestimate the importance of the Cold War events in Nicaragua. Though the Iran-Contra plot destroyed the hardliner coalition, it was the CIA's support for the anti-Sandinistas rebels that brought the resolute cold warriors Casey, Clarridge, and North together in the first place. It was also this same cause that inspired GA's Blue brothers to begin investment in unmanned aircraft. Whether the Blues officially formed part of the CIA's Contra mission is unknown, but what is clear is that without the brothers' plans to use UAVs against the Sandinista government, it is highly unlikely Karem's drones would have ever have made it to market following the cancelation of the Amber and LSI's consequent liquidation. So a coalition initially formed with the goal of overthrowing the Sandinistas went on to transform U.S. counterterrorism, laying much of the policy groundwork for what would eventually become the War on Terror and triggering the technological developments that would ultimately spawn the primary tool employed in that conflict, the lethal drone.

Evidence of the long-term impact of the failed hostage negotiations, and an acknowledgment of the immense pressure such terrorist actions can exert upon U.S. policy makers can be seen in the extent to which the United States has since consistently reaffirmed that it will make no concessions to individuals or groups holding U.S. nationals hostage. Officially the Obama administration's policy declared that the government's response to a hostage situation would include "diplomatic outreach, intelligence collection, and investigations in support of developing further options, recovery operations, and the use of any other lawful and appropriate tools," but in practice the U.S. government adopted the more aggressive, action-orientated approach to tackling hostage scenarios first endorsed by Clarridge before a lack of actionable intelligence paralyzed such plans.[169]

Obama first made this approach clear on April 10, 2009, just four months into his presidency, when he authorized operators from the U.S. Navy's SEAL Team Six to use lethal force to free the American captain Richard Phillips from his Somali pirate captors. Utilizing a ScanEagle drone launched from the USS *Bainbridge* to track the lifeboat the pirates had tried to abduct Phillips in, the SEALs killed the three pirates with sniper rifles, fired from the deck of the pursuing U.S. warship.[170] Three years later, in January 2012, members of the same SEAL unit successfully rescued the American aid worker Jessica Buchanan and her Danish colleague Poul Hagen Thisted from a compound in Somalia, killing nine hostage takers in the process. Emphasizing the perceived deterrent value of the raid, Obama called the action "yet another message to the world that the United States of America will stand strongly against any threats to our people."[171] And in May 2012, a Predator drone located the kidnapped British aid worker Helen Johnston and her three female colleagues in the desolate Badakhshan province of eastern Afghanistan, where a joint team of British and American Special Forces were then able to mount a successful rescue, killing eleven kidnappers in the process.[172]

Despite America's considerable success in countering hostage takers, the challenge of locating and safely liberating hostages and the risks of applying lethal force to such efforts have been tragically highlighted on a number of occasions. On October 8, 2010, Linda Norgrove, a British citizen kidnapped in Afghanistan while working for an American aid organization, was killed by the blast of a Navy SEAL's grenade during a rescue attempt.[173] A December 3, 2014, rescue effort in Yemen, authorized by Obama following intelligence that an American hostage, photojournalist Luke Somers, was in imminent danger, also ended in failure. Somers's location had reportedly been pinpointed by a combination of spy satellites, surveillance drones, and eavesdropping technology, but when the SEALs descended upon the site, Somers's captors, Al-Qaeda in the Arabian Peninsula, executed him and his fellow hostage, South African teacher Pierre Korkie. Despite the failure, Obama defended the use of military force in a statement: "As this and previous hostage rescue operations demonstrate, the United States will spare no effort to use all of its military, intelligence and diplomatic capabilities to bring Americans home safely, wherever they are located."[174] Finally, in a tragic combination of a technology introduced to help find hostages and neutralize terrorists, a CTC drone strike launched in the Pakistan border region on January 15, 2015, succeeded in killing the al-Qaeda leader Ahmed Farouq, but also accidently resulted in the deaths of two hostages. The presence of the American Warren Weinstein, kidnapped in 2011, and Italian Giovanni Lo Porto, who was seized in 2012, was, according to a government investigation, unknown to the CIA at the time.[175]

Though America's aggressive hostage policy produced mixed results, forcing the president to issue condolences for failed efforts and generating a degree of secrecy that infuriated the families of those held, it also allowed the Obama administration to avoid the political damage and high levels of public disapproval that both Carter and Reagan experienced.[176] What is more, by adopting the direct policy of non-negotiation, intelligence gathering, and confrontation that Clarridge had originally endorsed, it bypassed the risk of policy makers succumbing to the temptation to engage in negotiations and generating potentially larger political incidents. In the one instance that the Obama administration did engage in negotiation, engineering a prisoner exchange with the Taliban for the safe return of captured U.S. soldier Bowdrie "Bowe" Bergdahl (argued by officials to be a prisoner of war as opposed to a hostage), the deal proved highly unpopular. It is worth noting that the circumstances surrounding Bergdahl's capture, in which he was publically accused of desertion, doubtless helped contribute toward the criticism surrounding the settlement. But regardless of the specific details, it was the process of negotiation itself that prompted the most stinging domestic backlash.[177] For the American public, attitudes toward hostage taking are similar to those discussed in chapter 2 regarding counterterrorism action—Americans would rather see their government try and fail to free hostages by force than for hostages to be left to linger as diplomatic negotiations drag on.

A second factor in the lengthy timescale of drone development emerges from the fact that the unmanned aircraft technology envisioned by Clarridge was in its infancy. Though the CTC may have had a Radio Shack–style prototype working within a year, it would take more than a decade for GA-ASI and Big Safari's engineers to come up with a design reliable enough to be used on sensitive missions. In fact, despite being deployed sixteen years after the Eagle and Amber prototypes, the first armed Predator still encapsulated Clarridge's Radio Shack approach to technology. The jerry-rigged system featured a missile originally designed to be launched from low-altitude helicopters to destroy tanks, retooled to attack individuals from four miles up, fired from a flimsy unmanned aircraft developed to carry sensors not weapons, guided by an unfinished prototype sensor ball, and controlled with a largely untested data-connection system. The first armed Predator was, without a doubt, a work in progress, and even today, Predator UAVs have the highest accident rates of all the aircraft in the U.S. Air Force.[178]

The U.S. war logs from Afghanistan contain numerous accounts of drones suffering mechanical failures, computer glitches, or signal loss, usually resulting in a crash. For example, a report dated November 20, 2008, filed under the category "Equipment Failure" states: "1 x UAV (PREDATOR) with 1 x HELLFIRE on board, crashed on KAF [Kandahar Air Force base] near the JULIET Ramp."[179] Another report, dated December 27, 2008, revealed that a small Shadow drone had needed to make an emergency landing after its temperature had spiked and it had been unable to maintain its proper altitude.[180] The logs also reveal the concern over the potential loss of sensitive equipment each drone malfunction brings. Special units are deployed on salvage operations to try to ensure the remains of the aircraft do not end up being sold to the likes of Iran, China, Russia, or even Pakistan. For example, on September 4, 2009, a report logged that a Predator drone had "crashed due to suspected mechanical failure," and that "overwatch" of the crash site had been established. The report went on to add that a special unit had "taken lead in the recovery of sensitive materials," and that by "0050, all sensitive items [had been] recovered."[181] It would not be until the introduction of the MQ-9 Reaper in 2007, the successor to the Predator, that GA-ASI finally perfected the unmanned system its owners had sought since the mid-1980s. Capable of carrying fifteen times the ordnance of its predecessor and flying at three times the speed, the Reaper was, according to the DoD, the first "true hunter-killer" UAV.[182]

Despite the significant length of time it took for the CTC to effectively deploy the technology and embrace the techniques Clarridge had pushed in 1986, the first director of the CTC still managed to play a role in the downfall of the man who had inspired the creation of the Eagle program in the first place—Mu'ammar al-Gaddafi. Armed Predator drones were deployed to Libya to attack ground targets as part of the NATO support to the rebel forces that had risen up against the Libyan dictator in April 2011.[183] A study by the Bureau of Investigative Journalism has estimated that approximately two hundred drone strikes were launched by the United States in Libya by the time the conflict ended on October 31, 2011.[184] The most significant of these strikes came around 8:30 A.M. on October 20. According to a report from the *New York Times*, Gaddafi tried to slip out of his fortified compound in Surt but was observed by an American Predator drone. Before Gaddafi's convoy had traveled two miles, NATO officials reported that it was set upon by the Predator and a French warplane. The missile strikes forced the surviving vehicles in the convoy to detour and scatter. It is during the chaos following the strike that anti-Gaddafi fighters descended on the scene and captured the dictator as he sheltered for cover.[185] So, the very technology that Clarridge and the DST staff of the CTC had developed with the specific aim of dealing with terrorist sponsors such as Gaddafi eventually ended up playing a key role in both his downfall and his death. The capability of the drone to loiter, observe, and gather intelligence, consistent with Clarridge's original specifications, along with the ability to deliver lethal ordnance, made it impossible for the dictator to slip out of his stronghold undetected.

The Predator drone and its more advanced successor, the Reaper—the primary tools used by the CIA to wage is drone war against al-Qaeda and their affiliates—evolved directly from the DoD's investment in drone technology, which in turn emerged from, or at least in unison with, the CTC's Eagle program. The drone campaign itself is run by the CTC, the fusion center Clarridge created to ensure that the United States' counter-terrorism efforts were centralized, coordinated across all branches of government, and proactive in their pursuit of terrorists. It is global in reach, enabling the United States to attack terrorist safe havens in areas U.S. forces could not otherwise access. It is based upon the scrupulous analysis of multiple sources of intelligence, with hundreds of targeters and analysts compiling signals intelligence provided by the DoD's NSA and human intelligence drawn from the CIA's network of foreign assets, which officials argue makes the strikes precise and limits collateral damage as far as is possible. Finally, it enables the CIA to go on the offensive, preemptively targeting known and suspected terrorists before they can launch attacks against American targets and interests. It is striking, then, that the drone campaign is based upon the principles Clarridge sought to instill into U.S. counterterrorism and utilizes the same tool he sought to deploy, fittingly making the CTC's first director the father of the center's drone campaign.

4

"TALKING ABOUT CAPTURING BIN LADEN":
THE CLINTON ADMINISTRATION AND THE
LEGAL ARCHITECTURE OF LETHAL FORCE
IN COUNTERTERRORISM, 1993–2000

And the Clinton White House was only talking about capturing bin Laden,
not knocking him off.
—*CIA Counsel John Rizzo*

Upon assuming power, Clinton's administration quickly identified
the newly emergent threat of nonstate terrorism but struggled to develop
a solution. The fallout of the Iran-Contra scandal had seen Clarridge and
North removed from their positions and renewed congressional oversight
of the CIA. With Casey's passing, the emerging proactive counterterrorism
agenda had been left in tatters. By the time Osama bin Laden first declared
jihad against the United States in 1996, a culture of risk aversion had
emerged within the CIA, while the memory of the political damage caused
by Iran-Contra ensured neither Clinton nor the members of his NSC were
looking to sanction any risky covert operations. This lack of any proactive
method to pursue the terrorist threat left America vulnerable, giving its
enemy the time and space it needed to recruit, train, and dispatch fighters
to wage their global struggle against the "far enemy" of the United States.
As the threat heightened, the Clinton administration, drawing upon
Reagan-era legislation, cautiously began to establish a legal architecture
that authorized the CIA to take action against al-Qaeda. Though too late

to prevent the disastrous attacks of September 11, 2001, this preexisting architecture had formed the backbone of George W. Bush's War on Terror, and once strengthened with additional Congressional and presidential support, provided the legal sanction for the targeted killings undertaken by the CTC's drone campaign. This legal architecture proved remarkably stable in the first post-9/11 decade, empowering presidents to pursue a significantly more proactive counterterrorism strategy than their predecessors. It also served to demonstrate that American counterterrorism is significantly more effective and less prone to failure when operating within clear legal boundaries as opposed to working covertly outside them and running the risk of exposure and backlash.

The Covert Action Pendulum

Though there can be no doubt that much of al-Qaeda's success against the United States during the Clinton administration can be credited to the organization's exceptional planning, experience, and determination, the United States itself must accept a significant portion of the blame for failing to act in time to counter the threat. This is in spite of the effectiveness of the efforts of the U.S. counterterrorism community in identifying the newly emergent al-Qaeda and quickly recognizing it as a major threat. According to Richard Clarke, the veteran Washington bureaucrat brought in by the Clinton administration to serve as America's first national coordinator for counterterrorism, the administration was seized with the issue as early as 1994.[1] Much has already been written about why the Clinton administration failed in its efforts to kill Osama bin Laden and neutralize his nonstate Sunni Islamic terrorist group before, or even after, its successful attacks against American interests.[2] This chapter takes a different perspective. By arguing that one of the key factors that has enabled the CTC to be significantly more effective against al-Qaeda's leadership is the legal infrastructure within which it now operates, the chapter highlights how a lack of such clear legal sanction and authorization crippled Clinton's efforts to neutralize bin Laden. This illustrates how the laws that came to underpin the CIA's drone campaign have proven to be as vital to its operation as the technology itself.

The legal architecture that supports lethal counterterrorism actions undertaken by the United States intelligence community has taken decades to evolve. The transition between the Bush and Obama administrations was remarkably smooth with regard to the extent to which this architecture remained largely untouched. Despite the fact that certain counterterrorism policies introduced by the previous administration—such as the use of enhanced interrogation techniques—were heavily criticized and abandoned, there was what legal scholar Robert Chesney has described as "a remarkable degree of cross-branch and cross-party consensus manifested by legislation, judicial decisions, and consistency of policy across two very different presidential administrations."[3] This is especially true with regard to the use of lethal force in counterterrorism.

This stability certainly owes something to the policy commitment to confront al-Qaeda that followed the experience of 9/11. Prior to September 2001, the development of a policy and legal architecture addressing the use of lethal force to neutralize terrorist threats was much more hotly contested and evolved in fits and starts generally in the direction of increased constraint.[4] Rather than gradually building upon existing practice and legislation introduced by the counterterrorism hardliners within the Reagan administration, the responsibilities and accompanying legal authorities granted to the CIA lurched from one position to another in the decades between the creation of the CTC and 9/11, causing uncertainty and confusion among those responsible for carrying out America's aggressive covert operations. This state of affairs can be described as the covert action pendulum. The phrase denotes the way in which demand and support for aggressive covert action from the executive and Congress swings dramatically from one side to the other depending upon the political circumstances.

In times of need when presidents are struggling to come up with orthodox solutions to a foreign policy problem, they frequently turn to the CIA to provide a nonorthodox solution through aggressive covert action. These actions occur either with the approval of the appropriate congressional oversight committees or in the most controversial cases without any legislative oversight at all. Operation Cyclone, the CIA's support for anti-Soviet mujahideen in Afghanistan during the 1980s, serves as an example of a covert action supported by both the executive (in this case, first Carter then Reagan) and Congress. The Reagan administration's support for the Contras in Nicaragua and linked arms dealing with Iran provides an example of a covert action without congressional support.[5]

The CIA's drone campaign falls into the first camp, with the members of the House and Senate intelligence committees privy to most details and responsible for scrutinizing the highly classified program.[6] Obama stressed the necessity for this oversight in a speech on the future of U.S. counter-terrorism in May 2013, telling his audience: "The very precision of drone strikes, and the necessary secrecy involved in such actions can end up shielding our government from the public scrutiny that a troop deployment invites." The president also acknowledged the potentially seductive nature of an unregulated covert killing program, recognizing that if unchecked it could "lead a President and his team to view drone strikes as a cure-all for terrorism." Having highlighted the problem, Obama offered his solution to the audience: "For this reason I've insisted on strong oversight of all lethal action. After I took office, my Administration began briefing all strikes outside of Iraq and Afghanistan to the appropriate committees of Congress. Let me repeat that—not only did Congress authorize the use of force, it is briefed on every strike that America takes."[7]

To further increase oversight, the Obama administration explored the possibility of transferring responsibility for lethal drone strikes from the CIA to the Department of Defense. The move was opposed, however, by key lawmakers on the Senate Intelligence Committee, who were concerned that the military lacked the necessary intelligence-gathering capabilities to ensure the strikes would be as precise as possible. Influenced by a number of high-profile errors committed by JSOC forces, including an attack upon a wedding party in Afghanistan on July 1, 2002, and another wedding in Yemen on December 12, 2013, Senator Dianne Feinstein, the ranking Democrat on the committee at the time, argued that the CIA had proven itself capable of exercising "patience and discretion specifically to prevent collateral damage" and that she "would really have to be convinced that the military would carry it out that well."[8] To ensure that the CIA would not be made to cede responsibility for drone warfare, Congressional opponents of the shift inserted a secret provision into a massive January 2014 government spending bill that barred the use of any funds to facilitate the transfer of operations from Langley to the Pentagon, an unusually direct intervention on the part of Congress to preserve the structure of drone operations that had emerged, in a somewhat ad hoc manner, in the post-9/11 years.[9]

Revealing divisions among the attitudes of lawmakers toward the transparency of the CIA's drone campaign, there have been attempts by other members of Congress on both sides of the aisle to expand access to information about drone strikes beyond the intelligence committees, especially since Anwar al-Awlaki, an American citizen, was targeted and killed by a strike in Yemen on September 30, 2011. To date, these challenges have been unsuccessful. Responding to such pressure, and considering his legacy, Obama acted unilaterally in the final months of his presidency to further enhance the transparency of America's use of lethal force. On July 1, 2016, he signed an executive order committing future administrations to the annual release of basic statistics on counterterrorism strikes undertaken against terrorist targets outside areas of active hostilities. Furthermore, the order called for future reports to address any significant discrepancy between the post-strike assessments of the U.S. government and credible reporting from nongovernmental organizations, regarding noncombatant deaths. Finally, Obama used the order to set out a number of precautions and rules of best practice for future strikes, intended to reduce the likelihood of civilian casualties. While the move to enhance transparency and encourage greater caution was commended by human rights activists concerned about America's shadowy conflicts, this was offset by the fact that Obama's order further institutionalized and normalized lethal strikes outside conventional war zones as a routine part of U.S. counterterrorism policy. Moreover, the fact the reform came from an executive order revealed Congress's unwillingness or inability to legislate on the matter, as well as the fragile nature of the commitment.[10]

Historically, when Congress discovers that a covert action has occurred without its oversight, either through a leak (such as the NSA data-collection scandal exposed by the whistleblower Edward Snowden), through a conspicuous operational failure (for example, the capture of Eugene Hasenfus by Nicaraguan authorities while smuggling guns to Contra rebels), or through investigative journalism or congressional inquiries (as in Seymour Hersh's 1974 story about illegal CIA activity and the subsequent Church Committee hearings), the pendulum of political support for aggressive covert action swings the other way. The agency and those of its officers charged with planning and undertaking the aggressive covert actions become exposed to criticism and possible legal prosecution.[11] The

Church Committee hearings offer an illustrative case.[12] Originally established to investigate potential domestic abuses by the CIA, such as illegal wiretaps, the committee shifted its focus to lethal covert actions following an unguarded comment by President Gerald Ford to reporters suggesting the CIA had conspired to kill foreign political leaders.[13] Over the course of the hearings, various abuses were uncovered in which the CIA had operated outside its charter, with the use of assassination as an instrument of U.S. policy regarded as the worst.[14] "Central to the investigation," Kathryn Olmsted notes, "came the question, just who was to blame for these abuses?"[15]

Despite the CIA bearing the brunt of the blame for the abuses, Senator Church did acknowledge the complicity of the executive in the CIA's questionable activities during its first three decades of existence. The committee highlighted that while the CIA planned the coups and attempted to kill foreign leaders, these operations were always encouraged by those occupying the White House. As Mark Mazzetti notes, "The CIA offered secrecy, and secrecy had always seduced American presidents."[16] Senator Church's final report observed: "Once the capability for covert activity is established, the pressures brought to bear on the President to use it are immense," and went as far as to question whether the United States even needed a "regiment of cloak-and-dagger men" at the president's disposal.[17] Though the senator did not manage to see the CIA dismantled, he did succeed in ending what Loch K. Johnson has called the "Era of Trust" between Congress and the intelligence community, drawing the oversight committees more deeply into the making of intelligence policy.[18] The investigation also caused severe new restraints to be imposed on its covert actions. As a result of the hearings, President Ford signed EO 11905 on February 18, 1976, which, among other wide-ranging provisions affecting the intelligence community, stated: "No employee of the United States Government shall engage in, or conspire to engage in, political assassination."[19] Ford's order was as much about protecting his successors in the Oval Office from the seduction of the apparent quick fix offered by aggressive covert actions as it was about saving the CIA from future prosecution.[20] There were further swings of the covert action pendulum before al-Qaeda emerged as a significant threat to U.S. national security, the impact of which served to limit the options available to the U.S. counterterrorism community when called upon to neutralize the threat posed by the Sunni Islamic radical group.

William Casey, the director who oversaw the CIA's first foray into counterterrorism on the orders of President Reagan, arrived at Langley in the wake of the Church Committee recriminations. The new DCI acknowledged the executive's role in the agency's misfortunes during his first staff meeting by announcing to his assembled officers that "none of the sins the CIA had committed were born at Langley. They were carried out as the result of directives from above, including the wishes of American Presidents and secretaries of State, people who operated above the line while directing the dirty work to be done below it."[21] Despite Casey's reference to the agency's past actions as "sins," he dedicated his time at the CIA to trying to swing the pendulum back the other way, limiting consultation with Congress's intelligence oversight committees and keenly restoring a sense of risk taking and lethality to an agency he believed had been rendered weak and bureaucratic.[22] Under Casey, the CIA was to get back into the aggressive covert action business, with a new focus on counterterrorism.[23]

Rather than making the agency stronger, as Casey believed it would and his supporters have subsequently argued, the consequence of this push for risk taking and aggressive covert action was to leave the CIA vulnerable to another swing of the pendulum, weakening its counterterrorism capabilities still further. This is exactly what occurred when in late 1986 details of the Iran-Contra affair broke. As the public and Congress learned of the double-ended deal involving illegal arms sales to Iran aimed at securing the release of American hostages in Lebanon, and it also emerged that Nicaraguan rebels had been funded from the proceeds to aid in the overthrow of the Communist government, the agency found itself once again on trial.[24] Casey passed away as congressional investigations were progressing, and the nomination of the Reagan administration's chosen successor, Robert Gates, was opposed due to his links to those implicated in the scandal and his later efforts to cover it up by claiming ignorance, despite a paper trail linking him to knowledge of the affair.[25]

The White House's replacement nominee represented the mood of Congress at the time. William Webster, a former federal appeals court judge and a respected pillar of the Washington establishment, was just completing a nine-year tenure as director of the Federal Bureau of Investigation (FBI).

He was appointed with what John Rizzo, the CIA's counsel overseeing the Iran-Contra case, described as "a mandate from Congress to clean up the mess at Langley."[26] This cleanup was particularly damaging to the CTC. Having already lost its patron Casey, the CTC also lost one of its closest supporters, Clair George, head of the DO and a man described by Bob Woodward as "an old warhorse symbol of the CIA at its best and proudest."[27] George was asked to retire by Webster due to his links to Iran-Contra; he was eventually indicted on ten counts, including making false statements to Congress and obstruction.[28] Perhaps most damaging of all, the CTC also lost its founding member and leader, Duane Clarridge, who, along with several other lower-ranking officers, was forced into retirement, in what became known within the CIA as the "Holiday Party Massacre."[29] In addition, Stanley Sporkin, the CIA's chief counsel who had provided his legal backing to the CTC's aggressive counterterrorism policies, left the agency in December 1985, having been nominated to a seat on the U.S. District Court for the District of Columbia by Reagan.[30]

Despite the strong case against them, in an unprecedented step George, Clarridge, and several other Iran-Contra participants facing trial, including Caspar Weinberger and Robert McFarlane, were pardoned by the outgoing president George H. W. Bush on Christmas Eve, 1992.[31] Bush argued that the pardons were in the "healing tradition" set by previous presidents, and that the men—who had been motivated by patriotism—were the victims of "the criminalization of policy differences," adjudication on which belonged "in the voting booth, not the courtroom." Lawrence E. Walsh, the independent prosecutor of the Iran-Contra investigation, was furious, accusing the president of a misuse of executive power to complete the Iran-contra cover-up.[32] Accordingly, not one of the Iran-Contra participants was prosecuted for his role in the operation—the real cost was in the loss of their careers, credibility, and the ensuing damage to the agency's reputation.

Perhaps unsurprisingly, Clarridge is stinging in his criticism of Webster. He complained that the new DCI, though tasked with the job of tackling the threat of terrorism around the world, "didn't have the stomach for bold moves of any sort." Clarridge recounted his frustration at the rejection of a plan for the "fairly daring extraction of a dangerous terrorist." The plan was to snatch Mohammed Rashid, a well-known and skilled bomb maker wanted in connection with a number of airline attacks, from an

African country where he had been spotted by CTC assets.[33] The complex operation would require a C-130 landing in the desert, in-flight refueling, and considerable coordination with the U.S. military. Having worked out the details of the operation with administration officials just below cabinet level, Clarridge was convinced that it was "entirely feasible." Despite this, he was informed by Webster's office that the Joint Chiefs had rejected the plan and that Webster had terminated the operation.[34] Rather than considering the wider politics engulfing the CIA at the time, the former counterterrorism chief pinned the blame on the new DCI, arguing that the idea of a lawyer leading the CIA was antithetical. "All of his training as a lawyer and as a judge was that you didn't do illegal things," Clarridge complained. "He never could accept that this is *exactly* what the CIA does when it operates abroad. . . . It's why we're in business." In the view of the former counterterrorism chief, Webster "had an insurmountable problem with the raison d'etre of the organization he was brought in to run."[35] It is worth noting that Casey, Clarridge's champion and someone not opposed to rule breaking himself, was also a trained lawyer, suggesting Clarridge's grievance was with Webster himself or the attitude he personified rather than his training and previous profession.

Not all within the CIA shared Clarridge's perspective; those with a better sense of the vulnerabilities of the agency following the Iran-Contra scandal felt that Webster's cleanup was a necessary step. Though Clarridge railed that DCI Webster was "out of sync with the purpose and principles of the Agency," the reality is that, in the aftermath of Iran-Contra, it was Clarridge himself that was now out of sync with the changing principles of the CIA.[36] He may have believed it was in the nature of the CIA's work to break other countries' laws, but the CIA had been guilty of breaking its own country's laws. Furthermore, the agency had exposed American hypocrisy by flaunting the international arms embargo on Iran, which it had proactively been pushing other countries to observe.[37] At variance with Clarridge's view, under Webster both legislative and executive officials acknowledged the new value of congressional participation.[38] "On balance," remarked Webster, demonstrating a departure from Casey's adversarial attitude toward Congress, "the oversight process has clearly been useful and helpful."[39] Gates, having missed out on the chance to be director himself due to the scandal, clearly agreed, referring to the former judge as a "godsend," bringing "a reputation for integrity, honesty, and fidelity to the Constitution" to the CIA when it was still under a shadow.[40]

Whether the cleanup was inconsistent with the purpose of the CIA or a necessary step to save the agency from itself is not the main issue here, however. The point is that the violent swings between the CIA being "unleashed" (as Reagan had promised during his election campaign) and then "cleaned up" has had an extremely negative impact upon its effectiveness over time, especially in the field of counterterrorism.[41] First, the waves of prosecutions, even if nullified by presidential pardon, stripped out talent and ended the careers of some of the agency's most capable officers for following orders passed down from above. Second, the constantly shifting boundaries and expectations placed upon the CIA created a sense of confusion as to exactly what the agency's role was when it came to counterterrorism. Finally, and most crucially, the sense that its officers could be retrospectively prosecuted for following orders from their leaders in response to the demands of government officials created a sense of risk aversion. This risk aversion played a significant role in limiting the Clinton administration's options for dealing with bin Laden and the wider al-Qaeda network, and can be seen as a contributing factor to the United States' failure to neutralize the group before the 9/11 attacks occurred.

In the longer term, however, the CIA's risk aversion helped push U.S. counterterrorism toward the adoption of the Predator drone and the subsequent drafting of the legal sanctions that have accompanied its use. Ultimately the insistence upon greater legal authority for the use of lethal force enhanced America's counterterrorism capabilities by providing policy makers with the option of more aggressive covert action, such as targeted killings, while giving the CTC the authority and tools to deploy this force. Whether or not the availability of such an option is a positive development for the United States is still very much up for debate, but the roots of this policy can be found in the failures of the agency's past counterterror efforts.

The Weakness of the CIA's Counterterrorist Center

The CTC and the CIA more widely were still reeling from the effect of the Iran-Contra scandal and the Holiday Party Massacre when Clinton assumed office. The rapid collapse of the Soviet Union had created further challenges for an agency whose very raison d'être had been the winning of the Cold War. The resulting drop in anti-Western terrorist incidents that followed the loss of material support and sanctuary from collapsing Communist states put the CTC under even greater pressure, and it is telling that George H. W. Bush's administration never felt the need to introduce a counterterrorism policy of its own.[42] When Clinton entered the Oval Office, his initial focus was very much on the domestic economy. Not a single foreign policy objective was set out in the goals of his administration's first hundred days.[43] The Cold War was over, and the American public had voted for a president whose unofficial campaign motto had been "It's the economy, stupid!"[44] As a result of this domestic focus and the CIA's damaged reputation, it was singled out for significant budget cuts under Clinton, who had made a campaign promise to cut its funding by 25 percent ($7.5 billion) over five years.[45]

In contrast to the Clinton campaign narrative, which convinced voters that the first post–Cold War decade should be one of domestic regeneration over defense and intelligence spending, Clinton's pick for DCI, James Woolsey, sought to make the case that the CIA was as vital as ever in the post-Soviet security environment. Woolsey, a man who was commonly known to have had very little personal connection to the president, testified before the Senate Select Committee for Intelligence on February 2, 1993, and warned that the United States may have "slain a dragon" of Communism, but it now faced "a jungle filled with poisonous snakes." In many ways the dragon, Woolsey's colorful metaphor concluded, had been easier to keep track of.[46] Two terrorist incidents within eight weeks of Clinton's inauguration seemingly supported the new DCI's position, with analysts warning that the attacks were most likely linked to radical Islamist training camps in Afghanistan and Pakistan.[47] In spite of these warnings, the promised tranche of cuts went ahead.

Despite resistance from Clinton, the Democrat-controlled Congress followed the initial cuts with further budget reductions, eventually cutting $7.5 billion from the CIA budget in three years as opposed to the five proposed by the president. The CIA did recover from the budget cuts imposed during Clinton's first term as the threat posed by nonstate terrorism heightened. By 1997 the president sought to increase the agency's counterterrorism funding from the proposed $9.7 million to $41 million.

When Republicans gained control of Congress in 1998 Clinton supported their further budget increase to the CIA, which saw approximately $3 billion added to the agency's overall budget by 1999, the largest single increase to the intelligence budget in fifteen years.[48]

In spite of the agency's eventual recovery, the instability wrought by the budget cuts damaged its morale and capabilities. Deputy Counsel John Rizzo summed up the impact of the to and fro of the pendulum by reflecting nostalgically that the Reagan-Bush era had been "a period that began with the Agency reenergized and ascendant," but that the transition to the Clinton administration had left it "largely adrift and leaderless."[49] The budget cuts represented not only a loss of financial capital but also a loss of political capital, a decline in its standing, the primary currency of the CIA.[50] Hit particularly hard on both fronts was the agency's Directorate of Operations and covert action capability, held in particularly low regard due to its association with the Iran-Contra scandal. From a place of prominence during the anti-Communist actions of the Cold War, when covert activity may have accounted for up to 60 percent of the annual intelligence budget, by the early 1990s the CIA's clandestine mission had fallen into a state of near disregard, with funding leveling out at approximately 1 percent of the agency's stripped income.[51]

The purges of the agency continued through the 1990s. The hugely detrimental case of the double agent Aldrich Ames, a veteran case officer who was arrested on espionage charges in 1994 after spying for the Russians since 1985, saw the CIA further chastised by Congress and ridiculed in the media.[52] In the aftermath of the debacle Woolsey was forced to resign his post, despite the fact that Ames's crimes had been going on for a decade before he was appointed DCI. With volunteers for the post of director of the beleaguered agency thin on the ground, the Clinton administration eventually settled upon the reluctant John Deutch, who had been serving as deputy secretary of defense. Despite his stellar public-service credentials, Deutch proved to be extraordinarily unpopular at the CIA, having savaged the agency during his nomination testimony with some highly critical and tone-deaf comments, such as announcing it was time for "a new generation of leaders and managers at the CIA" to replace those who had "grown up fully in the Cold War."[53] The new DCI had also made no secret of his reluctance to take the job, or of the fact that he valued technical data collection over human spying and espionage, the core business of the DO. "I'm a technical guy," he told one interviewer. "I'm a satellite guy. I'm a SIGINT [signals intelligence] guy."[54]

Despite securing the post of DCI with an impressive Senate vote margin of seventeen to zero in committee and ninety-eight to zero on the floor, Deutch began his short time at the CIA disliked and mistrusted.[55] Packing the senior staff offices on Langley's seventh floor with his former DoD aides, Deutch's plans for agency reform, included mixing DO and DI officers to remove "stovepipes," building closer links with the FBI and the DoD's regional commanders in chief (CINCs), and a renewed focus upon technical intelligence, were remarkably unpopular and poorly sold.[56] The task of winning the support of his subordinates was made all but impossible for Deutch by his having to conduct what the *New York Times* referred to as a "cleanse" of "dirty assets." This cleansing was largely prompted by a report from the IOB submitted on June 28, 1996, concerning the CIA's conduct in Guatemala from 1984 to 1995, during which time funds were provided to military officers known to have been involved in the killing of one American and the spouse of another.[57] The report argued: "Our clandestine intelligence capability will clearly remain essential to our national security. However, though the conduct of clandestine intelligence collection at times requires dealing with unsavory individuals and organizations, the value of what we hope to gain in terms of our national interests must outweigh the costs of such unseemly relationships and be worth the risks always inherent in clandestine activity."[58]

As a result of the Oversight Board's feedback and Deutch's promise during his testimony to hold all members of staff personally accountable for their actions and any misdeeds that might occur as a result, the director fired Terry Ward and Fred Brugger, two senior DO officers with links to the Guatemala operation. According to the CIA lawyer John Rizzo, these officers had spotless records from decades of service and were enormously popular figures inside the agency.[59] Jeffrey Smith, Deutch's handpicked general counsel, worked closely with the new DCI to write what became known as the "Deutch rules."[60] These new guidelines severely narrowed the margins of moral tolerance to be observed by the agency when it went looking to recruit assets in the future.[61] Besides the practical constraints that the guidelines introduced with respect to recruiting assets linked to a group such as al-Qaeda, their introduction also had a devastating impact upon morale. "Our spirit was broken," said William Lofgren, chief of the Central Eurasian Division. "At the CIA, you have to be able to inspire people to take outrageous risks. Deutch didn't care about us at all." The result, according to Lofgren, was that "people retired in place or left."[62]

Deutch lasted less than two years as director of central intelligence, retiring from the post in December 1996. In a postmortem of his brief time at Langley, the *New York Times*'s Thomas Powers defined the abiding theme of his tenure as "a kind of ongoing guerrilla war between the DCI's office on the seventh floor and the clandestine folks," triggered by disrespect on Deutch's side and increasing dislike on the DO's.[63] By the time Deutch's deputy, George Tenet, was sworn in as director on July 31 1997, he inherited an agency that was in disarray. The constant turnover of directors had left its policies adrift, and its purpose unclear.[64] Moreover, the sackings of high-profile and popular staff had a tremendously detrimental impact upon morale, while the hefty budget cuts further drained off talent and spirit with early retirements and redundancies. The negative press associated with the CIA also caused recruitment to stall, with just twenty-five trainees becoming clandestine officers in 1995.[65]

For those who remained at the agency, a very real risk was that their actions could compromise one of the plethora of new rules and guidelines that had been introduced to curtail any morally ambiguous behavior. Worse still, their current actions could breach a guideline or break a rule which had not yet been introduced, but which could be applied retrospectively in the future, as had been the case with Ward and Brugger. Within this atmosphere, according to former CIA officer Melissa Boyle Mahle, a culture of risk aversion was "pervasive throughout the Agency."[66] As Mahle spent time prior to 9/11 working in the DO recruitment center, her criticism of the culture at Langley is particularly telling. So poor was the agency's standing that early in his tenure Tenet was invited by former president Ford to take part in a discussion panel entitled "Does America Need a CIA?" Though the debate was a purely academic exercise, the very existence of such a topic for discussion on a panel served to demonstrate the vulnerability of the agency.[67]

Speaking before the Senate Select Committee on Intelligence in the first hearing of his nomination, Tenet sought to tackle the area he saw as most responsible for the waning of the CIA's fortunes. He aimed to dramatically reduce the CIA's role in aggressive covert action and paramilitary activities, the very missions that caused the swinging of the pendulum for and against the agency. He described such activities as the "most controversial" of the CIA's major functions, as they exposed both the agency and its officers to unacceptable legal risk. "A succession of administrations," observed Tenet in his autobiography, "would tell [CIA officers] that they were expected to take risks and be aggressive. But if something went wrong, Agency officials faced disgrace, dismissal, and financial ruin."[68] Though being careful not to fully reject the use of covert force by promising to "sustain the infrastructure we need when the President directs us to act," Tenet argued that covert action, while "a critical instrument of U.S. foreign policy . . . should never be the last resort of a failed policy."[69] Tenet's concerns that the CIA often engaged in lethal covert actions because of a lack of other credible policy choices were well founded. It was exactly this scenario that had led to the CIA being put at the front line of America's first war against terrorism during the Reagan administration, when Casey was given the job of dealing with America's elusive enemies, a job that no other branch of government had wanted. A decade later, with its staff, budget, morale, and legal latitude all reduced, the agency was once again called upon to take on another such task, that of neutralizing the threat posed by Osama bin Laden and his al-Qaeda network.

The Importance of Correct Legal Sanction for Successful Covert Action

Faced with reports of the growing threat posed by the radical Sunni Islamic organization that had declared a jihad against the United States in 1996, Clinton turned to the CIA to provide a covert action solution. In theory, this was a positive development, as the agency had been looking for new challenges to make it relevant again after the Cold War.[70] As Richard Immerman observes, DCI Tenet recognized that "for the purpose of advancing the CIA's mission to enhance American security and restoring its reputation, budget and élan, the agency had to reorient its emphasis."[71] The mounting threat of the al-Qaeda network offered such an opportunity. Making the CIA the linchpin for a more aggressive and concerted effort to confront and counter terrorism significantly helped restore the agency's reputation and funding.[72] But placing the CIA in this position also presented new risks for the agency and its staff. As the acrimonious debates between the principals of the Reagan administration had previously revealed, neutralizing the threat posed by terrorist groups was fraught with legal and ethical challenges, not to mention physical dangers.

With its protagonists existing in a legally gray area between criminals and soldiers, there would always be questions about what level of force was proportionate for the agency to use in counterterrorism operations. Furthermore, terrorists frequently sheltered among civilians, which created additional risks and challenges when they were pursued. Neutralizing the threat posed by terrorism while remaining within the confines of domestic and international law was a challenge with which the CTC had already struggled at a time when its latitude for action had been considerably greater and its resources more plentiful. Risk averse from a decade under pressure, what the CTC required was a clear presidential Finding, a transparent Memorandum of Notification (MON), a direct Executive Order, or a detailed Presidential Decision Directive giving explicit legal authority to go after bin Laden and his followers with whatever force was deemed necessary. The CTC needed to be able to operate without fear of legal recriminations should the planned operation fail or result in wider casualties than planned. This is what the Reagan administration had sought to do with NSDD 138 and the accompanying Finding, endorsed by then CIA chief counsel Stanley Sporkin.[73]

Post-9/11, the Bush and Obama administrations both ensured the CTC had clear legal sanction to neutralize al-Qaeda members and affiliates when placed on the front line of the War on Terror. Bush's administration had initially held the same legal position as its predecessor, rejecting within its first month a CIA request for a more comprehensive and aggressive MON to target al-Qaeda.[74] Following the September 11 attacks, however, new legislation was quickly introduced that altered the conduct of U.S. counterterrorism. The most significant element was the Authorization for the Use of Military Force against Terrorists (AUMF), passed by Congress 420 to one and signed by Bush on September 18, 2001. Written as it was in the emotional days following the 9/11 attacks, the only U.S. politician to vote against the AUMF was the Democratic representative for California,

Barbara Lee. Her concerns were, quite rightly, that the wording of the authorization was a "blank check" for military action. "Let's step back for a moment," she argued; "let's just pause, just for a minute, and think through the implications of our actions today so that this does not spiral out of control." The AUMF was initially intended to give Bush the ability to retaliate against whoever orchestrated the attacks, but it has since become the primary legal justification for almost every covert American action in the War on Terror, including the drone campaign.[75] At just sixty words, the AUMF is suitably vague to license all manner of counterterrorism actions. It reads: "That the President is authorized to use all necessary and appropriate force against those nations, organizations, or persons he determines planned, authorized, committed, or aided the terrorist attacks that occurred on September 11, 2001, or harbored such organizations or persons, in order to prevent any future acts of international terrorism against the United States by such nations, organizations or persons."[76] "It's like a Christmas tree," said John Bellinger III, a lawyer who worked closely with Condoleezza Rice, first Bush's APNSA and later his secretary of state. "All sorts of things have been hung off those 60 words."[77]

Accompanying the AUMF was Resolution 1368, passed by the U.N. Security Council the day after the 9/11 attacks. The resolution declared al-Qaeda's actions a "threat to international peace and security" and called on "all States to work together . . . to bring to justice the perpetrators." The resolution also justified a focus upon the Taliban and any other group linked to al-Qaeda by adding, "Those responsible for aiding, supporting or harboring the perpetrators . . . will be held accountable."[78] This resolution has been significant in enabling the United States to validate its continued campaign against al-Qaeda and its affiliates as self-defense under Article 51 of the U.N. Charter, which guarantees a state's "inherent right of individual or collective self-defense if an armed attack occurs against a Member of the United Nations, until the Security Council has taken measures necessary to maintain international peace and security."[79]

The day before the AUMF was passed, Bush also signed a Finding explicitly authorizing the CIA to target and kill named terrorists.[80] Though media reports suggested Bush's order marked the first departure from America's ban on assassination since Ford had issued it, the order was actually based upon a Finding issued by Clinton in 1998. This Finding in turn had its roots in NSDD 138 and the accompanying Finding signed off on by Sporkin two decades earlier.[81] Those who argue that "9/11 changed everything," as Vice President Cheney put it, oversimplify the CIA's position.[82] In spite of the well-publicized examples of a gung-ho revenge-orientated attitude from members of the agency, such as then CTC chief Cofer Black's order to his agents to bring him "bin Laden's head in a box on dry ice," the agency did not immediately embrace targeted killing following Bush's Finding.[83] Even with the new authorities granted to the CIA, the *Washington Post* reported that concerns within the agency regarding legal protection still lingered. Agency lawyers felt the Bush Finding and the AUMF were too ambiguous. The agency was reluctant to accept a broad grant of authority to hunt and kill U.S. enemies at its discretion, but it was willing, and believed itself able, to take the lives of terrorists designated by the president.[84]

In the face of the CIA's apprehension, further legal cover was quickly added through the Bush administration's military order regarding the legal status of al-Qaeda and Taliban fighters, issued on November 13, 2001. The order designated the fighters of both groups as "unlawful combatants" rather than enemy soldiers or civilians, effectively denying them Geneva Convention protections.[85] This "in-between" category, as legal scholar Claire Finkelstein has described it, combined with Congress's authorization for the use of military force against al-Qaeda and the Taliban, eased the way for the CIA to undertake targeted killings without fear of legal reprisals.[86] This position was further bolstered on August 27, 2004, when Bush issued EO 13355, which superseded Reagan's EO 12333, the successor to Ford's original assassination ban. While the order did not overtly overturn the assassination ban, it did define "countering terrorism" as "a matter of the highest national security priority" and granted the intelligence community "the authority to translate the intelligence objectives and priorities approved by the President into specific guidance for the Intelligence Community."[87] With this legal architecture in place, the CTC's first lethal drone strike outside the official war zone of Afghanistan occurred in Yemen on November 3, 2002, when one of its Predators destroyed a jeep carrying six al-Qaeda militants.[88] Duane Clarridge's Eagle Program had come full circle, and the CTC was once again in possession of an armed remote-controlled aircraft.[89] Ironically, for a man who had argued that it was the role of the CIA to break laws, Clarridge's lethal vision for the CTC could only be realized once it was fully enshrined within laws.

A common criticism of the use of the CTC to lead the drone campaign, endorsed by many within the legal academy and the human rights community, has been that, despite the legal architecture, CIA drone operators are themselves unlawful combatants due to their civilian status.[90] The legal scholar Gary Solis—a veteran of the U.S. Marine Corps and author of *The Law of Armed Conflict*—has voiced such criticism, arguing that as fighters without uniforms or insignia, directly participating in hostilities, the CIA's agents were employing armed force contrary to the laws and customs of war, thus violating the requirement of distinction, a core concept of modern armed conflict.[91] Law scholar and former navy surface warfare officer David Glazier took Solis's argument further during a hearing before the House Committee on Oversight and Government Reform's national security and foreign affairs panel. "It is my opinion, as well as that of most other law-of-war scholars that I know," advised Glazier, "that those who participate in hostilities without the combatant's privilege . . . gain no immunity from domestic laws." Under this view, CIA drone pilots are liable to prosecution under the law of any jurisdiction where attacks occur for any injuries, deaths, or property damage they cause.[92] "One wonders," pondered Solis, "whether CIA civilians who are associated with armed drones appreciate their position in the law of armed conflict."[93]

Regardless of how conscious individual members of the CTC were in regarding the legality of drone strikes, it is something their superiors clearly paid attention to. A history of blowback against the agency, from the assassination efforts of the 1950s, the Iran-Contra affair in the 1980s, and the enhanced interrogation techniques scandal of the late 2000s served to demonstrate to the agency's senior staff that perspectives of wrongdoing for undertaking actions once fully endorsed by the White House could be applied retrospectively, and could be career-ending. A caution to avoid such fallout helps explain the vigorous legal defense that has been mounted against these arguments, and also explains the unorthodox chain of command that has since emerged under which drones actually function.

Charles Kels, an attorney for the Department of Homeland Security, dismissed criticism around the legality of CTC agents actively participating in drone warfare as "pernicious accusations" born of "an overly rigid and inaccurate reading of the law of war." International humanitarian law, Kels argued, does not prohibit the use of civilian personnel in combat, as long as they do not engage in "perfidy," or treachery, such as the feigning of civilian, noncombatant status to mount ambushes. Thus, Kels reasoned, while it would be unlawful for the CIA to "paint a drone in the insignia of a commercial airline carrier" to launch surprise attacks, there is nothing intrinsically wrong with civilians aiding combat operations.[94]

The second legal argument—that the CTC's drone pilots, as civilians, were not entitled to the same immunity accorded belligerents as active service personnel were—was rendered irrelevant with the revelation that the CTC's drones have in fact always been flown by regular air force pilots. Based upon the partnership first forged between the CIA and the air force when the agency acquired the GNAT-750-45 drones to aid the campaign over Bosnia, the operation of the CTC's drones was always a collaborative affair between the pilots who helped develop the technology, their GA-ASI partners, and the CIA.[95] Documents released to the National Security Archive by aviation historian Richard Whittle have revealed that on September 18, 2001, as America was still reeling from the 9/11 attacks, this collaboration continued. On the orders of Air Combat Command, the air force team members who had flown the Afghan Eyes reconnaissance missions over Afghanistan the previous year on the CTC's behalf were reassembled at the agency's headquarters in Langley. Based in a mobile home on the CIA campus, and operating under the name Detachment 1, Air Combat Command, Pentagon, the unit flew the agency's first armed drones over Afghanistan and elsewhere from ground control stations on the Langley campus. On May 29, 2002, the unit was officially disbanded and replaced by Detachment 1, 17th Reconnaissance Squadron, which was relocated from its temporary accommodation to Indian Springs Air Force Auxiliary Field, the small Nevada desert base where initial test flights of the Predator drone had been conducted.[96] Upgraded and renamed Creech Air Force Base in 2005, the airfield has since served as the primary location from which the CIA's drones are operated by the air force personnel posted there.[97]

The revelation that air force personnel operate the CIA's drones may have eliminated the legal concern relating to a civilian agency executing lethal strikes, but it raises an arguably more pressing issue with regard to the blurring of the civilian/military divide between the CIA and the USAF. Though possible to explain the initial chain of command governing the CTC's drone use through the historical structure of the Predator project, such reasoning does not explain why the collaborative approach remained when responsibility could have been transferred to either the CIA or the air force. The most likely explanation appears to be that the Bush and, subsequently, the Obama administration sought to keep the drone campaign associated with the CIA to use—or abuse—the civilian agency's authority to conduct covert action as set out in the National Security Act of 1947 (and later NSC directive 10/2), and its protection from needing to publicly disclose foreign activities under Title 50 of the U.S. Code.[98] This has enabled the air force, usually bound to declare its actions publically under the U.S. military's Title 10 code, to operate in a covert manner previously reserved for America's civilian spies, providing a level of deniability typically unavailable to the personnel of America's air force.

As the continuity of this unorthodox chain of command demonstrates, the Obama administration relied upon the same legal structures as the Bush administration to justify its significantly larger-scale targeted-killing campaign, with just two notable differences.[99] First, Obama abandoned the term "unlawful combatant" to describe members of al-Qaeda and their associates. This was a political decision rather than a legal one, as the term's association with the prisoners held in the controversial Guantánamo Bay detention facility meant it held negative political connotations. In practical terms the in-between category was retained, and Obama went to great lengths to ensure that his legal authority was seen to be drawn from and consistent with the laws of armed conflict and the principle of "just war" rather than purely from the AUMF. To this end Obama used his 2009 Nobel Peace Prize speech to reframe the rhetoric of Bush's War on Terror as a just war of self-defense against al-Qaeda, arguing that the United States' actions were "in self-defense," that "the force used is proportional," and that "whenever possible, civilians are spared from violence."[100] Other than these rhetorical changes, the Obama administration vigorously defended the legality of using lethal force against al-Qaeda and its associates, not just in the hot battlefield of Afghanistan but also in the geographically remote areas of Northwest Pakistan, Yemen, and Somalia, where al-Qaeda sought to establish safe havens. "Indeed," reflects Robert

Chesney, "the Obama administration quickly outstripped its predecessor in terms of the quantity and location of its airstrikes outside of Afghanistan, and it also greatly surpassed the Bush administration in its efforts to marshal public defenses of the legality of these actions."[101]

This continuation led to accusations of backtracking from many of those who interpreted Obama's 2008 campaign rhetoric of change to indicate a dramatic departure from the War on Terror.[102] A Reason-Rupe Poll published on September 10, 2013, revealed that 64 percent of Americans questioned said "President Obama's foreign policy is worse or [the] same as Bush's." It is worth noting, however, that this poll was conducted in the wake of the NSA scandal triggered by Edward Snowden's leaked revelations and while Congress was debating possible intervention in Syria, factors that helped to associate the president with the excesses of his predecessor. Trevor McCrisken argues that those who expected Obama to introduce wholesale changes to U.S. counterterrorism policy misread Obama's intentions. According to McCrisken, Obama always intended to deepen Bush's commitment to counterterrorism, with the change being a campaign more focused upon dismantling al-Qaeda than upon the wider distractions of regime change and democracy promotion, which the war in Iraq represented.[103] Throughout his 2008 election campaign, Obama made clear this intention to refocus on neutralizing al-Qaeda. For example, during his first debate with John McCain, Obama argued that, despite the United States having spent more than $600 billion on the war Iraq, sustaining four thousand military fatalities and a further thirty thousand wounded, Bush had neither captured bin Laden nor "put al-Qaeda to rest." In fact, Obama continued, the terrorist group was "resurgent, stronger now than at any time since 2001." The reason for this, the senator argued, was that the Bush administration had taken its "eye off the ball" by invading Iraq. Rather than ending with an antiwar message, however, Obama finished his answer by stating: "The lesson to be drawn is that we should never hesitate to use military force," but it should be used wisely. The invasion of Iraq had not been wise.[104]

During the same debate, Obama went on to set out his intentions for refocusing the War on Terror, introducing what would evolve into the CIA's drone campaign in Pakistan, by arguing that "every intelligence agency will acknowledge that al-Qaeda is the greatest threat against the United States," and that not enough had been done to "get rid of the safe havens . . . across the border in the northwest regions" of Pakistan. Indicating his willingness to act unilaterally against the terrorist group, Obama concluded his point by declaring: "If the United States has al-Qaeda, bin Laden, top-level lieutenants in our sights, and Pakistan is unable or unwilling to act, then we should take them out."[105] While some have argued that the continuation of certain policies under Obama, such as the failure to close the Guantánamo Bay detention facility, illustrate the structural limits to change within the U.S. system, when it comes to his policy against al-Qaeda, Obama followed through with what he proposed from the outset.[106] Hence, any misinterpretation of his intention to mount a more aggressive campaign against al-Qaeda on the part of his supporters was arguably an act of self-deception, not dishonesty on the part of the candidate.

The expansion of the drone campaign did force the Obama White House to strengthen the legal case in one area. The DoJ was pressured into publishing a white paper in order to provide further clarification on the administration's authority to target American citizens, after receiving heavy criticism from both the political left and the political right following the killing of Anwar al-Awlaki in a drone strike in Yemen on September 30, 2011.[107] Addressing the issue further in a speech on U.S. counterterrorism methods, delivered on May 23, 2013, Obama promised to "extend oversight of lethal actions outside of warzones that go beyond our reporting to Congress," while at the same time warning that such a course of action posed significant difficulties in practice and would therefore take time to implement.[108] As part of this promised extension of oversight, Attorney General Eric Holder had already submitted a letter to the Senate Judiciary Committee the day before the speech in which he explained that the decision to target Awlaki had been based on "generations-old legal principles and Supreme Court decisions handed down during World War II." The attorney general stated that for the small number who decide "to commit violent acts against their own country from abroad," merely being an American citizen does not make one "immune from being targeted." Holder went on to argue: "The Supreme Court has long 'made clear that a state of war is not a blank check for the President when it comes to the rights of the Nation's citizens.' . . . But the Court's case law and longstanding practice and principle also make clear that the Constitution does not prohibit the Government it establishes from taking action to protect the American people from threats posed by terrorists who hide in faraway countries and continually plan and launch plots against the U.S. homeland. The decision to target Anwar al-Aulaqi was lawful, it was considered, and it was just."[109]

Significantly, Awlaki was not the first American citizen killed in a drone strike and thus executed without a trial. On November 3, 2002, a Hellfire missile fired from a CIA Predator destroyed a car in Yemen that had been carrying a well-known al-Qaeda commander named Qaed Senan al-Harethi. Referred to as the "Godfather of Terror," Harethis was linked to the 2000 bombing of the USS *Cole,* which had killed seventeen U.S. sailors. Unbeknownst to the CIA at the time, one of Harethi's passengers in the incinerated vehicle had been Kemal Derwish, a naturalized U.S. citizen and leader of a seven-man al-Qaeda sleeper cell based in Lackawanna, New York. When questioned about the killing of the American by the *Washington Post,* the Bush administration denied knowing Derwish was in the vehicle— a statement later contested by the *Post*'s Dana Priest, who reported in January 2010 that the CIA had in fact been aware of the American's presence in the vehicle.[110] Despite the denial, an anonymous government official echoed the same principle set out in Attorney General Holder's letter a decade later, telling the reporters: "It would not have made a difference. If you're a terrorist, you're a terrorist."[111]

Despite Derwish's death not attracting anywhere near the level of publicity and protest Awlaki's did, the killing of an American citizen and the high-tech nature of its execution did bring some criticism. Alluding to the potentially seductive nature of the new remote method of targeted killing, an article published in the *Economist* later that year entitled "Assassination by Remote Control" predicted that "the use of pilotless aircraft to hunt down terrorists will become more appealing to America."[112] In typically bullish style and reflecting a period during which America's counterterrorism practices were under less scrutiny due to the relative proximity of 9/11, Bush defended the CIA's pursuit of al-Qaeda leaders through rendition and targeted killings in a speech delivered on the day of the Yemen strike. He argued: "The best way to secure this homeland, short-term and long-term, is to find those killers one at a time and bring them to justice." The struggle against al-Qaeda was a different kind of war, Bush explained. "It's a kind of war instead [*sic*] of trying to knock down airplanes and sink ships, we're looking in caves." The United States had to put "the spotlight on some of the dark corners of the world," because al-Qaeda's leaders hid there while they sent youngsters out to their suicidal deaths, the president continued. "The only way to treat them," Bush reasoned, "is what they are—international killers." And the only way to find them, the president added, was "to be patient and steadfast and hunt them down."[113]

The Clinton Administration's Lack of Appropriate Legal Architecture

Though the legal clarity that authorizes the CIA's lethal drone campaign may appear to be a product of the 9/11 attacks, the reality is that it is based upon a legal architecture that has its roots in the Reagan administration and gradually evolved during the Clinton administration. The 9/11 attacks accelerated the process and guaranteed greater public support, but the policy of targeted killing embraced by George W. Bush and to a greater extent by Obama has both its strategic and legal roots in the actions of the Reagan and Clinton administrations. For Clinton, however, in the pre-9/11 security environment in which the prevailing culture had been to constrain the CIA's covert paramilitary capabilities, the path to empowering the CTC to neutralize al-Qaeda's leadership was a cautious, at times faltering, and frequently frustrating one.

When the new wave of Sunni Islamic terrorism first revealed itself, initially with Mir Ami Kasi's attack upon CIA staff outside Langley on January 25, 1993, followed by Ramzi Yousef's bombing of the World Trade Center a month later, the unanimous view within the government at the time was that the law enforcement community was to take the lead in combatting the terrorists. The FBI headed the investigation, and Yousef and his fellow terrorists were to be indicted, arrested, and prosecuted in U.S. courts.[114] Some, such as the CIA's John Rizzo, have argued that, while such an approach presented relatively few legal challenges for the U.S., it left America vulnerable to the problem Shultz and his fellow counterterrorism hardliners had warned of two decades before, that a terrorist attack had to be carried out before any action could be taken against its perpetrators.[115]

This is not strictly true, however, as the FBI's TERRSTOP investigation showed. As a result of the investigation, Omar Abdel-Rahman, otherwise known as "the Blind Sheikh," and eleven of his followers were arrested, charged, and found guilty of planning a "Day of Terror," during which the group hoped to paralyze New York City by assassinating several political figures and bombing numerous landmarks simultaneously.[116] In defense of Rizzo's criticism, however, the records show that it is unlikely the FBI would have investigated the plot had it not been for the Yousef bombing, despite warnings from a well-placed informant, Emad Salem. Bureau documents quote Salem as having berated agents following the WTC attack, declaring in frustration: "It takes a bomb with you guys to wake up and start to move."[117]

Anthony Lake, Clinton's APNSA, described the WTC bombing as a "wake-up call."[118] The State Department's "Patterns of Global Terrorism" reports for 1994 and 1995 showed a significant increase in the number of Sunni-jihadist-linked terror attacks, and Paul Pillar, a counterterrorism analyst at the CIA, coined the phrase "ad hoc terrorism" to describe what he warned was "a watershed in global terrorism—the debut of a new generation of unaffiliated, religious and political violence."[119] Gradually, as intelligence reporting continued to warn of the growing transnational security threat posed to the United States by this Islamic extremism, both the Clinton administration and the CIA began to move toward the position previously held by the counterterrorism hardliners within the Reagan administration. In January 1996, the CTC focused its intelligence resources into an interdisciplinary virtual station staffed by CIA, NSA, FBI, and other officers to track an individual identified as "one of the most significant financial sponsors of Islamic extremist activities in the world."[120] That individual was Osama bin Laden, the leader of al-Qaeda, an Islamic extremist group formed in 1988 with the explicit aim of uniting the previously disparate Sunni Islamic radical groups together in a jihad against the "far enemy" of the United States.[121] The station's mission, according to DCI Tenet's testimony to the 9/11 Commission, was to "track him [bin Laden], collect intelligence on him, run operations against him, disrupt his finances and warn policymakers about his activities and intentions."[122] The creation of the Bin Laden Issue Station marked a significant return on the part of the CTC to what its founder, Duane Clarridge, had first envisioned. It was interdisciplinary, global in outlook, and focused on developing plans to neutralize terrorist threats. There was one noteworthy exception, however: its lack of lethal capability.

Despite Tenet informing the Clinton administration that "an al-Qaeda defector told us that UBL [bin Laden] was the head of a worldwide terrorist organization with a board of directors that included al-Zawahiri," and in spite of the terrorist leader issuing two public declarations of jihad against the United States in which he made clear his intent to kill Americans and their allies, the CTC had very limited legal authority to mount an operation to neutralize this threat.[123] The nature of the CTC's sources and evidence meant that initially, regardless of the significant intelligence attesting to his group's organizational abilities and determination to strike the United States, there was no legal indictment for bin Laden. The terrorist leader's use of virtually lawless safe havens made it impossible to investigate his group's activities in order to gather evidence of the quality to stand up in an American court. Furthermore, even after an indictment was eventually filed, following al-Qaeda's devastating bombing of American embassies in Kenya and Tanzania on August 7, 1998, bin Laden's sheltered locations made arrest almost impossible.[124]

Initially, bin Laden had based himself in Sudan, which under Hassan Turabi's rule had become what former CIA analyst Bruce Riedel described as "a safe haven for all kinds of radicals and dissidents."[125] The State Department defines a terrorist safe haven as "ungoverned, under-governed, or ill-governed physical areas where terrorists are able to organize, plan, raise funds, communicate, recruit, train, transit, and operate in relative security because of inadequate governance capacity, political will, or both." Sudan certainly met those criteria for al-Qaeda.[126] The United Nations did eventually exert political pressure on Turabi to expel known terrorists after an assassination attempt on Egyptian prime minster Hosni Mubarak was traced back Sudan.[127] As a result, bin Laden returned to Afghanistan, where he had fought in the anti-Soviet jihad a decade previously. According to the 9/11 Commission, upon his arrival Pakistani Inter-Services Intelligence (ISI) officers introduced him to the Taliban.[128] From here the link is well documented, with the Taliban sheltering bin Laden as their guest, having reportedly asked him not to carry out terrorist activities against other countries.[129] As U.S. diplomatic relations with the Taliban soured, any chance of prosecuting bin Laden through official legal and diplomatic channels dissolved.[130]

That bin Laden was a threat to the lives of U.S. citizens and the citizens of America's allies was clear, as was the fact that he was beyond the reach of both the U.S. judicial system and international law. The challenge of reaching terrorists who sheltered in safe havens around the world was one with which the Reagan administration had already struggled. The solution had been the CIA's creation of the CTC, backed by a presidential Finding, which had authorized its officers to undertake worldwide covert action against terrorism. After the fallout of the Iran-Contra scandal and a decade of budget cuts and legal constraints, however, the CTC was less well equipped to complete the mission than its founders had originally envisioned. The center quickly developed a plan that involved deploying the bin Laden unit's TRODPINT team (leftover assets from the agency's involvement with the Afghan mujahideen in the 1980s) to storm Tarnak Farm, a compound in which it was believed bin Laden was residing.[131]

Despite the fact that the Reagan-era Finding technically provided adequate legal authority for the CTC to begin moving against bin Laden, in the words of the 9/11 Commission the agency was "mindful of the old 'rogue elephant' charge," and senior CIA managers wanted something on paper to show that they were not acting on their own.[132] John Rizzo, who drafted both the original Reagan Finding and those that followed under Clinton, explained how significantly things had changed for the agency since the days of the Reagan counterterrorism hardliners, noting that "of all the officials who had been involved in the Finding back then, I was the only one who was still around in 1998." According to Rizzo's memoirs, he was uncomfortable with the agency acting simply on the basis of that Finding alone. "I wrote it in another era, to deal with terrorists from another era, long before there was anyone comparable to bin Laden." The CIA counsel also had doubts concerning the proposed operation. When the Clinton administration first ordered the CTC to develop a plan to capture bin Laden, the al-Qaeda leader was not actually facing criminal charges in the United States, nor was there another country that would have been willing to put him on trial on America's behalf. Furthermore, the CTC had reported that bin Laden was unlikely to come quietly, raising fears that any snatch operation undertaken by the Afghan tribals could result in a shoot-out causing the death of bin Laden, and perhaps the deaths of women and children around him. "And the Clinton White House was only talking about capturing bin Laden," noted Rizzo, "not knocking him off."[133]

In all, four MONs were drafted by the CIA for Clinton between February 1998 and February 1999. The first was for the planned tribal snatch plan, but the proposal never made it to the president's desk for approval. Michael Scheuer, the head of the CTC's Bin Laden station at the time, was strongly in favor of the operation, but senior managers at the agency had over-whelming doubts.[134] Former DCI Tenet provided his version of events in his memoir, explaining that he took Scheuer's recommendation "very seriously, but six senior CIA officers stood in the chain of command between Mike and me." Unlike Scheuer, who was an analyst untrained in paramili-tary operations, most of the senior staffers were seasoned operations officers. "Every one of those senior operations officers above Mike," Tenet noted, "recommended against undertaking the operation." The reasons given ranged from an outright belief that the operation could not succeed to a fear that too many innocent women and children were likely to be killed. Geoff O'Connell, director of the CTC at the time of the planning, allegedly told the former DCI that the proposed raid was "the 'best plan we had,' but that 'it simply wasn't good enough.' "[135]

Not only were the CIA managers concerned about the strategic elements of the plan, there was also apprehension concerning the reliability of the surrogates to conduct the operation within the boundaries of what would be acceptable. Tenet makes a specific point of the fact that no one in the Clinton administration gave the CIA permission to use elite American troops; instead, the agency had to rely upon its old tribal links. Having experienced first-hand from the failed Fadlallah mission in Beirut a decade earlier how difficult untrained surrogates can be to control in such situa-tions, and how the blowback of a failed operation can damage both the CIA and wider U.S. interests, the agency was determined to avoid collateral damage.[136] As Tenet explained, "[We] couldn't simply have our surrogates burst in, guns blazing, and hope for the best." Commenting on the shift in attitudes and latitude pre and post-9/11, the former DCI noted how "that sort of 'kill 'em and let God sort 'em out' approach" might have had a lot of appeal after the massacres of 9/11, but 1998 was a different envi-ronment, legally and otherwise.[137]

The second MON was signed in late August 1998, following al-Qaeda's bombing of U.S. embassies in Africa. With dramatic evidence that bin Laden's declarations of jihad against the United States had not been hollow threats, the new MON was meant to encapsulate the Clinton administration's increased determination to neutralize bin Laden. However, coupled with the strategic difficulty of planning an operation against bin Laden in a safe haven, this MON also highlighted the political difficulties involved in achieving this objective. The use of the intelligence services to undertake assassination was still banned by executive order. Though Stanley Sporkin, the CIA counsel under the Reagan administration, may have been willing to argue that this ban extended only to heads of state and not terrorists, both Rizzo and Tenet have identified the 1990s as a very different legal environment. In an effort to avoid any overt language linking the White House to something that could be interpreted as an assassination plot, Rizzo was instructed by staff within Clinton's NSC to add language to the MON authorizing the tribals to use force only in "self-defense." In Rizzo's words, "It was a phrase I did not remember ever including in a Finding or MON before and one more suited for a judge's jury instructions in a criminal trial than as a covert-action directive."[138] It was language that added a layer of complexity to an already difficult mission, as all planning had to be conducted with the exfiltration of bin Laden in mind.

As time dragged on and the CTC proved unable to develop a workable snatch plan, the CIA pushed for language that incorporated an acknowledgment on the part of the president that any effort to capture bin Laden was most likely going to end up with him killed in a gunfight. Clinton signed yet another MON in December 1998 to this effect, but once again political self-interest interfered with the language. Somewhere up the NSC chain on the way to the president a phrase was added that shifted the responsibility for bin Laden's death, should it occur, firmly to the CIA. The edited MON stated that the CIA could authorize the tribals to kill bin Laden, but only if capture was "not feasible." Again, Rizzo observed that this was another term he had previously never seen in a covert-action authorization.[139] There was good reason for this. The phrase was unnervingly vague for those accountable for signing off on the mission within the agency. It put responsibility for bin Laden's life or death squarely on the CIA, despite the fact that the orders to neutralize the threat came from the White House. Who was to judge whether capture was feasible? The DCI, the DO, the tribals' handlers in the CTC? Or was responsibility to be handed to the tribals themselves? At what point was the decision meant to be made? Was this something that could be decided and relayed to the tribals during the planning stage, or was it a decision the tribals were meant to make in the heat of the moment during the operation itself? Worst of all, as bitter past experience had taught the senior managers of the CIA, the judgment could even be made retrospectively by a member of the executive branch or a Senate Select Committee.

One final MON was introduced in February 1999. As doubts lingered about the TRODPINT team's likelihood of ever being up to the task of successfully snatching bin Laden, the CIA developed a plan to use the better trained and more capable forces of Ahmed Shah Massoud.[140] Massoud was arguably Afghanistan's most capable warlord and leader of the Taliban's bitter foes, the Northern Alliance. The CIA's hope was that Massoud would be able to accomplish what was being asked of the CTC within the boundaries of what Clinton's White House had imposed. While still uncomfortable with the feasibility aspect of the December 1998 MON, the CIA had greater confidence in the ability of Massoud's more professional and disciplined forces to conduct the operation cleanly and efficiently than they had with respect to the tribal assets. Though the Massoud plan was sound in theory, its chances of success were again undermined by political concerns surrounding assassination and the language of the MON. This time, it was Clinton himself who annotated the MON. The president reinserted the qualification from the August 1998 MON which stated that lethal action could only be taken in self-defense. Ultimately, this meant that members of the less capable TRODPINT group had the most latitude with their raid: they would be able to use their own judgment as to the feasibility of getting bin Laden out alive and act accordingly, whereas Massoud's more capable forces had stricter limitations, only able to use force in self-defense. The ambiguity of the American position was not lost on Massoud and his men. As an American officer responsible for liaising with the warlord's forces recalled, "The Northern Alliance thought, 'Oh, okay, you want us to capture him. Right. You crazy white guys.'"[141]

Garry Schroen, the CIA's main liaison with Massoud, whose links dated back to the Afghan jihad of the 1980s, believed that the capture proviso did not actually stop Massoud planning an operation against bin Laden. It did, however, inhibit those plans.[142] During his testimony before the 9/11 Commission five years later, Clinton claimed to have no recollection of editing the MON and was unable to explain why he did so.[143] Despite this, in his memoirs Clinton makes it clear that he felt he had given the CIA all of the authority it needed to neutralize the al-Qaeda leader, remarking that he signed several Memorandums of Notification authorizing the CIA to use lethal force to apprehend bin Laden. When the CTC officers and Afghan tribals were uncertain as to whether they had to try to capture bin Laden before they used deadly force, "I made it clear they did not," he claims unequivocally. Continuing to defend the extent to which he granted the CIA the necessary authorization to act, Clinton added that within a few months of signing the first MONs he had extended the lethal force authorization by extending the list of targeted bin Laden associates and the circumstances under which they could be attacked.[144]

Richard Clarke, Clinton's National Coordinator for Counterterrorism, agrees with the president's viewpoint, asserting that the CIA had all the legal authority it needed to carry out the operation against bin Laden. Expressing disbelief at what he regards as the CIA proving "pathetically unable to accomplish the mission," Clarke complains that he cannot understand "why it was impossible for the United States to find a competent group of Afghans, Americans, third country nationals, or some combination who could locate bin Laden in Afghanistan and kill him." Responding to criticism that the lethal authorizations were convoluted and that the people in the field did not know what they could do, Clarke argues that "every time such an objection was raised during those years, an additional authorization was drafted with the involvement of all the concerned agencies, and approved by the President's signature." Justifying the caution surrounding the language of the MONs, the former counterterrorism tsar described how the principals and the president sought to avoid sparking a cycle of violence similar to that which the Israelis found themselves locked into following their broad assassination policy introduced in retaliation for the 1972 Munich Olympics terror attacks, but he adds that "the President's intent was very clear: kill bin Laden." Any claims otherwise, Clarke protests, are excuses to cover up the agency's failings.[145]

Unsurprisingly, many within the CTC disagree with Clinton and Clarke's assessments. Scheuer, the first head of the Bin Laden station, blames what he saw as obstructionism, shortsightedness, and even cowardice by senior management in both the CIA and the executive branch for the failure to neutralize bin Laden. Shortly after the formation of the Bin Laden station, Scheuer had categorically stated that the destruction of bin Laden and al-Qaeda "could be accomplished only with the direct intervention of some type of force on the ground."[146] In Scheuer's view, senior staff lacked the will to do what was necessary to neutralize al-Qaeda. When relieved of command of the Bin Laden station in May 1999 for a series of recriminatory e-mails, Scheuer's response summed up his frustrations. Rather than retiring as was expected, he reported for work the following Monday and took a seat at a desk in the library, "waiting for the agency to come to him when it was ready to kill, not dither."[147] Following the 9/11 attacks, Scheuer was indeed taken back on and served as a special adviser to the chief of the Bin Laden unit from September 2001 to November 2004.[148]

In his book *Imperial Hubris,* first published anonymously the year he left the advisory post, Scheuer reasserted his case for the use of lethal force against al-Qaeda, heavily criticizing the defensive mentality America's government had adopted as "cowardice" cloaked by "words about international comity, civilized norms, and high moral standards." "Such words," Scheuer provocatively claimed, "are proper only in a suicide note for the nation." Instead, the former CIA analyst called for the engagement of "whatever martial behavior is needed" to remove the threat posed by al-Qaeda.[149] Scheuer argued that large-scale lethal force, with the accompanying risk to civilian lives, would be necessary to destroy al-Qaeda. "Progress will be measured by the pace of killing and, yes, by body counts." These body counts, Scheuer advised, would need to be precise and would likely run into "extremely large numbers." "The piles of dead will include as many or more civilians as combatants," warned the former head of the Bin Laden unit ominously, "because our enemies wear no uniforms."[150] Though Scheuer's forecast that civilian casualties would match those of America's enemies has proven unduly pessimistic, in many ways his prediction that effectively combating al-Qaeda would require a much more significant deployment of lethal force, precise body counts, and a greater tolerance of collateral damage has proven correct through the expansion of the CTC's drone campaign.

Scheuer's frustrations at the lack of lethal action were shared by other agents, who complained that politics had become more important to the agency than action. Robert Baer, a twenty-one-year veteran of the DO, described how changes to the CIA left him with "a growing rage that [he] was having more and more trouble containing." Nostalgically referring back to his days working under Clarridge in the CTC, Baer explained that "back then, the CIA was expected to operate on the edge, do things no other government agency would consider." By the time he retired in 1997, however, Baer complained that "politics had seeped down to its [the CIA's] lowest levels, into operations, where I worked."[151] Explaining his reasons for leaving the agency he loved, Baer continued his criticism: "Whether it was Osama bin Laden, Yasir Arafat, Iranian terrorism, Saddam Hussein, or any of the other evils that threaten the world, the Clinton administration seemed determined to sweep them all under the carpet. Ronald Reagan and George Bush before Clinton were not much better. The mantra at 1600 Pennsylvania Avenue seemed to be: Get through the term. Keep the bad news from the newspapers. Dump the naysayers. Gather money for the next election—gobs and gobs of it—and let some other administration down the line deal with it all. Worst of all, my CIA had decided to go along for the ride."[152] Gary Berntsen, another experienced member of the DO and the CTC, shared similar grievances in his memoirs. For Berntsen, a combination of Clinton, who "d[idn't] have it in his DNA to act . . . boldly," and "risk aversion on the seventh floor [the DCI's office]" made the necessary action against al-Qaeda "a non-starter."[153] Even Tenet's declaration of war against al-Qaeda following the embassy bombings in East Africa was rejected by Berntsen as nothing more than a rhetorical flourish. To Berntsen, the statement was not backed up by actions and was undermined by the ongoing series of budget cuts and early retirements: "I thought, Great, let's get rid of our most senior and qualified officers, tie those who are left in red tape, and then declare war. . . . In George Tenet's CIA the conduct of operations was less important than Beltway politics and networking on the seventh floor."[154]

Paul Pillar, who served as the deputy chief of the CTC in the late 1990s, provided a more nuanced but equally critical account of the mixed messages the agency received from the executive branch in a December 2013 interview with the investigative journalist Chris Woods: "It was probably understood [at the CTC] that there were at least some circumstances, involving resistance by the target or difficulty in capturing him, where lethal force would be permitted." Just what those extenuating circumstances would be, however, were never fully clarified by the White House, which Pillar suspects "did not want to put clearly on paper anything that would be seen as an authorization to assassinate," preferring instead to give "more of a wink-and-nod to killing bin Laden."[155]

In its efforts to apportion blame, the 9/11 Commission, having heard the testimony from all involved, decided the failure to neutralize bin Laden was not due to a lack of will on the part of the president and his national security team. Nor did it blame a lack of competence or imagination on the part of the CIA. Instead, the commission identified a lack of clear legal sanction for the use of lethal force as the problem. There was no clear bipartisan support or creation of the necessary legal architecture to empower the CIA to act without fear of prosecution and institutional damage. The commission's report concluded that "the President's intent regarding covert action against bin Laden was clear: he wanted him dead." That intent, however, was never well communicated to or understood by the CIA. Senior managers, operators, and lawyers within the agency all confirmed this unfortunate disconnect to the commission, with a former chief of the Bin Laden unit telling its members, "We always talked about how much easier it would have been to kill him [bin Laden]."[156]

Despite demonstrating sympathy toward the CIA, the commission still concluded that the agency's approach to paramilitary operations had been "muddled." Apparently ignorant of the constraints that had been placed on the CIA over the previous decade, the commission members had been struck by what they saw as the agency's unwillingness to take risks and reluctance "to pull the trigger when opportunities were presented." The biggest problem, in their view, was that "the Pentagon had the capabilities for hunting-and-killing operations, but the CIA had the authorities."[157] The commission's conclusion that the CIA had all the legal authorities it needed prior to 9/11 ignores the fact that it was not until Bush issued an additional Finding on September 17 and the AUMF was passed the following day that the CIA undertook its first targeted killing against al-Qaeda.

Factors Acting to Constrain the Clinton Administration's Use of Lethal Force

Despite Scheuer, Baer, and Berntsen's complaints (though to an extent acknowledged by Pillar), Beltway politics were a necessary concern as the White House considered the legal sanction to use lethal force against bin Laden and al-Qaeda. On the most basic level, regardless of his orders for the CIA to deal with the threat of al-Qaeda, Clinton was very conscious that he did not want to fall foul of the U.S. ban on assassination. Despite the fact that the Bush and Obama administrations made it clear that they regarded CIA drones to be undertaking targeted killings rather than assassinations, the dividing line between the two is a fine one. The concept and practice of assassination has proven hard to define, and many critics of the drone campaign, such as the investigative journalists Jeremy Scahill and Glenn Greenwald, have consciously chosen to refer to the practice as assassination rather than use what they describe as "the euphemistic term" of targeted killing, which "the government wants [them] to use" in order to imply accuracy and legitimacy.[158] In basic terms, an assassination is "the killing of an individual or group of individuals for purely political or ideological reasons," whereas a targeted killing is "the killing of an individual or group of individuals without regard for politics or ideology, but rather *exclusively* for reasons of state self-defense."[159] The state of war that the United States has argued has existed between it and al-Qaeda since 9/11, legally supported by the wording of the 2001 AUMF, has made it significantly easier for the post-9/11 administrations to define their actions as being in self-defense.[160] For Clinton, however, this was much more difficult. There was no precedent for the United States being in a state of armed conflict with a nonstate terrorist group. During the Reagan administration, CIA counsel Stanley Sporkin may have ruled that targeting terrorists did not constitute assassination, but the terrorist groups in question were state sponsored. Moreover, as already illustrated, the 1990s were a very different environment, both legally and politically, for the executive and the CIA. Hence, Clinton and the members of his NSC were at pains to keep the wording of the MONs as far away as possible from anything that could be construed as an order to kill an individual for political or ideological reasons.

Almost certainly, this sense of caution also extended to the potential political fallout should the CIA's mission against bin Laden go wrong. As President Ford had stated, one of the reasons he introduced the ban on assassination was to protect future presidents from the seduction of covert operations, which, if exposed or unsuccessful, had the power to embarrass, weaken, or even bring down presidencies. Clinton's childhood hero, John F. Kennedy, was a president who suffered significant political embarrassment when the CIA's covert operation to back the Bay of Pigs invasion of Cuba failed miserably. Kennedy had placed much of the blame for the failure on the CIA, confessing that he had wanted to "splinter [it] in a thousand pieces and scatter it to the winds."[161] The last Democrat to occupy the White House before Clinton had also suffered costly political humiliation when his risky operation to rescue the U.S. hostages in Iran failed.[162] And it was not just the CIA that was ravaged by the breaking of the Iran-Contra scandal. The reputation of President Reagan was also adversely affected: his approval ratings sank by twenty-one points, and for a time his administration struggled to articulate its messages over the din of the media circus generated by the scandal.[163] Though Reagan recovered his popularity, and to a lesser extent his credibility, later in his second term, there is no doubt the scandal blighted his legacy.[164] Scheuer and company may have deemed Clinton's caution as cowardly, but it is a responsibility of the president to maintain not just his or her own political position but also the public's faith in the office, and little undermines that faith more than a covert operation gone wrong, exposing the darker side of politics to the public.

Clinton was not immune to scandal, and he did not exert the same caution he showed with the wording of the MONs in his relationship with White House intern Monica Lewinsky. There are commentators who have suggested that the ensuing sex scandal and threat of impeachment undermined the president's authority and withered his political capital at exactly the time he needed it. Those closest to the president at this time have rejected these accusations, though it is clear that it was an issue when it came to building the case for the use of lethal force against al-Qaeda. "I feared that the timing of the President's interrogation about the scandal . . . would get in the way of our hitting . . . al-Qaeda," confessed Clarke, adding: "It did not. Clinton made clear that we had to give him our best national security advice, without regard to his personal problems."[165] CIA Director Tenet agrees with Clarke, giving assurances that he "never saw any evidence that Clinton's personal problems distracted him from focusing on his official duties."[166] While it may not have distracted Clinton, however, Tenet does suggest that the wider loss of credibility may have impacted Clinton's policy-making options. "Perhaps they [Clinton's personal problems] circumscribed the range of actions he could take—he was, after all, losing political capital by the hour."[167]

Clinton's secretary of state, Madeleine Albright, hints at a wider problem that may have served to limit Clinton's ability to act—a lack of congressional, media, and public support for lethal action against al-Qaeda. Reflecting upon the reaction to Operation Infinite Reach on August 20, 1998, in which Clinton authorized the use of cruise missiles against al-Qaeda in retaliation for the terrorist group's bombing of the U.S. embassies in Africa, Albright reports that "most members of Congress applauded the mission" but that "the public reaction was muted." Though al-Qaeda may have hit American embassies, it had not yet launched a domestic attack upon the United States. This was a vital distinction, and, as Clinton's secretary of state notes, there were "even those who thought we had overreacted and questioned the President's motives." Citing the fortuitously timed 1997 black comedy *Wag the Dog* in which the White House fakes a war to distract from a sex scandal, partisan commentators attacked the administration's retaliation against bin Laden, suggesting it was an effort to shift the media focus from Clinton's problems with the special prosecutor. "These allegations were groundless and repugnant," stated Albright, adding that they did, however, contribute "to an atmosphere in which there were virtually no calls from Congress or the media to take further military action against al-Qaeda."[168]

This lack of bipartisan political support, coupled with media cynicism, is significantly different to the wider support given to the counterterrorism policies of the Bush and Obama administrations in the wake of 9/11.[169] However, the fact that Clinton was acting prior to a terrorist attack upon the U.S. homeland does not excuse his inability to build a national consensus around the need to neutralize the threat posed by al-Qaeda. Rallying public support against threats is a vital part of a president's national security responsibilities. Harry Truman had to win both congressional and public support in 1947 when articulating the doctrine of containment, and though the threat Clinton was addressing with al-Qaeda was not as great or immediate in the minds of Americans, nor was the commitment required to combat it.[170] The administration did make some effort to sell the policy of targeting al-Qaeda to the American public. Clinton delivered an address from the Rose Garden on the day of the strikes highlighting the threat posed by Osama bin Laden, whom he described as the "preeminent organizer and financier of international terrorism," whose mission was to "wage a terrorist war against America." The president went on to suggest America "must be prepared to do all that [it] can" and warned that inaction would only embolden the terrorist group.[171] To further push the case for targeting bin Laden following Clinton's address, Sandy Berger, Clinton's APNSA, provided a detailed press briefing, and Albright took to the cable networks to answer interviewers' questions.[172] Despite this, so great was the circus Clinton's infidelity had created that the administration never succeeded in directing political or public attention to the issue. This muted response undermined Clinton's mandate to direct the CIA to take politically risky action, and the president lacked the necessary political capital to win support from such a hostile Congress.

Another significant political difference between Clinton's administration and the administrations of his successors is that Clinton's attorney general, Janet Reno, did not support the CIA's use of lethal force against al-Qaeda. Attorneys general are always kept in the loop with regard to new Findings or MONs in order to provide legal oversight and guidance for the president. According to John Rizzo, when it came to the MONs relating to bin Laden, Reno's behavior was "often erratic and alarming." The CIA counsel claims the attorney general and her staff would scrutinize each word and agonize over every detail.[173] Such action in itself is not so surprising, as attention to detail with respect to such a sensitive subject is a necessary quality in an attorney general. Reno's attitude towards the MONs, however, provides further explanation for Clinton's caution. Despite the fact that the DoJ did not provide any specific legal objection to the MONs, on two of the three occasions Reno's final judgment was accompanied by a separate memo in which she expressed her concern and opposition in policy terms rather than legal terms. She was concerned about blowback. Reno warned of possible retaliation, which could include the targeting of U.S. officials should any operation be construed as an assassination attempt.[174] This lack of legal support undermined the Clinton administration's confidence with regard to explicit authorization for the CIA to use lethal force.

Tenet and Rizzo have also made claims that Reno went further than her written cautions, orally warning the DCI and Geoff O'Connell, then director of the CTC, that she would consider any action intended solely to kill bin Laden as illegal. Tenet interpreted this warning to mean anyone involved in a mission that ended with bin Laden dead ran the risk of facing a murder charge. Coupled with the already ambiguous language of the MONs, this discouraged both the DCI and the director of the CTC to push for aggressive action against bin Laden, which could see members of the agency on murder charges.[175] Reno would later make no mention of these policy memos or her oral warnings during her testimony to the 9/11 Commission.[176]

The fear of targeted killings undertaken by an intelligence agency setting off a cycle of violence was not unprecedented. The disastrous consequences of the Mossad's campaign against those behind the 1972 Munich Olympics massacre has served as an important lesson on the risks of unilateral, state-led targeted killings. Israel's retaliatory missions may have successfully killed most members of Black September, the Palestinian organization responsible for the deaths of eleven Israeli athletes and officials, but it also resulted in one of the greatest disasters in the history of the Mossad.[177] On July 21, 1973, in Lillehammer, Norway, Mossad agents killed an innocent Moroccan waiter named Ahmed Bouchiki, having mistaken him for Black September's chief of operations, Ali Hassan Salameh. Norwegian authorities subsequently arrested many of the Mossad team who had participated in the attack, some of whom were sentenced to jail. As well as involving the death of the innocent Bouchiki, the disastrous mission exposed Israel's secret intelligence network to public scrutiny and prompted the deterioration in its international prestige, generating significant political fallout.[178] More important, the targeted killing failed to neutralize the terrorist threat against Israel, and though Black September may have ceased to exist, the exposure of Israel's campaign triggered retaliatory attacks from Palestinian militant groups, with the cycle of violence continuing to this day.

While the attorney general was right to be cautious about the potential political fallout of targeted killing campaigns, her judgment of the situation regarding al-Qaeda underestimated the determination of the organization's leaders. Bin Laden's declarations of jihad had already made clear that al-Qaeda held the United States responsible for what it perceived as a wide range of crimes against Muslims. Showing restraint against bin Laden was not going to make any difference, as Al-Qaeda already had all the motivation it needed. Evidence of this was recovered in the Shomali compound near Kabul, Afghanistan, in 2002. Documents and videos dating back to the mid-1990s were recovered, which included notes on presidential protective details and what vulnerabilities to look for, as well as sketches of Secret Service procedures and documents disclosing which streets or open areas were the best locations to stage a presidential assassination.[179] It is highly likely that al-Qaeda had been plotting to assassinate Clinton before any lethal action was launched against bin Laden. The U.S. embassy in Islamabad, Pakistan, had also received a warning late in 1997, again before the United States had launched an attack on bin Laden, that the al-Qaeda leader had ordered the assassination of senator Hank Brown, then chair of the Senate Foreign Relations Subcommittee on the Near East and South Asia.[180]

The final factor that can help explain the Clinton administration's caution in providing legal sanction to the CIA to undertake the targeted killing of bin Laden relates to the wider political environment of the time, and the strategic outlook the Clinton administration had for the United States' role in the post–Cold War order. Following the collapse of the Soviet Union, the United States was the world's sole superpower, what some commentators referred to as a "hyperpower."[181] How the United States conducted itself and used this position was a matter of significant political debate. Within George H. W. Bush's administration, those classed as assertive nationalists, such as then secretary of defense Dick Cheney and his neoconservative assistant secretary Paul Wolfowitz, believed that having achieved such power, the United States no longer needed to compromise its goals in pursuit of allied support. In the minds of these assertive nationalists, the United States was free to act unilaterally in its own interests, focusing upon the prevention of the emergence of a new rival, and no longer had to bind itself to organizations such as the United Nations and other multilateral institutions, which they regarded as inefficient and often at odds with U.S. interests.[182] Instead, in their view of the new world order, the United States was to maintain a position of a dominant hegemonic leader. Alliances were to be limited to the world's leading democracies, which collectively would use their power and influence to consolidate the spread of democracy and free market economies, systems they were convinced would bring global prosperity and stability.[183]

However, Clinton's victory in the 1992 presidential election brought a significant change to the strategic outlook for this new world order. Clinton was a liberal internationalist. Far from favoring a unilateralist and dominant approach, liberal internationalists believed that the best way to ensure a peaceful transition from superpower competition to post–Cold War stability was to firmly enmesh the United States at the center of the global community. Hence, liberal internationalists favor the expansion of trade ties, military agreements, and alliances and full participation in multilateral institutions, in the belief that the mutual dependence such a global community creates closely parallels stable domestic societies and ensures that overt rivalry and competition does not develop and escalate into military confrontation.[184] Clinton summed up this thinking in his first inaugural address, creating an image of "a new generation assum[ing] new responsibilities in a world warmed by the sunshine of freedom," a generation with a "universal ambition for a better life" driven by "peaceful competition across the Earth."[185]

Such an outlook, however, inevitably created inherent contradictions between U.S. national security interests and the international laws and treaties that held this global community together.[186] The intention to snatch or kill Osama bin Laden was one such contradiction. When first discussing the practice of seizing terrorist subjects from foreign countries in a meeting in early 1993, White House Counsel Lloyd Cutler warned Clinton that such action was a clear violation of international law. Vice President Gore, who was also present during the discussion, was less cautious. "Of course it's a violation of international law, that's why it's a covert action."[187] For Clinton, the risk of undermining the international legal system while exposing the United States to charges of hypocrisy added a further risk to be considered when contemplating the use of covert action against bin Laden. This having been said, the Clinton administration did not always stick to this internationalist approach, and as his presidency progressed Clinton adopted a more unilateral stance, refusing to ratify the 1998 Statute of Rome, which established the International Criminal Court, endorsing National Missile Defense in 1999 in spite of its violation of the existing Anti-Ballistic Missile Treaty, and intervening without U.N. Security Council support against Iraq and in the former Yugoslavia.[188] Clinton's use of cruise missiles against bin Laden in 1998 can be seen as part of this increasingly assertive trend, but the overall mentality of trying to keep the United States within the bounds of international law wherever possible certainly helps explain his initial reluctance to sanction unilateral covert action.

The combination of cruise missiles and the Clinton administration's desire to honor international law and treaty obligations threw the very legality of armed drones into question, and partially helps explain a costly delay in the mobilization of the armed Predator drone following Clarke's request for a "see it/shoot it" option against al-Qaeda in his November 2000 strategy paper. The Intermediate-Range Nuclear Forces Treaty (INF), signed by Reagan and Gorbachev on December 8, 1987, had required both the United States and the Soviet Union to disassemble their ground-launched ballistic and cruise missiles with ranges of between three hundred and thirty-three hundred miles. Despite the fact that the Soviet Union had since collapsed, the agreement between Washington and Moscow remained in force as the air force sought to weaponize the Predator drone in late 2000. In the wording of the treaty, a ground-launched cruise missile that is a weapon-delivery system was defined as "an unmanned, self-propelled vehicle that sustains flight through the use of aerodynamic lift over most of its flight path."[189] The initial legal opinion of the State Department's lawyers was that, under these terms, an armed Predator could constitute a cruise missile, causing an INF Treaty violation. Seeing the threat the State Department's initial legal judgment posed to the Predator project, and to future UAV research, the air force scrambled to the drone's defense, and months of bureaucratic stalemate followed.[190]

A resolution to the legal deadlock over the armed drone was eventually reached once Clarke, keen to see the armed drone deployed to Afghanistan as quickly as possible, weighed in. Drawing upon his prior experience as deputy assistant secretary of state for intelligence when the INF Treaty was negotiated, Clarke pointed out that, by definition, a cruise missile had a warhead, something the Predator did not. The weaponized drone was still just a platform; an aircraft complete with landing gear and pilot, even if it was flying remotely, designed to return to base after a mission. Based upon these terms, government treaty experts agreed the armed drone was permissible under the treaty terms, and development resumed on January 2, 2001, after a five-month delay.[191]

Aside from the impact of 9/11, the willingness of the Bush administration to authorize the CIA to undertake its unilateral targeted killing campaign reflected its own rejection of liberal internationalist ideas. During the 2000 election campaign, the Bush camp had mounted strong attacks upon the liberal internationalist outlook, which Al Gore had maintained as a central tenet of his proposed foreign policy agenda. Bush derided the limits of diplomacy in a speech entitled "A Distinctly American Internationalism," while his future APNSA and secretary of state Condoleezza Rice argued in a *Foreign Affairs* essay that, under the principles of liberal institutionalism, "the 'national interest' [had been] replaced with 'humanitarian interests.' "[192] The Obama administration sought to present its drone campaign as sitting somewhere between the two approaches, putting American national security first but doing so within the confines of the international legal system. The May 2010 National Security Strategy indicated that when the use of force was deemed necessary the Obama administration would "seek broad international support, working with institutions such as NATO and the UN Security Council." Yet the same document immediately went on to state: "[We reserve] the right to act unilaterally if necessary to defend our nation and our interests." It did, however, seek to distance the United States from the worst extrajudicial actions of the Bush presidency by adding that when acting unilaterally America would "seek to adhere to standards that govern the use of force."[193] This position of adherence to international standards in its war against al-Qaeda, including drone deployment, lies behind Obama's reference to America's meeting the criteria of a just war in his 2009 Nobel Peace Prize speech.[194]

Conclusion

The political analyst Graham Fuller, a former CIA operations officer with twenty years' experience, described terrorism and counterterrorism as "a very rough game"; "sometimes to win it, you have to use very rough tactics." The reason for this, Fuller explained, was that your opponents were "not nice people, these are not citizens of anyone's country, they live in a gray area."[195] This nonstate gray area, somewhere between criminality and war, was responsible for causing the Clinton administration and the CIA so much difficulty during the 1990s and into 2000, and left the United States much more vulnerable than a country with a $17 billion counterterrorism budget ever should have been.[196] What al-Qaeda represented was a new kind of post–Cold War threat, unlike anything the United States, or any other country, had experienced before. Al-Qaeda was, and still is, a transnational threat capable of waging a sustained guerrilla war against a superpower and its allies. And yet, even following the devastation of the 9/11 attacks, lawyers were divided on how the group should be treated.

Andrew Orr, when serving as editor in chief of the *Cornell International Law Journal*, expressed the view held by the U.S. government in arguing that it seems "almost absurd to argue that terrorism is a law enforcement matter," and that the "scale, sophistication and complexity of the al-Qaeda threat has long-since evolved into something more substantial."[197] Writing in February 2014, the terrorism expert J. M. Berger agreed with this position, arguing that America was fighting al-Qaeda like a terrorist group, while al-Qaeda was engaging the United States as an army. Berger suggested the reference to al-Qaeda and affiliates as a terrorist group was increasingly misleading, instead suggesting the vast weight of jihadist manpower and funding combined with the objective to capture and govern territory makes America's enemies "warfighting jihadists," an analysis proven all the more accurate by the emergence of the Islamic State as a significant military force.[198] Conversely, Christopher Greenwood, a judge in the International Court of Justice, has challenged such a perspective, contending that "in the language of international law there is no basis for speaking of a war on al-Qaeda or any other terrorist group." To Greenwood, irrespective of their scale and lethality, al-Qaeda and its ilk are "a band of criminals," and to treat al-Qaeda as anything else "risks distorting the law while giving the group a status which to some implies a degree of legitimacy."[199] Berger's own argument concedes that "becoming a warfighting jihadist is a much more appealing moral choice than terrorism."[200]

This debate continues today and helps to explain the indecision within the Clinton administration surrounding the granting of the legal authority to the CIA to neutralize bin Laden and his followers. Clinton's national security staffers were quick to identify the threat of al-Qaeda, but they were too slow to grant the appropriate legal sanction necessary to deal with it. In the CTC, designed by Clarridge a decade earlier to meet a different but equally transnational terrorist threat, the Clinton administration already had the perfect organization to counter the new danger, but until the center was granted the appropriate authority to act through the Findings and MONs it was powerless to stop bin Laden.

Currently, the CTC's lethal drone campaign functions within its own legal architecture. Targeted killings are undertaken through precision strikes against enemy combatants with whom the United States regards itself as being at war.[201] These combatants, made up of al-Qaeda and its associated forces, are logged onto a "Disposition Matrix"—a next-generation capture/kill list with a more sanitized name—with the president himself or herself making the final judgment as to whether or not those listed should be killed.[202] This approach is not without its critics. For example, journalist Glenn Greenwald has argued that the Matrix has established "simultaneously a surveillance state and a secretive, unaccountable judicial body that analyses who you are and then decrees what should be done with you, how you should be 'disposed' of, beyond the reach of any minimal accountability or transparency."[203] While debate over what level of accountability and transparency is appropriate will likely remain ongoing between the executive, Congress, and concerned members of the public, what is clear is that for better or worse the United States has developed the legal sanctions and associated bureaucracy that empower its intelligence community to deploy lethal force against al-Qaeda and its associates without fear of legal reprisal. Comparatively, one can see how the Clinton administration's vague language of self-defense and concern with the feasibility of capture lacked the necessary conviction and legal authority to empower the CTC to do what the president actually wanted it to do—kill his enemy. Just as the CIA was transformed from an agency that was "chugging along" into "one hell of a killing machine" by the post-9/11 legislation, so too was the executive's language on counterterrorism transformed from ambiguous and woolly phrases to distinct authorizations to hunt and kill.[204]

The CIA was entirely right to push for strong legal clarification from the Clinton administration on what exactly it was and was not authorized to do in order to deal with bin Laden. Historically it had been called upon time and time again by Eisenhower, Kennedy, and Johnson to engage in similar covert activities during their terms, from attempts on Castro's life to the larger-scale Phoenix Program in Vietnam.[205] These missions had resulted in the agency being publicly castigated by the Church Committee and its sibling hearings, had seen officers removed from their posts and legislation introduced to prevent the CIA from violating what were described as the "moral precepts fundamental to [the American] way of life."[206] The reforms that were introduced, including Ford's ban on assassination, were intended to protect the potential targets of the CIA, but also agency officers themselves and administration officials from overstepping their authority in the future.[207]

When, however, it was felt that this legislation had held the CIA back in its mission against the Soviet Union, the Reagan administration set about "unleashing it" once more.[208] The prosecutions that followed the Iran-Contra scandal ended the careers of a number of officers who simply saw themselves as following orders from their president and his NSC. The ensuing period of atrophy and loss of confidence within the agency served al-Qaeda well. The renowned scholar of CIA history Rhodri Jeffreys-Jones reflects upon the impact of the post–Cold War years on the Agency, remarking that "it is plain that the CIA's standing was impaired, and to a measurable degree." As Jeffreys-Jones goes on to argue in *The CIA and American Democracy,* the agency can only be as good as its standing and influence.[209] The high turnover of directors, low morale, and lack of presidential patronage weakened the CIA just at a time when the United States was struggling to find an effective response to the rising threat of nonstate terrorism. Existing outside the realms of law and order, but also not warranting a full military response, the threat of al-Qaeda in the 1990s was exactly the kind of nonconventional threat the CIA should have proven capable of dealing with, had it not been suffering the fallout of the two previous periods of adventurism and overextension. The covert action pendulum swings resulted in an unstable agency and, as a result, unstable national security.

Following 9/11, this pendulum swing and counter-swing can be seen to have occurred once more, with the CIA initially being granted carte blanche to attack al-Qaeda and to gather intelligence on the terror network and its plans for future attacks.[210] It is these orders that led to the CIA's use of black sites and enhanced interrogation techniques under the Bush administration, until the pendulum swung back and the agency found itself condemned at home and abroad for what was deemed unconstitutional and morally unacceptable conduct.[211] Despite the lack of any formal prosecutions by the U.S. government for the use of enhanced interrogation, its legacy still serves as a black mark against the CIA and the officers who followed the orders given, with President Obama, Attorney General Holder, and even the CIA Director Leon Panetta publicly criticizing the policy.[212]

Doubtless a similar nervousness exists within the agency with respect to its conduct of its lethal drone campaign. Though in theory supported by the authority granted to the president by 2001's AUMF and Bush's Finding, Obama's use of this legal architecture to justify the expansion of the campaign to include strikes against the Yemen-based al-Qaeda in the Arabian Peninsula (AQAP)—a group that formed almost a decade after 9/11, in 2009—and the Islamic State in both Iraq and Syria—a rival offshoot rather than affiliate of al-Qaeda's core—stretches the legal remit. The authors of a joint Stanford Law School–NYU School of Law report on America's use of drones have expressed skepticism that killings carried out around the world today can still be justified by 2001's AUMF, as well as taking issue with the use of the U.N. Charter's Article 51 to justify the strikes as self-defense.[213] Thus, if the scholars' analysis is correct, the expansion of the CIA's drone war over the past decade, along with other strikes conducted by JSOC and the USAF reveals an exploitation of the terms of the AUMF, which has relied upon the willful cross-branch, cross-party ignorance of the majority of lawmakers. This poses a significant question with regard to the expanded drone campaign's actual constitutional authority, meaning its legal architecture is therefore built upon unstable foundations and the legislature's corroboration, something future presidents—and the CIA—may not always be able to count upon.

What the history of the pendulum swings illustrates is that the CIA is less effective if used to excess beyond what is legally and morally acceptable at that given time, as there will be a backlash when circumstances change. Such backlashes cause the agency to be diminished. What this highlights is the importance of a clear and transparent legal architecture. After the first decade and a half post-9/11 the U.S. government has settled upon what it deems to be acceptable boundaries for the CIA to function within. A covert intelligence agency, its boundaries brush right up against—and even challenge—what is acceptable in international law, as revealed by the criticism of some legal scholars. However, the agency's actions have been condoned by policy makers and lawmakers, making them, in theory at least, as accountable as the agents who follow their orders. Should there come a time when an intelligence leak, change in administration, or discovery of fresh evidence casts a new light upon the policy of drone warfare, leading to fresh criticism being made of the CIA's conduct, it must be remembered that the CIA's drone operatives were functioning within the boundaries set out by the executive and overseen by Congress in order to guarantee the national security of the United States. Though the likes of Clarridge, Baer, and Berntsen may have come to see such legal boundaries, oversight, and transparency as anathema to the way the agency used to operate, one only need to look at the impact of the pendulum swings on their own careers to see that clear boundaries and regular review will serve both the CIA and the United States better. Greater transparency, solid oversight, and clear legal boundaries, though a long way from the principles the CTC was originally founded upon, have proven the most effective way of ensuring both a powerful and a responsible counterterrorism capability.

5

"NINJA GUYS IN BLACK SUITS": ALTERNATIVE COUNTERTERRORISM TOOLS, 1993–2008

> What I think would scare the shit out of these al-Qaeda guys more than any cruise missiles . . . would be the sight of U.S. commandos, Ninja guys in black suits . . .
> —*President William J. Clinton*

Counterterrorism works on multiple levels. At its most basic level it includes defensive measures designed to protect citizens and make it difficult for terrorist groups to function. This includes such actions as building diplomatic pressure against states that sponsor or harbor known terrorists, tracking and seizing illegal assets and money transfers relating to terrorist activities, surveillance and disruption of communications between known and suspected terrorists, and maintaining security checks at airports and other sensitive locations. Such things are standard practice and low profile and pass along from one administration to the next, constantly being undertaken to undermine terrorist threats. In addition to these are the more proactive policies intended to take the fight to terrorist groups by neutralizing their members and ultimately destroying the groups. Unlike the more consistent defensive measures, these policies have changed from administration to administration over the past three decades. When al-Qaeda emerged as a national security threat, the Clinton administration sought to use the practice of extraordinary rendition to bring the group's leader, Osama bin Laden, to justice before adopting the cruise missile as its weapon of choice in an effort to kill the elusive leader. Empowered by the fallout from 9/11, Clinton's successor, George W. Bush, instigated renditions and interrogation on a mass scale, detaining hundreds of suspected al-Qaeda and Taliban members while dramatically expanding the DoD's counterterrorism role. Critical of this approach, Obama scaled back the rendition program and instead embraced targeted killings executed by the CTC's drones as his primary counterterrorism tool, authorizing more strikes in his first year in office than his predecessor had launched during the entirety of his two terms. Far from being kneejerk changes, the different policies adopted from president to president are influenced by and built upon the successes and failures of the counterterrorism policies of their predecessors. The Obama administration's adoption of drone warfare was no different, evolving from what had previously proven successful and rejecting what had not.

Countering Terrorism through Extraordinary Rendition

The policy of extraordinary rendition is the counterterrorism method most intrinsically linked to George W. Bush's time in the White House. Extraordinary rendition is legally defined as the transfer—without legal process—of a detainee to the custody of a foreign government for the purposes of detention and interrogation, a practice that became a core element of Bush's War on Terror.[1] Reeling from the impact of 9/11 and fully expecting further attacks, the Bush administration charged the CIA with finding out what other al-Qaeda plots existed against the United States, where other terror cells might exist, and how they operated. For the CTC, that meant not only apprehending key members of al-Qaeda but also getting them to talk. Less than a week after the September 11 attacks, Bush provided the CIA with written authorization to capture, render, and interrogate terrorists in agency-run prisons (referred to as "black sites") in an operation dubbed GREYSTONE.[2] The full scale and scope of the GREYSTONE program remains unknown, as does the extent of foreign governments' cooperation. Amrit Singh's study on the CIA's use of rendition and detention under the Bush administration has identified 136 individuals who were reportedly subjected to extraordinary rendition, and fifty-four governments who purportedly participated in these operations in various ways.[3] It should also be noted that this figure does not include those transfers conducted by the DoD and its confinement of suspected terrorists in Guantánamo Bay or other military detention facilities. A list released by the DoD on May 15, 2006, included the names of 759 individuals who were or are held at the U.S. detention facility for war-on-terrorism detainees.[4]

Despite its association with the Bush administration, the rendition of terror suspects to face trial and incarceration through the American justice system was a practice established well before Bush assumed power and sought a response to the shock of the 9/11 attacks. In the wake of the Iranian hostage crisis, with incidents of international terrorism, in particular hijackings, rising dramatically, Congress added a section to the Comprehensive Crime Act of 1984 detailing how the United States was to deal with hostage situations involving American citizens.[5] The new section made federal crimes of air piracy and terrorist attacks upon U.S. citizens abroad, and specifically gave the FBI jurisdiction over any terrorist acts in which Americans were taken hostage—no matter where the acts occurred.[6] This authority was further strengthened and extra resources for counterterrorism operations made available when Reagan signed the Omnibus Diplomatic Security and Antiterrorism Act in August 1986.[7] Stressing the global nature of America's security commitment, he used the occasion of the signing of the act to send a message to terrorists planning attacks involving Americans: "This act once again puts those who would instigate acts of terrorism against U.S. citizens or property on notice that we will not be deterred from carrying out our obligations throughout the world. I am committed to ensuring the safety of our diplomats, servicemen, and citizens wherever they may be."[8] In many ways the introduction of these two acts marked a positive development for U.S. counterterrorism. They showed that Congress recognized international terrorism was an issue that required action, and that tackling terrorists' use of safe havens was a vital part of reducing the terrorist threat. Moreover, the passing of the acts revealed an understanding between the executive and America's lawmakers, with Reagan promising to "continue to work with Congress to identify legislative gaps in our ability to combat terrorism."[9] However, the acts also highlighted one of the most significant divisions in Reagan's administration concerning counterterrorism. The Omnibus Act was passed shortly after the creation of the CIA's CTC. The center's founders intended it to operate as the central hub through which all of America's counterterrorism operations would be organized and coordinated. But Congress's two acts revealed the prevailing consensus, especially among lawmakers, that terrorism was a criminal activity, not an act of war, therefore falling within the FBI's jurisdiction. Despite some successful cooperation, the competition between the CIA and the FBI over who led America's counterterrorism agenda remained, eventually being identified by the 9/11 Commission as a significant factor in America's failure to prevent the September 11 attacks.

Even with its new legal authority the Reagan administration's first attempt to prosecute international terrorists for hostage taking proved unsuccessful when Italian authorities intervened following the hijacking of the *Achille Lauro*.[10] However, buoyed by the positive domestic reaction to its efforts to seize the cruise liner's hijackers, the government continued in its efforts to subject international terrorists to the justice of U.S. courts. On September 13, 1987, America undertook its first successful terrorist rendition, capturing a Lebanese national named Fawaz Younis. Younis had been implicated in the hijacking of Alia Royal Jordanian Airlines flight 402 on June 11, 1985, as it left Beirut for Amman. The aircraft's crew and more than seventy passengers, including four American citizens, were held hostage for more than thirty hours as the hijackers forced the aircraft's pilots to fly them to Cyprus, then Sicily, and finally Beirut. Once there the hijackers released the hostages, held a hastily arranged press conference to denounce a resolution adopted by the Arab League, blew up the plane on the tarmac, and successfully fled.[11] The suspected hijacker Younis was captured through a complex joint FBI-CIA sting operation code-named "Goldenrod." Younis was lured onto an FBI-owned yacht off the coast of Cyprus with the promise of drugs. Once aboard he was taken into international waters, where the FBI arrested him.[12] The prisoner was then moved to the USS *Butte,* which steamed for four days to the western Mediterranean, where it transferred him to the American aircraft carrier USS *Saratoga,* and he was flown to Andrews Air Force base on the U.S. mainland. He was tried in a U.S. court and found guilty of conspiracy, aircraft piracy, and hostage taking, for which he was sentenced to thirty years in prison in October 1989.[13] In a direct challenge to the calls for lethal operations against terrorists that had dominated the counterterrorism agenda in the years preceding the Iran-Contra scandal, Edwin Meese, Reagan's attorney general, drew upon the Younis operation to stress the efficacy of a law enforcement response. Speaking after the Lebanese terrorist's successful prosecution, Meese affirmed that "acts of terrorism are criminal acts, pure and simple." He urged others to follow suit: "The world must deal with them as criminal acts and utilize the rule of law and order to combat this very serious threat to the lives and well-being of citizens of every country."[14]

The rejection of the Reagan hardliners' calls to treat terrorism as an act of war and renewed emphasis upon handling it as a criminal act is reflected in the United States' approach to counterterrorism when Clinton entered the White House. Following Mir Ami Kasi's attack outside the CIA's Langley headquarters and Ramzi Yousef's bombing of the WTC in 1993, the Clinton administration set out to bring both men back to the United States to face trial in federal court.[15] In the case of Yousef's rendition, the CIA and FBI cooperated closely with Pakistan's ISI to help find and arrest the fugitive. Following a tip-off, on February 7, 1995, ISI officers, accompanied by special agents from the U.S. Diplomatic Security Service (DSS) raided a room in the Su Casa guesthouse in Islamabad and arrested Yousef inside. The following day, the Pakistani government waived formal extradition proceedings and immediately handed the wanted man over to members of the FBI's Joint Terrorism Task Force (JTTF), who were waiting in a U.S. military transport plane at Islamabad airport, having scrambled over from Washington, D.C., the night before.[16] Yousef was flown to the United States, where he faced two separate trials in federal court in New York. On January 7, 1998, Yousef, described by the U.S. District Court judge as a "virus that must be locked away," was sentenced to a total of life in prison plus 240 years.[17]

Following the success of the Yousef operation, the Clinton administration was assured of the efficacy of rendition as a counterterrorism tool. On June 21, 1995, it introduced Presidential Decision Directive 39 (PDD 39), entitled "US Policy on Counterterrorism." This set out new guidelines to deter and respond to terrorist acts, as well as measures to reduce U.S. vulnerabilities to terrorism. In addition, a provision on preventing terrorist organizations acquiring or developing weapons of mass destruction (WMD) was inserted, following an incident on March 20, 1995, in which a religious cult named Aum Shinrikyo released liquid sarin into the Tokyo subway during the morning rush hour, killing twelve and injuring a further 3,769.[18] The directive reaffirmed rendition as a key component of U.S. counterterrorism, both as a response to terrorist acts and as a deterrent to future attacks. Acknowledging the tendency of terrorists to exploit safe havens around the world, PDD 39 provided additional legal authority for those charged with seizing indicted terror suspects, confirming that the United States would act unilaterally if it did not receive adequate cooperation from a state that harbors a terrorist.[19]

The Clinton administration exercised this self-assigned freedom to act unilaterally shortly after PDD 39 was introduced. Despite the efficiency in locating the higher-profile and more active Yousef, by early 1997 the CIA was no closer to finding Mir Ami Kasi. A combination of his low profile, the nature of the sprawling, unregulated territory of the AfPak border lands he was hiding in, and a deteriorating relationship with the ISI following America's introduction of sanctions against Pakistan's burgeoning nuclear program meant that the agency had no clear leads to act upon.[20] In order to help locate and capture this terrorist fugitive, the CTC recruited former Afghan mujahideen assets, who had received formal military training during the 1980s, to act as bounty hunters. The tribal fighters, codenamed TRODPINT, the same group who would later be called upon to develop a plan to capture bin Laden, were given hundreds of thousands of dollars in cash, weapons, motorcycles, trucks, and secure communications equipment and were ordered to try to locate Kasi.[21] One of the pieces of equipment with which the CTC provided the TRODPINT team was a mobile GPS beacon. Should the team locate Kasi, it could use the beacon to send a signal to U.S. satellites overhead to provide the exact location of the target. "The technology," reported Steve Coll, "would allow an American counterterrorism team to swarm an obscure location quickly once it was lit up by the Afghan agents."[22] Despite this investment in equipment, the TRODPINT team never actually located Kasi. In the end it was an informant from Kasi's own region who led the CIA to its target, and on June 15, 1997, he was seized by a joint FBI-CIA team supported by the Pakistani ISI. After a day in a holding facility in Islamabad he was transferred to the United States to face trial, receiving a death sentence for capital murder on February 4, 1998.[23]

The TRODPINT team may not have captured Kasi, but its creation did establish the blueprint for intelligence gathering and targeting in the AfPak border territory that was to have significant repercussions for U.S. counterterrorism. The agency created Afghan Counterterrorism Pursuit Teams (CTPT), evolved from the TRODPINT team, which serve as a vital link in the intelligence chain for the CTC's drone campaign in the AfPak region.[24] With reportedly more than three thousand men in these teams, their primary mission has been described by an anonymous U.S. official familiar with their operations as "intelligence collection." The official added that the teams do not engage in lethal action when crossing over into Pakistan but had made "major contributions to stability and security."[25] According to Shuja Nawaz, a Pakistani journalist, the teams "see traffic coming and going from the fortress homes of tribal leaders associated with foreign elements [in the FATA region of Pakistan], and they pass the information along, placing GPS locators on the relevant homes and vehicles." Once the reports are corroborated through American surveillance, "someone pops a couple of hundred-pound bombs at the house."[26]

The importance of CTPTs is further highlighted by journalist Bob Woodward, who provides an account of an early transition meeting between Bush and Obama, during which the outgoing president told his successor that the drones were little more than "flying high-resolution video cameras armed with missiles," and that the "only meaningful way to point a drone towards a target was to have spies on the ground telling the CIA where to look, hunt and kill."[27] It took the CIA five years to develop its network of spies and informants in the AfPak region, a timescale that helps explain the significant gap between the agency first proving its armed drones worked, in November 2002, and actually launching strikes on regular basis. The CTC's technological capacity was ahead of its human intelligence resources.[28] Once the CTC had an effective network providing targeting information, strikes in Pakistan surged from four in 2007 to thirty-six in 2008 and fifty-four in 2009. While the change in administration and a shift in strategy no doubt contributed to this rise in strikes, it was due also to the fact that the CIA's long-term investment in pursuit teams on the ground was beginning to pay off.[29]

Why the Obama Administration Abandoned Extraordinary Rendition as a Primary Counterterrorism Tool

Within the first week of his presidency Obama signed Executive Order 13491, revoking Bush's previous order and effectively ceasing the CIA's use of black sites as prisons. The order did not ban the actual practice of rendition, nor did it prohibit the short-term detention of terrorist suspects for interrogation before handing them over for prosecution.[30] As a result, renditions continued under the Obama administration.[31] For example, on October 5, 2013, in a predawn raid in Libya, Delta Force operatives assisted by FBI agents and CIA officers captured Abu Anas al-Libi, a terrorist under indictment in the United States for his role in the 1998 African embassy bombings. Libi was held in custody and interrogated on board the USS *San Antonio* in the Mediterranean Sea before being transported back to the United States for trial in a federal court in New York.[32] Libi's rendition, however, is far from the norm for the Obama administration's counterterrorism practice. Despite the secrecy surrounding the practice of rendition, it is clear that the numbers seized and detained under Obama represent a fraction of those taken by the previous administration. So why did the Obama administration move away from the GREYSTONE policy that had been embraced so fervently by the Bush administration and instead adopt drone strikes as its primary tool for neutralizing al-Qaeda and its Taliban allies?

One of the primary motivations behind the Bush administration's desire to capture al-Qaeda members was the desire to gather intelligence on the organization and its future plans. This information was to be gained through the controversial enhanced interrogation techniques (EITs). Unsurprisingly, those most closely associated with authorizing the techniques have strongly defended the benefits of their use. In his memoirs, President Bush claimed that the techniques proved highly effective, and that those interrogated revealed large amounts of information on al-Qaeda's structure and operations. Ultimately, Bush declared, "the CIA interrogation program saved lives."[33] Bush's vice president, Dick Cheney, perhaps the most vocal proponent of the use of EITs, rejected criticism of the practice as "recklessness cloaked in righteousness" in a high-profile speech he delivered at the American Enterprise Institute on May 21, 2009.[34] Cheney used his memoirs to further defend the ethics of the EITs, arguing that they provided invaluable intelligence that enabled the government to prevent attacks and save American lives.[35] The most senior account from the CIA on the issue of the EITs comes from Jose Rodriguez Jr., who was appointed head of the CTC following 9/11. As the center's first post-9/11 director, Rodriguez was responsible for gathering intelligence on al-Qaeda's network, and ultimately for advocating the use of enhanced interrogation. In his memoirs, Rodriquez stated categorically that he was certain, beyond any doubt, that these techniques, which had been approved by the highest levels of the U.S. government, certified as legal by the Department of Justice, and briefed to and supported by bipartisan leadership of congressional oversight committees, "shielded the people of the United States from harm and led to the capture and killing of Usama bin Laden."[36]

The extent to which the EITs helped save American lives, and whether or not intelligence gathered from them led to Osama bin Laden, is a hotly debated topic. A controversial five-hundred-page report, composed of the declassified portion of a sixty-seven-hundred-page government study on EITs, was released by the SSCI on December 3, 2014, challenging the assertions above. In its extensive criticism of the agency's conduct, the report suggested that the CIA's program was more brutal than policy makers were initially led to believe: detainees were kept awake for up to a hundred and eighty hours, usually standing or in stress positions, at times with their hands shackled above their heads, subjected to involuntary rectal feeding and rectal hydration without medical necessity, and in the case of Khalid Sheik Mohammed, waterboarded on 183 separate occasions with no new intelligence being gleaned. Despite such extreme methods, the report's key findings concluded that the CIA's use of EITs was not an effective means of acquiring intelligence or gaining cooperation from detainees.[37]

The CIA issued a lengthy 136-page response to the SSCI's findings and posted a summary fact sheet on its website. Acknowledging a number of the serious mistakes raised by the report, CIA Director John Brennan conceded that EITs were "an inappropriate method for obtaining intelligence" and affirmed his resolute intention "never to allow any Agency officer to participate in any interrogation activity" in which EITs would be involved.[38] Reflectively, the agency conceded that the program's primary shortcomings stemmed from the fact that the agency was "unprepared and lacked the core competencies required to undertake an unprecedented program of detaining and interrogating suspected terrorists around the world."[39] By highlighting its fundamental unsuitability to act as global jailer and interrogator, the CIA indirectly sought to shift criticism from its officers—who Brennan attested had made many sacrifices to prevent terrorist attacks—to President Bush and his wider administration, who had given the agency the unsuitable mission in the first place.[40]

In spite of the CIA's admission of errors, the agency did seek to rebut certain aspects of the Senate's report, in particular challenging what it described as "the Study's unqualified assertions that the overall detention and interrogation program did not produce unique intelligence that led terrorist plots to be disrupted, terrorists captured, or lives to be saved."[41] Contrary to the SSCI's findings, the CIA maintained that most of the twenty case studies cited in the Senate report "remain valid examples of the program's effectiveness." In particular, the CIA's response maintained Rodriquez's pre-report argument that information obtained from detainees played a role, in combination with other streams of intelligence, in finding bin Ladin.[42] Many have characterized the proclamation that EITs helped locate the al-Qaeda leader as unduly simplistic at best and a flagrant lie at worst, accusing its advocates of massively overplaying the contribution of the CIA's program while downplaying the extensive intelligence work the other branches of the agency undertook.[43] According to Peter Bergen, finding bin Laden was the result of the painstaking and cumulative assemblage of information from multiple detainee interviews, including noncoercive FBI interviews, from thousands of al-Qaeda documents recovered on the battlefield or following an arrest, and from the scouring of open-source reporting about bin Laden. This time-consuming and diligent analysis helped to assemble a picture of who his associates were and how and where he was likely to be living.[44]

In addition to the EIT program's political controversy and debatable efficacy, revelations from the NSA files leaked by the former contractor Edward Snowden have provided further evidence as to why the capture and interrogation of terror suspects assumed less importance by the time Obama took office, enabling him to all but discontinue the practice. The files revealed that the NSA's electronic surveillance capabilities had increased significantly in the years since the War on Terror began. The DoD's signals intelligence specialists had developed the ability to track patterns of movement based upon an individual's digital footprint, a practice that has produced significantly more valuable leads in the pursuit and targeting of specific al-Qaeda and Taliban members than enhanced interrogation ever did.[45] Glenn Greenwald, the investigative journalist to whom Snowden passed the files, and his colleague Jeremy Scahill have raised concerns as to the accuracy of targeting individuals for drone strikes by their metadata and electronic footprint, a practice referred to within the NSA as "We Track 'Em, You Whack 'Em."[46] Conversely, Joshua Foust, a fellow at the nonpartisan American Security Project think tank, has argued that the revelations show the NSA to be "an invaluable part of the intelligence chain leading to successful strikes against known (and very dangerous) figures within al-Qaeda." Rather than exposing the CTC as a "heartless killing machine intent on harming children—a caricature often employed in opposition to drone strikes"—Foust argued that the leaked files reveal a careful and methodical program, resulting in "a terrorist organization deeply afraid of an effective weapon."[47]

Even with the controversy regarding the collection of citizens' metadata, the NSA's approach has presented significantly fewer legal challenges than rendition and interrogation, and has the added benefit of disrupting al-Qaeda's communication and networking capabilities.[48] Though actual details of how the NSA supports the CIA are scarce, one specific example provided in James Bamford's book *The Shadow Factory* reveals how an NSA listening post in the U.S. military's Camp Doha in Kuwait intercepted a phone call from a satellite phone associated with Qaed Salim Sinan al-Harethi on November 3, 2002. Al-Harethi was one of the masterminds of the October 2000 attack upon the USS *Cole*. The intercept allowed the NSA to provide positive identification of al-Harethi and guide a CIA Predator from the nearby base in Djibouti, East Africa, to his exact coordinates. The Predator destroyed al-Harethi's vehicle, killing him and five associates.[49]

Perhaps even more significant in the Obama administration's rejection of rendition and enhanced interrogation as a core counterterrorism tool is the extent to which it became a politically poisonous issue. For many both within and outside the United States, the association of America with the practice of torture created intense hostility and anti-Americanism. When on June 12, 2008, the U.S. Supreme Court ruled against the Bush administration, judging that the detainees at Guantánamo Bay had a constitutional right to habeas corpus—to challenge their detention before a neutral judge in a real court—it was clear that opposition to extraordinary rendition and the associated interrogations was at a critical level.[50] During his election campaign, Senator Obama strongly opposed both the mass detention of suspected al-Qaeda members and the use of EITs, which he described as torture. The Nobel Committee's decision to award Obama the Nobel Peace Prize in December 2009 echoed the relief the international community felt when the newly elected president laid America's most controversial counterterrorism policy to rest. Upon receiving the award, Obama summed up his reasoning for prohibiting torture and attempting to shut the prison at Guantánamo Bay, telling the assembled audience: "The United States of America must remain a standard bearer in the conduct of war." Abiding by the Geneva Conventions when confronting "a vicious adversary that abides by no rules" was what made America different from those it fought. "We lose ourselves when we compromise the very ideals that we fight to defend," stated Obama in reference to the deep domestic divisions the use of EITs had produced.[51]

Furthermore, while rendition may have been effective when the United States was seizing individual terrorists in the wake of attacks that had already been undertaken and were therefore followed by legal indictments, this was no longer the case once America entered a state of armed conflict with al-Qaeda and a legally declared war with the Taliban. The number of suspects the United States was dealing with made a law enforcement approach impossible. One has only to look at the difficulty the Obama administration had trying to prosecute or release those held in the Guantánamo facility to understand why the U.S. government is not keen to add to this number. Following 9/11, the nature of the United States' conflict with al-Qaeda shifted from the defensive, retaliatory approach adopted during the 1990s to a preemptive strategy. President Bush articulated this change of approach in a speech delivered at West Point on June 1, 2002. "If we wait for threats to fully materialize, we will have waited for too long. . . . Our security will require all Americans to be forward-looking and resolute, to be ready for pre-emptive action when necessary to defend our liberty and to defend our lives."[52] This position was formalized three months later with the release of the 2002 National Security Strategy. Widely referred to as the Bush Doctrine, the strategy stated: "The greater the threat, the greater is the risk of inaction—and the more compelling the case for taking anticipatory action to defend ourselves, even if uncertainty remains as to the time and place of the enemy's attack. To forestall or prevent such hostile acts by our adversaries, the United States will, if necessary, act pre-emptively."[53]

Though the case for preemptively deploying force against nations when "uncertainty remains" was critically undermined by the debacle of Iraq's nonexistent stockpiles of WMDs, the concept is still central to the CIA's drone program. The U.S. government assigns the status of "enemy combatant" to anyone who travels to an al-Qaeda training camp or associates with known al-Qaeda, Taliban, or affiliated group members. While critics of the drone campaign argue that these individuals do not meet the imminent threat requirement to trigger legal self-defense, the United States' position is that terrorists represent continuous and ongoing threats of unlawful aggression. As law professor Russell Christopher explains, under this definition it does not matter whether or not the terrorist is actually being aggressive at the time of the strike: "It is their very *status* as a terrorist which qualifies as the *conduct* of posing an imminent threat."[54] The philosopher Jeff McMahan explains this position further, arguing that "even while terrorists are sleeping or eating dinner or doing some other innocuous activity, they do not lose their status as terrorists and thus are continuously and invariably constituting imminent threats."[55] Despite the lack of legal consensus within international law, the United States is clear in its own legal justifications. Therefore, while the U.S. government considers it legal to attack these individuals as known enemy combatants in the prevailing state of armed conflict between America and al-Qaeda, its affiliates, and the Taliban, there might be little or no evidence to prosecute these individuals should they be seized and held in U.S. custody. The awkward truth is that militarily, legally, and politically it has proven much easier for the United States government to kill known and suspected terrorists than to arrest and take them through the justice system.[56]

Why Military Force from the Department of Defense Was Not an Option

While the legal and political fallout of the GREYSTONE program may help clarify why extraordinary rendition was abandoned as a primary counter-terrorism tool, it does not explain why the CTC's drone campaign took its place. If lethal action was the United States' preferred approach to counterterrorism, there were other more obvious tools to which the president had access. When the Clinton administration eventually abandoned the cautious restraint it had shown in the face of al-Qaeda's threats and launched a lethal operation, it chose to use cruise missiles to try to neutralize the al-Qaeda leader. This was not the first time cruise missiles had been deployed for counterterrorism purposes. In June 1993 the Kuwait police claimed to have foiled an Iraqi-sponsored assassination plot targeting former president George H. W. Bush during a visit to the Middle Eastern ally. Secretary of State Warren Christopher equated the plot to kill a former U.S. president with "an attack upon our nation," prompting Defense Secretary Les Aspin to argue that the United States could exercise of the right of self-defense under Article 51 of the U.N. Charter.[57] The Clinton administration opted to retaliate unilaterally by using cruise missiles against Saddam Hussein's intelligence headquarters in central Baghdad, where it was suspected the assassination plot had originated (though an investigation by Seymour Hersh later questioned the Iraq government's complicity in the plot).[58]

On June 26, 1993, the cruise missiles were launched at the Iraqi intelligence building from a U.S. destroyer and cruiser stationed in the Red Sea. "Well, when will we get the pictures from the missiles?" Clinton asked his special assistant, Richard Clarke. Clarke, who confesses he was almost floored by the question, replied that the missiles had no cameras, but that they would learn the degree of the damage from satellites the following morning. Showing the technological foresight worthy of a man labeled the globalization president, Clinton asked: "Why don't the missiles have cameras in them? . . . I'm going on TV in an hour to say we blew up this building, I want to know that we did." "Well, if the missiles communicated," explained Clarke, "someone might see them coming or interfere with them."[59] Clarke's explanation was accurate. In 2009 it was discovered that Iraqi insurgents were hacking the video feed from U.S. drones over Iraq using a piece of software named "SkyGrabber," an illegal Russian program designed for stealing satellite TV feeds.[60] This explanation, however, was not enough for Clinton. "We can't communicate with the missiles? What if I wanted to turn them back?" he asked. Clarke, who was flustered and stammering by this point, asked: "You don't want to sir, do you? . . . because you can't . . . there is no mechanism to . . ."[61]

Twenty of the twenty-three cruise missiles launched hit their target, leveling the intelligence headquarters. Three, however, went astray, killing Layla al-Attar, a revered Arab artist and director of the Iraqi National Art Museum. Her husband was also killed in the strike, and her daughter lost her sight.[62] Unaware of the collateral damage but insistent upon knowing whether or not the missiles had hit their target before addressing the American public, Clinton pushed Clarke. The adviser notified Clinton's APNSA, Anthony Lake, who in turn called Admiral William Studeman, the deputy DCI, only to be told that there was no option but to wait. Unsatisfied, Clinton horrified his national security team by telephoning CNN and asking if they had any journalists in Baghdad who could see whether the target had been destroyed. The cable network did not have any reporters on the ground in the city, but it did manage to reach the cousin of one of its Jordanian bureau cameramen, who happened to be able to see the intelligence headquarters from his downtown Baghdad apartment and confirmed that "the whole place blew up."[63] Finally confident in the results of the strike, Clinton addressed the American people, informing them that the United States could not let the assassination plot "go unanswered," because America's enemies would "repeatedly resort to terrorism or aggression if left unchecked." Clinton finished by assuring the public that every effort had been taken to minimize the loss of innocent life.[64]

Clinton had no intention of turning the missiles around, but his expectation of direct command and control over lethal operations and immediate feedback on the results goes some way toward explaining how Obama came to be at the center of a killing network, giving personal authorization for strikes with access to live feeds of ongoing missions. For the commander in chief there is clearly a felt need to have direct control over the operations ordered. In many ways the availability of a live feed from lethal drones circling targets around the world has made this a reality. While a live feed may provide the immediate feedback that Clinton sought, it also runs the risk of encouraging micromanagement, creating too close a personal relationship between the president and his authorized targeted killings. The well-established drone policy critic Tom Engelhardt refers to Obama's role at the center of the CIA's targeted killing campaign as a marker of an "imperial presidency." Engelhardt's fellow TomDispatch blogger Peter Van Buren has gone further, suggesting the direct control over drone operations led to the "transformation of the White House into a killing machine" headed by "the assassin-in-chief."[65] Though both Engelhardt and Van Buren's polemics perhaps overplay the direct level of control those in the executive have over actual operations, Clinton clearly felt too detached from the mission he was ordering—something that is no longer the case.

The records provide a number of examples of proposed cruise missile strikes against bin Laden that were rejected due to concerns over the lack of reliable intelligence and too great a risk of collateral damage. Clarke describes a strike planned in February 1999, when an informant advised that bin Laden was visiting royals from the United Arab Emirates, who had set up a falconers' camp in the desert south of Kandahar, Afghanistan. Between the single source of human intelligence on the ground and the satellite photography, Clinton's NSC was not confident enough bin Laden was visiting. Even if it could confirm his presence in the camp it did not have any real-time intelligence that could confirm what part of the camp should be targeted. Ultimately the risk of dead princes from an Arab ally was enough to ensure the strike did not go ahead.[66] Another account provided by the 9/11 Commission describes a strike proposed in Kandahar on December 20, 1999. Intelligence suggested bin Laden would be staying overnight at the governor's residence. The NSC principals are reported to have again considered launching cruise missiles, but once more concerns about collateral damage were raised. Marine Corps General Anthony Zinni, who had been responsible for planning the December 1998 strategic bombing of Iraq's WMD capacity in Operation Desert Fox, projected the potential casualties at more than two hundred, and also estimated that significant damage would be caused to a local mosque.[67] A separate projection from a senior intelligence officer attached to the Joint Staff was significantly more conservative with regard to casualties, and predicted there would be no damage to the mosque. The uncertainty, however, was enough to see the strike called off.[68]

On the occasion the United States did launch cruise missiles against bin Laden a number of other factors reduced the effectiveness of the action. August 1998's Operation Infinite Reach was initiated in retaliation for al-Qaeda's bombing of the U.S. embassies in Africa. It involved the launching of seventy-nine cruise missiles at al-Qaeda-linked sites in Afghanistan and Sudan. Intelligence on the ground had suggested bin Laden would be hosting a conference to discuss their next attacks with his organization's senior leadership, at the Zhawar Kili al-Badr camp, near the eastern town of Khost. The conference site and three other training camps associated with al-Qaeda were selected as targets.[69] Not for the first time, the geography of Afghanistan hampered the U.S. mission. The cruise missiles had to fly hundreds of miles from U.S. warships and submarines stationed in the Arabian Sea to landlocked Afghanistan. In addition, the missile guidance systems had to be prepared and programmed with target coordinates. So, even at the speeds the missiles flew, there was an interval of several hours between Clinton's launch order and their arrival at the target.[70] This gave plenty of opportunity for a highly mobile target such as bin Laden to move out of danger. Furthermore, the United States had to warn Pakistani authorities of the launch due to the missiles' flight path over Pakistan's air space. As the attacks were taking place at a time of increased tensions between the Pakistani and Indian governments, it was feared the Pakistanis might interpret the missiles as an Indian strike and retaliate. So, shortly before the missiles entered Pakistan's airspace, the vice-chairman of the Joint Chiefs, General Joe Ralston, was dispatched to alert Pakistani officials of the U.S. operation. There was significant concern that this information would likely be passed from sympathetic elements in the Pakistani government to the Taliban, especially given the close links between that organization and members of Pakistan's ISI. It was likely that should the Taliban, as hosts to bin Laden, have learned about the strikes, they would have warned the al-Qaeda leader, giving him time to flee the scene.[71]

Whether bin Laden was tipped off or was just fortunate enough to have moved on before the missiles arrived is unclear. Peter Bergen suggests the evacuation of foreigners from Kabul shortly before the strikes may have alerted al-Qaeda's observant leadership.[72] Mohammed Odeh, an al-Qaeda operative arrested by the FBI for his role in the 1998 embassy bombings, suggested to his bureau interrogators that bin Laden predicted the United States would retaliate with cruise missiles rather than risk casualties from either a commando or aircraft bombing raid, and had warned his people to evacuate.[73] What is clear is that the missiles missed their mark, and in doing so helped promote bin Laden's status further. The world's super-power had tried to kill the terrorist leader using hundreds of millions of dollars' worth of equipment and had failed.[74] A number of planned follow-up strikes were prevented by the extremely limited intelligence resources the United States had on the ground in Afghanistan. Just as the lack of American eyes on the target had prevented Clarridge from being able to launch a mission to rescue the American hostages from Beirut, so too the inability to corroborate intelligence supplied by Afghan sources with American eyes prevented the Clinton administration from taking action. Clarke reflected upon the situation in his "Strategy for Eliminating the Threat from the Jihadist Networks of al-Qida," which he completed shortly before Clinton left office, noting that "follow-on attacks were considered and military assets deployed on three occasions when al Qida commanders were located in Afghanistan by Humint [human intelligence] sources," but on all three occasions the attacks were called off because the single source was not deemed sufficiently reliable to justify U.S. military action.[75]

During the Clinton administration's final year Clarke sought to see the use of cruise missiles in America's pursuit of bin Laden replaced by the deployment of armed drones. Clinton's counterterrorism adviser drew up a strategy for eliminating al-Qaeda in which he referred to the September 2000 reconnaissance flights the CTC had conducted over Afghanistan using the Predator drone. The document noted that bin Laden had been located on two separate occasions by the unmanned aircraft, and it called for the flights to recommence in spring 2001, but that in future the Predator should have "a new capability" to launch Hellfire missiles at ground targets. Clarke reported that this would give the United States a "see it/shoot it" option, significantly reducing the kill chain from the president's order to the impact of the missile.[76]

For counterterrorism purposes, which required the pursuit of highly mobile targets who frequently hid among civilians, the cruise missile was just too blunt and inflexible an instrument. It was a Cold War–era weapon designed for leveling buildings, not killing individuals, and its singular deployment against bin Laden exposed how ill equipped the DoD was to engage in counterterrorism. Since adopting the use of armed drones, thus providing the CTC with the see it/shoot it option, the United States has launched hundreds of drone strikes against known and suspected al-Qaeda-related targets. While much of the reasoning behind the increased number of strikes can be put down to America's adoption of a war footing against al-Qaeda and the Taliban following 9/11, the role played by the drone itself in increasing the number of strikes should not be ignored. Phillip Alston, the U.N. special rapporteur on extrajudicial killing, warned in his May 2010 report that the perceived accuracy of drones combined with the greater level of direct control and lack of immediate risk to American personnel could lower the threshold for leaders to order the use of lethal force.[77] The increased willingness of Clinton's successors to unleash lethal strikes adds validity to Alston's case.

The likes of Richard Clarke knew the limitations of the cruise missile all too well and set in motion the process of getting the armed Predator drone deployed. While this may not have been complete by the time Clinton left office, it ensured that the weapon was ready for the Bush administration to deploy in the aftermath of 9/11. The development of a weapon system that significantly shortened the timescale between ordering to launching the strike, provided real-time intelligence on the target, and delivered a significantly smaller blast radius that limited the risks of collateral damage was specifically driven by the need to hit bin Laden in his safe haven of Afghanistan.[78] The armed Predator drone was custom made for lethal counterterrorism operations, and it is clear to see why the Obama administration adopted this weapon over other military options.

Rejection of the "Boots on the Ground" Option

The final alternative for Obama to wage his war against al-Qaeda in the AfPak region was to make more use of the U.S. military, primarily in the form of raids conducted by the counterterrorism operators within JSOC.[79] As Mark Mazzetti reveals in *The Way of the Knife*, Bush's secretary of defense, Donald Rumsfeld, had significantly bolstered the DoD's counterterrorist capabilities in the wake of 9/11. Following a visit to JSOC's headquarters at Fort Bragg, North Carolina, in November 2001, the secretary more than doubled the special operators' budget, reaching nearly $8 billion by 2007.[80] Coupled with Rumsfeld's signing of the Al-Qaeda Network Executive Order in spring 2004, which granted the military the authority to spy on, capture, or kill al-Qaeda members anywhere in the world, the DoD's counterterrorism capability was transformed. "No longer was JSOC capable merely of twenty-four-hour hostage-rescue missions," observes Mazzetti. "It could run wars of its own."[81] Jeremy Scahill's exhaustive investigative journalism, published in his book *Dirty Wars* and featured in the Academy Award–nominated documentary of the same name, reveals how accurate Mazzetti's observation is. Scahill's work exposed the scale of U.S. Special Forces counterterrorism operations sanctioned under Bush, and the extent to which these were continued and even expanded by the Obama administration.[82] These missions ranged from Afghanistan to Yemen, Somalia, and beyond, and they illustrate what journalist David Sanger calls Obama's "surprising use of American power" in his "secret wars."[83] In their most high-profile mission to date, it was the JSOC, backed by CIA intelligence, that undertook the raid into bin Laden's Abbottabad hideout.

In adopting this approach, Rumsfeld took terrorism more seriously than any secretary of defense before him, and he significantly expanded the options a president has for using lethal force against al-Qaeda and its affiliates. Whereas his predecessors, such as Weinberger, Aspin, Perry, and Cohen, had fought to keep the DoD out of counterterrorism, Rumsfeld sought to challenge the CIA in this field and wrestle command of lethal counterterrorism operations away from the agency, as revealed in an October 2001 memorandum. "Does the fact that the Defense Department can't do anything on the ground in Afghanistan until the CIA people go in first to prepare the way suggest that the Defense Department is lacking the capability we need?" Rumsfeld asked. "Given the nature of our world," he continued, "isn't it conceivable that the Department ought not to be in a position of near total dependence on CIA in situations such as this?"[84] Through his reference to "the nature of the world," Rumsfeld indicated that counterterrorism was going to be a top national security concern for the foreseeable future, and that to remain influential within U.S. policy making the DoD had to end its reluctance to involve itself in counterterrorism operations.

In order to allow the DoD to take the lead in U.S. counterterrorism, Rumsfeld sent a proposal to DCI Tenet for the Pentagon to create its own counterterrorism cell. It would be similar to the CTC, only bigger and more suited to the new, global war against terrorism. He wrote, "From everything I hear, CTC is too small to do a 24–7 job." Rumsfeld outlined his plan for the creation of an entirely new counterterrorism body based in the Pentagon, named the Joint Intelligence Task Force for Combating Terrorism (JITF-CT), which would "establish a single focused effort" in the newly launched War on Terror.[85] It was Rumsfeld's hope that this focused effort would put the Pentagon, not the CIA, at the heart of the War on Terror. In contrast to previous secretaries of defense, who saw counterterrorism as a risky and messy distraction from the core business of the DoD, Rumsfeld saw it as the vehicle to exert American dominance.[86]

An impression of the scale of Rumsfeld's plans can be gleaned from the strategy memo he sent to the president within weeks of the 9/11 attacks: "The USG [U.S. Government] should envision a goal along these lines: New regimes in Afghanistan and another key State (or two) that supports terrorism (To strengthen political and military efforts to change policies elsewhere)." He added, "If the war does not significantly change the world's political map the U.S. will not achieve its aim."[87] Just as anti-Communism had been used to justify the expansion of American influence around the world during the Cold War, counterterrorism would provide the motivation for America to shape the post-9/11 world. Rumsfeld's JITF-CT was created by the Defense Intelligence Agency (DIA) in 2007 and has since served as the DoD's primary counterterrorism analysis center.[88] Despite the creation of JITF-CT and the significant expansion of JSOC, it is the CIA and its drone campaign that has remained America's primary counterterrorism tool in Northwest Pakistan. Why was Rumsfeld's DoD so supportive of the CIA when it came to engaging al-Qaeda?

Before 9/11, the DoD's attitude toward counterterrorism was shaped by a number of unsuccessful interventions that had driven the Joint Chiefs and defense secretaries to steer clear of such operations. As already discussed in chapter 3, there was significant collateral damage and the loss of two airmen following Reagan's use of airpower against Gaddafi.[89] Chapter 1 detailed the failure of Delta Force's commando raid in the disastrous Operation Eagle Claw. Not only did this cost Carter dearly in a political sense, it also ensured that the Pentagon's commanders, both civilian and military, sought to avoid any similar commitments throughout the Reagan years.[90]

This reluctance to engage in smaller-scale operations was challenged in the autumn of 1992. Following the collapse of the Soviet Union, President George H. W. Bush had announced the emergence of a new, American-led world order in which "brutality will go unrewarded and aggression will meet collective resistance." Addressing Congress during his 1991 State of the Union address, the president told the assembled lawmakers that America would bear a major share of leadership in this new global system because, "among the nations of the world, only the United States of America has both the moral standing and the means to back it up." The United States was, Bush declared, "the only nation on this Earth that could assemble the forces of peace," making America a "beacon of freedom in a searching world."[91] Though the primary focus of Bush's speech was Saddam Hussein's aggression in Kuwait, the wider message of an assumption of American responsibility was not lost on the international community, in particular the recently elected U.N. secretary general, Boutros Boutros-Ghali. He challenged Bush to support his lofty rhetoric of a post–Cold War "new world order" with action. A civil war had erupted between clans in the East African nation of Somalia in 1988, destroying the fragile economy. With hundreds of thousands of Somalis facing starvation, the United Nations appealed to the American president to provide assistance to an ongoing U.N. Operation in Somalia (UNOSOM I) peacekeeping mission. The United States agreed to the United Nations' request, which, according to Bush's APNSA, Brent Scowcroft, would demonstrate that "the United States was not afraid to intervene abroad," and might even form a model for future peacekeeping missions under joint U.S.-U.N. auspices for the future.[92]

Unfortunately, in reality the operation was a calamitous failure. The United Nations and the United States had divergent views as to the goal of American intervention in Somalia. Despite the deployment of twenty-five thousand troops, the United States saw its role as purely humanitarian—to ensure food reached those in need, and nothing more. For Boutros Boutros-Ghali, however, the mission was one of nation building, with Somalia to be stabilized through the disarming of the warlords. A tug of war ensued between Boutros-Ghali and the United States, which resulted in the worst of both worlds.[93] Keen to maintain his liberal internationalist vision of the post–Cold War world, the recently elected Clinton agreed to keep a limited U.S. military force in the country to assist with the U.N. mission, but he reduced the size of the force significantly, leaving a contingent of fifteen hundred troops and four hundred Special Forces commandos to accomplish an objective that had already looked beyond the capabilities of the full force. When four American servicemen were killed by soldiers loyal to the Somali warlord Mohamed Farah Aideed, the U.N. Security Council responded by ordering his capture and detention. "Failure to take action," warned Clinton's secretary of state, Madeleine Albright, in an op-ed in the *New York Times,* "would have signaled to other clan leaders that the UN is not serious."[94]

The snatch mission fell to the elite Delta Force and Army Rangers that made up the U.S. Quick Reaction Force (QRF), who deployed to the capital, Mogadishu, on October 3, 1993, to apprehend Aideed.[95] Due to a combination of bad luck, poor planning, the lack of necessary equipment, and what the United Nations later described as a superficial understanding of Somalian clan politics, the snatch mission was a disaster.[96] In what became known as the Battle of Mogadishu, a gun battle ensued with Aideed's forces, during which three American Black Hawk helicopters were shot down, eighteen American soldiers were killed, seventy-four were wounded, and more than five hundred Somalis died. In the aftermath, the body of one of the slain Black Hawk crew members was filmed being dragged through the streets of the Somali capital and beaten by an angry crowd.[97] Al-Qaeda later claimed to have played a prominent role in the downing of the American helicopters, with bin Laden boasting that they had taught the Somali militia to aim for the tail rotors with rocket-propelled grenades. The leader of the Somali al-Qaeda affiliate al-Shabaab has also described how members of al-Qaeda played important combat and support roles during the battle itself.[98]

The impact of the Somalia episode on the attitudes of those within the Pentagon and the Clinton administration should not be underestimated, and helps explain the collective hesitancy to authorize an armed snatch mission against bin Laden during Clinton's second term.[99] In his memoirs, Clinton compared the debacle to his political hero John F. Kennedy's most humiliating experience, the Bay of Pigs invasion.[100] The parallels go beyond the military fiasco, as both failures were partially caused by faulty intelligence and poor planning and could be traced back to commitments and decisions made by the previous administration. It is highly likely that this feeling of having been boxed in by the previous administration contributed to Clinton's decision to hold off from any retaliation against al-Qaeda following the bombing of the USS *Cole* in the closing months of his administration, preferring instead to pass on the details during the transition meetings with Bush, leaving the choice of action to the incoming president.[101]

The effect of this military failure was perhaps felt even more acutely within the DoD. The month before the snatch mission, Colin Powell, then chairman of the Joint Chiefs, had requested that Secretary of Defense Les Aspin approve a request filed by William F. Garrison, the U.S. commander in Somalia, for tanks, armored vehicles, and AC-130 Spectre gunships to support his heavily depleted forces. This request was in line with Powell's preferred strategic approach, which had been dubbed the Powell Doctrine during the buildup to the 1990–1991 Gulf War. Drawing heavily on the Weinberger Doctrine, the Powell Doctrine emphasized that should military force become necessary it ought to be overwhelming in scale and deployed only with substantial public support. Powell advocated the utilization of "every resource and tool" in order to "achieve decisive force against the enemy, minimizing U.S. casualties and ending the conflict quickly by forcing the weaker force to capitulate."[102] To military officers like Powell and Garrison such requests were perfectly reasonable and offered the best way of achieving the mission while limiting U.S. casualties. To civilian policy makers such as Aspin, however, the doctrine created an all-or-nothing approach to American military action. Despite almost certainly limiting American military casualties and preventing policy makers from rashly deploying force, it also served to establish a prohibitively high benchmark for military action. Aspin rejected the request, leaving Garrison's force poorly equipped to succeed in its mission. Following the debacle in Mogadishu, Aspin was forced to answer to a congressional committee on what had gone wrong. After an unconvincing performance by Aspin, several members of Congress demanded his resignation; in December 1993 Clinton announced that his secretary of defense was retiring after less than a year in his post.[103]

Aspin's replacement, William J. Perry, and other senior Pentagon leaders absorbed the lessons of Somalia. Both Operation Eagle Claw and now the mission in Somalia had, in the eyes of the military, been doomed to failure by political errors. The result of this was to reinforce the tenets of the Powell Doctrine. Problematically for Clinton, the Powell Doctrine, like the Weinberger Doctrine before it, limited the role of the military to huge-scale interventions. Unfortunately, this meant when it came to looking for a solution to a small but professional terrorist group hiding out in the difficult-to-reach terrain of Afghanistan, the Pentagon was completely unwilling to be involved. Clarke recalls a discussion between Clinton and Joint Staff Chairman Hugh Shelton in the Cabinet Room in the aftermath of the 1998 embassy bombings. The president asked the former Special Forces commander: "Hugh, what I think would scare the shit out of these al-Qaeda guys more than any cruise missiles . . . would be the sight of U.S. commandos, Ninja guys in black suits, jumping out of helicopters into their camps, spraying machine guns. Even if we don't get the big guys, it will have a good effect."[104] According to Clarke's account, Shelton "looked pained." His explanation for why such an action was not feasible was that al-Qaeda's camps were a long way away from anywhere the United States could launch a helicopter raid. Despite this, Clinton's top military officer did agree to "look into it," but no plan for a Special Forces raid was ever presented to the president.[105] Shelton must not have looked very hard, because it was only a year later that the CIA secured basing rights for the Predator reconnaissance flights into Afghanistan from neighboring Uzbekistan.[106] The reality was that even with the potential for basing in Uzbekistan, staff at the Pentagon had no desire to organize a mission that combined the long-distance helicopter raid elements of Operation Eagle Claw with the snatch operation of the Somalia raid. Despite high-tech equipment and well-trained Special Forces, the challenge of flying across hundreds of miles of hostile territory to apprehend or kill an individual belonging to a highly motivated and well-trained terrorist organization, based upon limited intelligence, was so rife with risks and potential negative fallout that the Pentagon resisted any pressure to engage. In accordance with the tenets of the Powell Doctrine, if the U.S. military was to combat al-Qaeda, it would need to be on a massive scale where superior equipment and numbers could be brought to bear to ensure force protection and success, something the U.S. public, media, and most important Congress had no interest in supporting.[107] Thus small-scale raids were out of the question for Clinton. The Pentagon's resistance to the mission against al-Qaeda is perhaps best demonstrated by the fact that immediately following 9/11, when the United Nations granted the United States permission to invade Afghanistan to track down al-Qaeda and overthrow its Taliban hosts, the American military had no prearranged plan for attacking the country.[108]

The public humiliation of these failures was eventually reversed with the success of the high-risk raid against bin Laden's Abbottabad hideout, deep in Pakistani territory, on May 2, 2011. Despite the enormous publicity that was guaranteed to accompany the raid whatever the outcome, the Pentagon's JSOC felt confident to take on the mission. Why was Obama able to order a raid when Clinton's proposal for such an operation had been resisted by his military chiefs? Unlike Eagle Claw, which was the first such operation undertaken by America's new Delta Force, and the Somalia raid, which was unfamiliar territory for the SEALs and Rangers, by the time the CIA pinpointed bin Laden's location, JSOC's special operators had, according to a senior Defense Department official, conducted more than two thousand similar raids in Afghanistan and Iraq. They had become such common practice on the front lines of America's War on Terror that the same official described them as "mowing the lawn," adding that on the night of the bin Laden raid a further twelve missions were conducted in Afghanistan, resulting in the killing or capturing of fifteen to twenty targets.[109] Even with this extensive experience, the crash landing of one of the custom-built stealth Black Hawk helicopters in the courtyard of bin Laden's compound emphasized the inherent risks associated with such missions, no matter how well planned and prepared.[110]

The political fallout of the raid goes further to demonstrate why the United States' use of Special Forces against al-Qaeda and the Taliban in Pakistan could not be a mainstay strategy. Even before the raid, Obama indicated he was aware that there would be huge political ramifications.[111]

Though drones had been operating in Pakistani territory for years before the raid, there was a significant difference between the fallout from the unmanned and semiauthorized incursions and actual American boots on the ground in Pakistan. Humiliated by its inability both to locate bin Laden and to detect or react to the American forces, the Pakistani government lashed out at the United States. In the immediate aftermath Salman Bashir, the Pakistani foreign minister, described the raid as a "violation of sovereignty," which raised "legal questions in terms of the UN Charter."[112] Opinion polls commissioned by CNN shortly after the raid revealed that 63 percent of Pakistanis expressed disapproval of America's actions, with 51 percent saying they felt it would adversely affect their country's relations with the United States.[113]

Despite the apparent backlash against the bin Laden raid, polling by Pew Research revealed that though the al-Qaeda leader was not well liked in the years before the raid, only 14 percent of Pakistanis felt his death was a good thing, suggesting any action that had resulted in his killing would be looked upon unfavorably. Moreover, even though the Pakistani public was generally critical of the U.S. action, America's overall favorability rating among Pakistanis was not worsened any further by the raid.[114] The report of Pakistan's Abbottabad Commission further highlighted Pakistanis' attitude towards the raid, describing it as an act of war by the United States against Pakistan. The report declared that the raid demonstrated Washington's "contemptuous disregard of Pakistan's sovereignty, independence and territorial integrity in the arrogant certainty of . . . unmatched military might."[115] As political retaliation, Pakistani members of Parliament voted to temporarily end NATO transit convoys taking supplies to forces in Afghanistan.[116] While this was only a short-term measure, it was nonetheless costly and highly disruptive to operations in neighboring Afghanistan.

In addition to the vitriol aimed at the United States, the fallout of the raid caused significant damage to the already vulnerable Pakistani government, headed by President Asif Ali Zardari and his fellow Pakistan Peoples Party (PPP) prime minister, Yousaf Raza Gilani. The Abbottabad Commission's report described the U.S. action as the "greatest humiliation" suffered by Pakistan since the loss of Bangladesh in 1971, and it issued a damning indictment of the authorities' "culpable negligence and incompetence" in failing to pick up on the clues of bin Laden's whereabouts. The report even went on to suggest that there was a possibility that current and former officials provided "plausibly deniable support" for bin Laden, though it did ultimately clear the Pakistani establishment of any involvement in sheltering the al-Qaeda leader.[117] In polling following the raid, only 11 percent of Pakistanis had a favorable view of Zardari, while his prime minister's ratings had dropped from 59 percent in 2010 to 37 percent. It is worth noting, however, that Zardari's government was already deeply unpopular before the raid due to a range of factors, such as rising prices, unemployment, and political corruption, with the fallout of the American attack serving to solidify rather than create the image of an incompetent administration.[118] What the episode of the Abbottabad raid did reveal is that the presence of American soldiers in Pakistan is significantly more controversial than that of drones. U.S. Special Forces may have honed their skills in undertaking such raids over the previous decade of the War on Terror, but such actions could still only be used sparingly and could not serve as the primary tool for combatting al-Qaeda and its Taliban allies in the Northwest provinces of Pakistan.

The fallout from the Abbottabad operation revealed that a larger presence of American soldiers in Pakistani territory was not an option for the Obama administration. The Pakistani public's anger at the undermining of Pakistan's sovereignty through the presence of even a small number of American soldiers indicated it would not tolerate larger incursions. Furthermore, the large-scale deployments under Bush had not proven an effective method of destroying al-Qaeda. From the outset, bin Laden and Zawahiri had sought to suck the United States into a protracted, full-scale war through which they could drain the American economy and illustrate to other Muslims that there was a clash of civilizations between the "Zionist Crusaders" of the West and Islam.[119] While bin Laden had not managed to trigger a global clash, the presence of American forces in Iraq had certainly acted as a recruitment tool for al-Qaeda, breathing new life into the organization and attracting recruits from all over the Arab world. The Sinjar Records, retrieved by U.S. forces from the small Iraq-Syria border town of the same name in October 2007, revealed the extent to which foreign fighters were flocking to Iraq to fight Americans. The records provided details of 606 foreign fighters who had entered Iraq via Sinjar since August

2006. Of these fighters, 41 percent were Saudi, 19 percent were Libyan, and the others came from across the Middle East, including Syria, Kuwait, and Jordan, as well as African nations, such as Algeria and Sudan. Of the 389 fighters who designated what their work was, more than half stated martyr.[120] Though Obama did opt to increase the number of troops on the ground in Afghanistan, to commit similar forces elsewhere in the pursuit of al-Qaeda—in particular into an area such as the North-West Frontier of Pakistan—would have invited significant anti-American attacks. The remote nature of the drones has served to downplay the crusader image of the United States al-Qaeda has sought to spread, and there is no evidence of large numbers of foreign fighters flocking to the Federally Administered Tribal Areas (FATA) to repel the infidel, as was the case with Iraq.[121]

The final option open to the Obama administration, and one it did initially pursue, was to exert pressure on the Pakistani military to act against al-Qaeda and the Taliban by undertaking military action in the FATA region. This approach faced two main problems. The first was similar to the risks posed by the presence of U.S. boots on the ground in the region: namely, the presence of the Pakistani military could inadvertently escalate the conflict rather than suppress it. When the Pakistani military did engage in hostilities against the Taliban in the FATA in 2004, it exposed the fragility of the government's control and the intensely divided nature of Pakistan. Influential clerics at the Lal Masjid mosque in Islamabad called for the people of South Waziristan to resist. Furthermore, the Pakistani military lacked the precision munitions and intensive counterterrorism training of the U.S. Special Forces. Instead, the Pakistani army bludgeoned the villages of the FATA. As Mark Mazzetti describes, the Pakistanis used helicopter gunships and heavy artillery, destroying hundreds of compounds belonging to tribesman suspected of harboring militants, looting civilian homes, and even shelling civilians fleeing in a convoy of trucks. Despite its extreme use of force, the Pakistani military still took heavy casualties and had more success in turning the local residents of the region against it than in actually harming the militants.[122] On April 24, 2004, the Pakistani government signed a peace treaty with the Pakistani Taliban.[123] The document itself was a mockery, and soon the Pakistani state was under siege from the militants in its frontier provinces. However, the Pakistani government's willingness to sign the treaty revealed that it did not have the equipment, training, or motivation to try to dismantle the militant groups who had conquered their border territories.

Conclusion

The United States has significantly expanded both its lethal and nonlethal counterterrorist capabilities since the Reagan administration's limited options for retaliation led to the bombing of Libya in 1986. In theory, Barack Obama entered the White House with more options of how to deal with al-Qaeda than any president who had preceded him. These included renditions backed by a global network of CIA black sites, Special Forces raids undertaken by JSOC with larger military interventions led by the DoD, and increasingly sophisticated and technologically advanced drones supported by the NSA's electronic surveillance and extensive human intelligence provided by the networks of informants in the AfPak border lands. Despite this apparent plethora of options, Obama's choices were actually curtailed by a range of factors.

First, the United States' finances were already strained from a decade of war in Afghanistan and Iraq. The economic crash of 2008 had also had a devastating impact upon the nation's economy, putting pressure upon the new president to develop a strategy that was relatively modest in its costs compared to the extravagant expenditure of his predecessor's War on Terror. Second, the domestic pressure to end what was regarded as the CIA's torture program and the growing worldwide embarrassment caused by the Guantánamo Bay detention facility made the continuation of large-scale rendition a political impossibility. Furthermore, political opposition to expanding U.S. troop deployments and intense anti-Americanism in countries such as Pakistan ensured that any counterterrorism efforts in the AfPak border region would need to be handled covertly. Finally, Obama's key objective in his more focused war against al-Qaeda—the decapitation of the organization's core leadership in the AfPak region and denial of safe haven—was supplemented by a desire, with respect to Afghanistan, to reverse the Taliban's momentum and weaken the insurgency to a level that could be contained by the Afghan National Security Force.

With Obama's core objectives in mind, expanding the drone campaign was a sensible choice for the new president. The CTC's program enabled a sustained and relatively precise campaign against al-Qaeda's leadership. It also empowered the United States to penetrate the Taliban's previously safe haven of the FATA and undertake the strikes necessary to weaken the group and at least reduce, if not reverse, its momentum in Afghanistan. At the same time, the campaign was relatively cheap, in U.S. lives as well as cash, in comparison to the large-scale military operations of the Bush administration, and despite organized protests in parts of Pakistan, generated considerably less anti-Americanism than the physical presence of U.S. boots on the ground had done in Afghanistan and even more so in Iraq. That is not to say that the Obama administration was wholly reliant upon drones. Obama made extensive use of America's other counterterrorism tools, deploying JSOC forces on missions all around the world, authorizing extensive electronic surveillance, and rendering a number of high-profile targets back to the United States for trial by military tribunal. These activities, however, were largely reserved for the other battlefields of America's war against al-Qaeda. When it came to the AfPak region the CTC's drones were undeniably the primary tool to help the Obama administration achieve its core objectives on both sides of the border.

6

"THE ONLY GAME IN TOWN": THE STRATEGY AND EFFECTIVENESS OF THE CIA'S LETHAL DRONE CAMPAIGN, 2009–2012

Very frankly, it's the only game in town in terms of confronting or trying to disrupt the al-Qaeda leadership.
—*Director CIA Leon Panetta*

During his 2008 election campaign Senator Barack Obama rejected the concept of the War on Terror that President Bush had declared in the aftermath of the 9/11 attacks. There were those who believed Obama's rejection of the notion of fighting an open-ended global conflict with all terrorism meant that he would repeal the military authority granted to the president by Congress's passing of the 2001 AUMF and set about dismantling the legal architecture that had evolved over the previous two decades, leading to increased domestic surveillance, the CIA's network of detention centers, and the preemptive deployment of unilateral force in breach of international law. Those who anticipated this were quickly disappointed. In discarding the War on Terror concept, Obama was rejecting the scale of Bush's conflict, not the actual use of force for counterterrorism purposes. Under Obama the war would continue, but it would be refocused. No longer would the United States seek to change the regimes of the so-called Axis of Evil states or preemptively attack nations with a WMD capacity, real or perceived. Instead, what Obama promised was a slimmed-down, intensive campaign directed against those responsible for 9/11—the core of the ever-expanding al-Qaeda franchise—and those who harbored them.

Working in close collaboration with his counterterrorism adviser John Brennan, Obama declared the AfPak region as the front line of the war with al-Qaeda, and set out a three-part strategy for destroying the terrorist group. First, the United States would aim to decapitate al-Qaeda's leadership; second, it would deny safe haven for both al-Qaeda and its Taliban allies; and, finally, it would degrade the Taliban insurgency in order to aid the NATO mission in Afghanistan and ensure the group could not find the time and space to recover. Principal responsibility for achieving these objectives fell to the CIA's Counterterrorism Center, and the primary tool that was to enable the CTC to hunt and kill America's terrorist enemies was its ever-evolving unmanned drone fleet, a prototype of which the center had introduced three decades previously. How effective the drone campaign has been in achieving these aims is a hotly debated topic, the answer to which has significant implications for the future direction of U.S. counterterrorism.

The Architect of the Obama Administration's Drone Campaign

A former CIA analyst with an impressive résumé, John Brennan served as adviser to the then senator Obama on intelligence matters during his 2008 election campaign, having been personally recommended to the Democratic candidate by both Anthony Lake, who served as Clinton's national security adviser, and the previous DCI, George Tenet. The erstwhile CIA bureaucrat and political realist quickly won Obama's trust and respect, building a close rapport with the victorious candidate based upon a mutual agreement of the problems with the Bush administration's conduct of the War on Terror—a phrase both men agreed was ridiculous.[1] Having had a distinguished career within the intelligence community, Brennan initially hoped the president-elect would nominate him as the next director of central intelligence, something Obama strongly considered, having opposed the serving D/CIA Michael Hayden's nomination back in 2006 (ironically, for Hayden's role in establishing the NSA's surveillance program, something Obama would later come to exploit extensively).[2] Controversially, however, Brennan had served as deputy executive director of the CIA from 2001 to 2003, during which time the agency established its extraordinary rendition and EIT program. Though not personally connected to the practices, Brennan had publically defended the agency's actions on a number of occasions since, telling the *New Yorker*'s Jane Meyer in 2007 that "the U.S. would be handicapped if the CIA was not . . . able to carry out these types of detention and debriefing activities."[3] When an open letter signed by two hundred psychologists and academics was published on November 24, 2008, urging Obama not to nominate a "torture apologist" as CIA director, it was clear Brennan did not represent the clean break from the abuses of the Bush administration those who had supported the newly elected president expected, and he promptly withdrew from consideration.[4]

Despite the removal of Brennan from contention for D/CIA, Obama's desire to have his closest national security adviser develop his administration's counterterrorism strategy was revealed by the president-elect's decision to offer Brennan an alternative post, which did not require Senate confirmation but still granted as much, if not more, power to shape counterterrorism policy than the Langley post. He was appointed assistant to the president for homeland security and counterterrorism, a senior position within the National Security Council similar to that held by Richard Clarke within the Clinton administration, and given a bunker-like office just one floor down from the Oval Office and an equal number of steps from the Situation Room. This quite literally placed Brennan at the heart of the national security infrastructure of the United States.[5] To avoid a possible turf war between his counterterrorism strategist and the director of the CIA, Obama surprised many, including his nominee, by offering the leadership of the agency to Leon Panetta.[6] An experienced Washington insider who had previously served as Clinton's chief of staff, Panetta was "a total blank slate on intelligence issues," as John Rizzo, one of those responsible for preparing the nominee's initial briefings, put it.[7] This suited Obama perfectly. He did not need a D/CIA who would seek to bring a new strategic vision to the CIA's counterterrorism mission—Langley's highest priority at that time—as Brennan already had that. Instead, the president needed a director who had the political clout, respect, and Washington connections to enable this new strategy to be implemented. The close cooperation Obama expected between his counterterrorism chief and D/ CIA was demonstrated by the fact the two shared an eight-by-ten-foot transition team office during the last months of 2008 and early 2009, with Panetta later noting that "he was grateful that [Brennan] would be serving on Obama's team as the administration's chief advisor on counterterrorism," a position for which, Panetta added, "he was eminently well suited."[8]

During Obama's first term, Brennan took what was a disparate collection of counterterrorism tactics and turned them into a White House–centered strategy with him at its core.[9] In doing so, his influence was so great that some military officials reportedly began referring to him as "deputy president."[10] Hence, Brennan is rightly reputed to be the main architect and proponent of the CTC's drone campaign in Pakistan, a position that won him numerous supporters and critics alike.[11] Yet the relative novelty of the drones used to implement the strategy belied how closely Brennan's approach followed that espoused by the counterterrorism hardliners of the Reagan administration three decades earlier. The connection is no coincidence, with Brennan's career lineage revealing the long-term impact of the hardliners' influence on the agency. From 1990 to 1992, Brennan led terrorism analysis in the CTC, arriving shortly after Clarridge had been forced to retire but clearly absorbing the lessons the counterterrorism hardliner had sought to instill within the center.[12]

As well as exhibiting the hardliners' belief that lethal force should be employed to preemptively neutralize terrorist threats, Brennan's career also demonstrated an understanding of the importance Clarridge had placed upon interdepartmental cooperation and intelligence collection in counterterrorism. In March 2003 Brennan was appointed chief of the agency's newly founded Terrorist Threat Integration Center (TTIC), a center intended to enable full integration of U.S. government terrorist-threat-related information and analysis across departmental lines.[13] Furthermore, from 2004 to 2005 he served as interim director of the newly established National Counterterrorism Center (NCC), a clearinghouse for all the government's various terrorist databases. Congruently, Obama's 2008 election campaign had made extensive use of databases built up via data mining and microtargeting to target prospective voters.[14] Accordingly, both men entered the White House with a firm understanding of the potential that technological developments in the field generally referred to as big data offered in creating a new approach to tracking and targeting America's elusive terrorist enemies. The Disposition Matrix was born—a next-generation kill list populated with thousands of names, biographies, and biometric data acquired by the U.S. intelligence, military, and law enforcement communities, and with it, the full realization of the CTC's data-driven approach to counterterrorism first conceptualized by Clarridge upon its founding.[15]

Brennan's adoption of drone warfare may have been built upon principles established by his predecessors, but the astute counterterrorism chief also took full advantage of what, in technological terms, is referred to as connectedness—the sudden conjunction of a range of technologies that had developed independently of one another. GA-ASI's advances in drone technology meant its 2007 Reaper variant could carry fifteen times the ordinance at three times the speed of its predecessor, the Predator; Raytheon's introduction of the Griffin missile in 2008, a precision munition weighing a third of the weight of a Hellfire and designed specifically to eliminate personnel rather than tanks, reduced both the risk of collateral damage and the cost of each strike; new digital techniques were employed by the National Geospatial Agency (NGA) to map out the physically inaccessible territory occupied by America's enemies; the air force further improved its satellite guidance; and third-party advances in laser and infrared sensor technology, coupled with the increased feed quality from Raytheon's Multi-Spectral Targeting System, promised greater precision. Finally, the creation of machine-learning algorithms that can learn from and make predictions on data, coupled with increased computing power and near-infinite data storage capacity, allowed the NSA to conduct surveillance on a hitherto unknown scale, capturing massive amounts of data that could be used to build a picture of known and suspected terrorists' patterns of life.[16] Thus, "the entire apparatus of the United States government," observed the investigative journalist Chris Woods, was "bent toward the process of targeted killing over the past decade."[17] The expansion of the drone program was a product not just of the CIA but of the U.S. government in its greatest sense.

The Obama Administration's Use of Drones

During Barack Obama's first year in office, the CIA's Counterterrorist Center undertook more lethal drone strikes than it had in the previous eight years of the Bush administration. The New American Foundation

(NAF) has estimated that, by the end of Obama's first term, the CTC had launched 294 lethal drone strikes in Pakistan, compared to forty-six during Bush's time in office. NAF's assessment suggests these first-term strikes killed between 1,250 and 2,118 militants, 124 to 154 civilians, and a further 130 to 222 individuals whose status was categorized as unknown. The Obama administration's second term saw a significant reduction in the number of these strikes, with sixty-one launched by June 24, 2016. In tandem with stricter targeting criteria and increased safeguards, the down-scaled second-term campaign, according to the NAF, resulted in the deaths of approximately 309 to 382 militants, between five and seven civilians, and no casualties of the unknown category. Of those killed in the first-term, 15 percent were civilians or of unknown status; that figure fell to 2 percent during the second term. Confident in the accuracy of the drone campaigns he had unleashed, and motivated by a desire to reduce suspicion and bolster support for a tool he believed would continue to be necessary, Obama ordered the release of an official summary of the counterterrorism strikes launched outside the designated areas of active hostilities (Afghanistan, Iraq, and Syria). Released on July 1, 2016, and covering January 2009 to December 2015, the first official figures revealed that Obama had authorized 473 strikes, killing an estimated 2,372 to 2,581 combatants and between sixty-four to 116 civilians across Pakistan, Yemen, Somalia, and Libya. While human rights advocates welcomed the move toward greater transparency, many challenged the discrepancies between the government's figures and those of nongovernmental assessments. James R. Clapper, director of national intelligence, argued that the administration had access to superior intelligence before, during, and after a strike, including video observations, human sources and assets, signals and geospatial intelligence, accounts from local officials on the ground, and open source reporting, providing unique insights that were unavailable to nongovernmental agencies.[18] Despite this apparent accuracy, the use of the range of estimated combatant and civilian deaths in its official data underscored that even the U.S. govern-ment did not always know the identity or affiliations of those killed in its drone strikes.

Journalist Daniel Klaidman provides an account of a discussion between Obama, his deputy national security adviser, Tom Donilon, and the D/ CIA, Michael Hayden, following a misjudged drone strike at the very start of Obama's first term, on January 23, 2009.[19] Rather than destroying the home of a leading Taliban commander, the strike killed Malik Gulistan Khan, a prominent tribal elder in the FATA who was a member of a pro-government peace committee. The missile also killed four members of his family, including two children. Reacting to Hayden's description of the CIA's "signature strikes," a technique authorized by Bush in which the drones could fire on groups of military-aged men who bear certain signatures associated with terrorist activity, but whose identities are not necessarily known, Obama is reported to have said, "That's not good enough for me." Despite Hayden's efforts to defend the approach, Donilon is described as having told the DCI: "We have to review how we do this stuff. I'm not sure I'm comfortable with how we're doing it."[20] Given that the casualties under the Bush administration's drone campaign were between 205 and 348 militants at a cost of 126 to 154 civilians, with a further forty-six to fifty-six unknown casualties, it appears to be the case that the CIA's experience, combined with additional safeguards and what Clapper described as a methodology that has been "refined and honed over the years," has made drone strikes significantly more accurate.[21]

Strategic Objective 1: Decapitation of the Al-Qaeda Leadership

Though the CIA itself does not comment upon the drone campaign, an understanding of what its main strategic objectives were can be gleaned from the comments of administration officials after Obama came to office. Ultimately, the drone campaign had three primary purposes. The first aim was publicly set out by John Brennan in a speech delivered at the Paul H. Nitze School of Advanced International Studies in Washington, D.C., on June 29, 2011. In the speech, which never specifically referred to the classified drone campaign, Brennan stated that the Obama administration's counterterrorism aim in the AfPak region was to wage a "broad, sustained integrated and relentless campaign" against al-Qaeda's "core leadership in the tribal regions of Pakistan" with the aim of "dismantling" it.[22] In counterterrorism circles, targeting high-value leadership is referred to as a decapitation strategy. Such a strategy is intended to reduce a terrorist group's operational capability through the systematic elimination of its most valuable members, the disruption of its organizational routine by deterring other members from assuming leadership roles, and potentially even spurring organizational collapse through a power vacuum.[23] It is an approach that has generated significant debate about its effectiveness among terrorism and security experts. Many scholars, such as the political scientist Robert Pape, have argued that targeting enemy leaders within terrorist groups has met with meager success. Pape's view is supported by the recent studies of international affairs scholar Jenna Jordan, among others, which conclude that leadership decapitation is a misguided strategy, ineffective at best and possibly counterproductive.[24]

Audrey Kurth Cronin, one of the most noted scholars studying how terrorist groups end, has suggested that the study of the impact of decapitation is far from complete. She states that past experience with the decapitation of terrorist groups is just beginning to be studied in a systematic way. As a result, the relationship between decapitation and a group's demise is not straightforward, nor fully understood.[25] This point is supported by two extremely detailed empirical studies that have recently challenged the current consensus that decapitation is ineffective. Rand's Patrick B. Johnston has criticized the methodologies of previous studies and has argued that his own data-driven approach has enabled a more accurate measurement of successful and failed decapitation campaigns dating back to the mid-1970s. Johnston concludes that neutralizing insurgent leaders has been shown to have a substantively large and statistically significant effect on numerous metrics of countermilitancy effectiveness. Specifically, he claims that his results show that eliminating insurgent leaders increases governments' chances of defeating insurgencies, reduces insurrectionary attacks, and diminishes overall levels of violence.[26]

Bryan C. Price's separate study, conducted using what Price refers to as the largest and most comprehensive database of its kind, argues that terrorist groups are actually highly susceptible to decapitation because they have unique organizational characteristics (being violent, clandestine, and values-based organizations) that amplify the importance of leaders and make leadership succession difficult.[27] The accuracy of Price's further conclusions are substantiated by the consequences of the CIA's campaign against al-Qaeda. For example, Price's study suggests that decapitation increases the mortality rate of terrorists, as less experienced members are forced into senior positions, increasing their chances of making mistakes and being killed or captured, an argument supported by bin Laden's own warnings to his deputies. "It is important to have the leadership in a faraway location," the al-Qaeda leader cautioned in a letter sent to one of his lieutenants in October 2010. "When this experienced leadership dies, this would lead to the rise of lower leaders who are not as experienced as the former leaders and this would lead to the repeat of mistakes."[28] The notoriously frequent turnover of the third-in-command position in al-Qaeda supports Price's point and validates bin Laden's concerns.[29] The role, which essentially functions as a field commander, exposes the al-Qaeda member to significantly greater risk by requiring him to conduct operations in the open. Since 9/11 the United States has killed or captured those charged with this role with increasing frequency, including: Mohammed Atef in November 2001; Abu Zubaida in March 2002; Ramzi bin al-Shibh in September 2002; Khalid Sheikh Mohammed in March 2003; Abu Faraj al-Libbi in May 2005; Abu Hamza Rabia in November 2005; Abu Laith al-Libi in January 2008; Mustafa Abu al-Yazid in June 2010; and Abu Yahya al-Libi in June 2012.[30]

Yet despite the high mortality rate among its field commanders, al-Qaeda has been able to routinely replace its casualties with leaders capable of maintaining its operations. This supports the core argument advanced by Jordan, that due to al-Qaeda being an older, larger, and more established group that benefits from a multilayered bureaucratic structure, leadership roles are institutionalized, meaning the terrorist group is capable of resisting destabilization in the face of leadership attacks. In addition, the group has been able to sustain significant enough recruitment to replenish its losses while preserving its bureaucratic structure, suggesting a consistent degree of popular support. Bin Laden was, as Rohan Gunaratna notes, an ideologue as well as an activist, and following the founding of al-Qaeda, his most significant contribution to the group, in collusion with Ayman al-Zawahiri, was the development of an appealing narrative and set of beliefs, wrapped in an anti-Western message, with which al-Qaeda could attract a broad base of support.[31] This support has been maintained in the face of, and arguably fueled by, America's assault. As terrorism scholar Mia Bloom has warned, what are perceived as heavy-handed counterterrorism tactics, such as preemptive attacks on the supporters of terrorism, could easily backfire and mobilize greater support for terror.[32] Such backing benefits the group through the provision of safe haven as well as financial and material supplies. "The resonance of al-Qaeda's beliefs within local communities," Jordan contends, "has increased the organization's ability to withstand leadership targeting."[33]

Contrary to Jordan and Bloom's arguments, both Price and Johnston's studies offer more positive appraisals of the prospects of a decapitation strategy, though both maintain that such an approach is not a silver bullet to end terrorist groups and not likely to be successful in isolation. Instead, they independently reach a very similar statistical conclusion that a successful decapitation campaign creates a 25 to 30 percent greater chance of a government successfully reducing violence and ending the conflict with the terrorist group, a sizable benefit, which partially helps to explain why the United States has come to rely so heavily on this approach.[34] The available evidence suggests a substantial payoff with respect to al-Qaeda in the AfPak region. Early into the Obama administration's first term, Panetta told those assembled at the Pacific Council on International Policy in Los Angeles that the CTC's drone strikes had been "very effective" against al-Qaeda, adding that they were "the only game in town in terms of confronting or trying to disrupt the al Qaeda leadership."[35] In its September 2011 Quarterly Report to Congress on Afghanistan and Pakistan, the Obama administration continued to extoll the campaign, declaring that it had had "significant success" in the region, having "taken out more than half of al-Qa'ida's leadership."[36] Reflecting back on two terms of intense drone warfare, Michael Morell, who served as the CIA deputy director from May 2010 to August 2013, reiterated the campaign's success in the AfPak region, remarking in his 2015 memoir that the CTC's drone strikes had "decimated al Qa'ida's core leadership in South Asia," where "multiple al Qa'ida leaders there have been removed from the battlefield."[37]

Official American claims on the efficacy of the drone campaign are supported by al-Qaeda sources, with a June 2009 publication written by Abu Yahya al-Libi, at the time the terror group's third in command, lamenting the damage the drone strikes have done to al-Qaeda, specifically mentioning the high casualty rate among members of the core leadership: "The harm is alarming, the matter is very grave. So many brave commanders have been snatched away by the hands of the enemies. So many homes have been levelled with their people inside them by the planes that are unheard, unseen and unknown."[38] The al-Qaeda document provided photographs of the small GPS homing beacons, referred to as *pathrai* (a Pashto term for a metal device), with which the CTC equips its agents. The "spies painstakingly transport [pathrai] to the targets they are assigned by their infidel patrons," noted al-Libi, resulting in "the firing of the murderous and destructive missiles whose wrath is inflicted on the Mujahedeen."[39] Though there is a risk of the CTC's spies mistakenly placing the homing beacons on the wrong house or vehicle, or deliberately misusing them, both the tone of al-Qaeda's document and the behavior of its members, along with that of their Taliban hosts in the region of the strikes, suggests that the drones have been finding their targets. As paranoia gripped the jihadists, they launched hunts throughout the FATA, creating what Gul Rafay Jan, a Waziri from the border town of Miranshah, referred to as "a climate of fear." "Sometimes we see a body a day lying by the road-side, . . . they've got signs around their necks saying they were spies planting chips," the tribesman told the *Times*. "Sometimes they have been tortured to make confession videos," Jan continued, "by having rods pushed through their arms or stomachs, or being suspended over fire."[40] The Pakistani Taliban have even created their own force whose only purpose is to identify, capture, and execute people allegedly working for the web of local spies created by the CIA.[41] The Lashkar al Khorasan (Army of Greater Afghanistan), as it is known, operates in North Waziristan, arresting and publicly executing tribesman in all manner of brutal ways, primarily, local tribal elders have said, just to terrorize ordinary tribesman and discourage further spying.[42]

Documents recovered from Osama bin Laden's Abbottabad compound following the U.S. Navy SEAL raid in 2011, revealed that the consequences of drone strikes for al-Qaeda's senior members also troubled the organization's leader, who had urged his "brother leaders, especially the ones that have media exposure," to "exit Waziristan." In a letter to Libyan-born Atiyah Abd al-Rahman (also known as Shaykh Mahmud), who was one of bin Laden's top deputies and a key link to the outside world, al-Qaeda's isolated leader stressed that his senior deputies should "choose distant locations . . . away from aircraft, photography and bombardment while taking all security precautions."[43]

Price and Johnston's findings suggest that the removal of senior leaders can cause interorganizational rifts, with power vacuums causing even disciplined groups such as al-Qaeda to fracture and split, leading to the organization's demise. Though the elimination of key al-Qaeda personnel through drone strikes has not triggered complete organizational collapse, the persistence of the CTC's drones, able to loiter for days thanks to their high-endurance design, combined with the risks posed by electronic communications due to the NSA's surveillance, forced al-Qaeda's core leadership into a constant state of concealment. In order to survive in the face of relentless drone pressure, the organization had to change its operating methods, diverting its limited resources and time from generating plots and training new recruits to protecting its leaders.[44] By geographically and electronically isolating them from the jihadi brethren they seek to lead, the drone campaign has had a hugely detrimental effect upon the legitimacy and influence of the members of al-Qaeda's core leadership, alienating them from the global jihadist movement.

Evidence of the impact of targeting the leadership on the decline of al-Qaeda's core leadership came in the form of an embarrassingly public split with the organization's errant former affiliate Islamic State of Iraq and al-Sham (ISIS) in June 2013. The descent of Iraq's neighboring state into civil war in March 2011 attracted the attention of the Islamic State of Iraq's (ISI) leadership, who had maintained extensive recruitment networks and funneled fighters—with the tacit support of a Syrian government keen to keep the American-backed Iraqi regime off-balance—through the country into Iraq after the U.S. invasion in 2003.[45] ISI's recently appointed leader, Abu Bakr al-Baghdadi al-Husseini al-Qurashi (a nom de guerre drawing upon Islamic tradition), who had assumed control of the terror group on May 10, 2010, following the death of his predecessor, dispatched his operations chief, Abu Muhammad al-Jowlani, to Syria with instructions to establish a new ISI front.[46] The move was endorsed by Zawahiri, who sought a new affiliate at the heart of the burgeoning jihadist movement in Syria, having previously doubted ISI's loyalty and wisdom.[47]

Zawahiri's mistrust of ISI was spawned by the fact that, since the death of the group's Jordanian founder, Abu Musab al-Zarqawi, from a JSOC-coordinated airstrike on June 7, 2006, ISI leaders had not pledged *bay'a* (allegiance) to bin Laden, or Zawahiri. Instead, Baghdadi owed his position to ISI's new senior leadership, made up of former Ba'ath Party members who had run Iraq for decades under Saddam Hussein and had come to dominate ISI following JSOC's purge of the group's original leadership. Thirty-four of the forty-two most senior ISI jihadists had been eliminated by mid-2009 as a result of General Stanley McChrystal's high-tempo, high-value target raids.[48] On the verge of collapse, ISI bolstered its ranks by lifting its ban on admitting former Ba'athists, who, though not sharing the jihadists' extensive knowledge of and dedication to the Qur'an, possessed invaluable insights into the vulnerabilities of the Iraqi military, held intelligence on the powerbrokers of every tribe across the country's cities, towns, and villages, and knew how to exploit black markets to export oil, arms, and antiques in exchange for the resources that would help ISI rebuild itself.[49] It was these men, argues Raheem, a former personal aide to Zarqawi, who helped found the organization that eventually evolved into ISIS, who nominated Baghdadi, a "quiet and uncharismatic man" who was "regarded as a minor figure" with "no military experience and [with] scholarship of little note," despite his Ph.D. in Islamic studies. Following Baghdadi's appointment, ISI's inner circle reportedly closed ranks around its chosen leader, discontinuing all but the most cursory communication with the isolated al-Qaeda central command, as the few remaining senior figures considered loyal to Zawahiri were either deliberately sidelined or killed off on the battlefield.[50]

As al-Jowlani's Syrian front, named Jabhat al-Nusra (JAN), began to make significant progress in the country's civil war, ISI's leadership grew concerned that al-Qaeda would declare the Nusra Front—as it was popularly known—the official al-Qaeda affiliate in the region, undermining ISI and denying it a role in the expanding Syrian conflict. In an effort to bring his former subordinate back under ISI's control, Baghdadi issued an audio statement on April 9, 2013, declaring that al-Nusra was a branch of ISI and henceforth would be subsumed into the expanded Islamic State of Iraq and al-Sham (ISIS). Al-Jowlani rejected the demand, instead pledging bay'a directly to Zawahiri as a way of maintaining JAN's independence under the looser franchise system of al-Qaeda central. The high-profile clash between two affiliates of a global jihadist movement of which al-Qaeda purported to be the leader was an existential threat to the organization, which challenged not only Zawahiri's personal authority but also the founding purpose of al-Qaeda to act as a spearhead behind which disparate jihadi groups across the globe could unite with common purpose against the perceived Western crusade against the ummah.[51]

Zawahiri sought to intervene in the jihadists' dispute, but the persistence of the CTC's drone coverage left him limited to dispatching handwritten letters via a trusted emissary, Abu Khalid al-Suri, in whom he vested the power to resolve the dispute. "Sheikh Abou Bakr al-Baghdadi was wrong when he announced the Islamic State in Iraq and the Levant without asking permission or receiving advice from us and even without notifying us," berated the al-Qaeda leader in his communiqué dispatched on May 23, 2013, instructing that ISIS was to "be dissolved, while Islamic State in Iraq is to continue its work."[52] Zawahiri's diminished authority was brutally revealed when the newly declared ISIS responded, murdering Zawahiri's loyal servant with a suicide bomb attack on February 21, 2014, leading to a state of civil war between the two groups.[53]

ISIS completed its coup against al-Qaeda's leadership of the jihadi movement on June 30, 2014, with a bold, if absurd, claim to religious authority when it announced by audio recording the establishment of a caliphate, declaring Baghdadi caliph and thus leader of Muslims everywhere.[54] An official document released online alongside the declaration specifically highlighted the difference between the isolated Zawahiri and the new caliph, describing Baghdadi as "the sheikh, the fighter, the scholar who practices what he preaches, the worshipper, the leader, the warrior."[55] "The legality of all emirates, groups, states and organizations becomes null by the expansion of the caliph's authority and the arrival of its troops to their areas," said the group's spokesman, Abu Mohamed al-Adnani. To reflect its new, self-awarded status, the group changed its name to Islamic State (IS), announcing that its territory stretched from Iraq's Diyala province to Syria's Aleppo.[56] Though Zawahiri's Syrian affiliate JAN and the Yemen-based franchise al-Qaeda in the Arabian Peninsula (AQAP) remained loyal, other jihadi groups, such as the Nigerian-based Boko Haram and Egyptian Ansar Bayt al-Maqdis, swiftly shifted their allegiance to al-Qaeda's new rival.[57] By July 2015, approximately three dozen jihadist groups, across at least eighteen nations, had pledged support or allegiance to the Islamic State, shattering the previously united jihadist movement.[58]

"Such impudent behavior," grumbled Abu Muhammad al-Maqdisi of IS's rebellion, "would never have been accepted in the days when Bin Laden was alive." Maqdisi, a radical Jordanian cleric regarded as one of al-Qaeda's most significant intellectual founding fathers, and a personal tutor to the founder of the group that evolved into IS, Abu Musab al-Zarqawi, has poured scorn on the conduct of the Islamic State's leadership. "No one used to speak against him," Maqdisi reminisced. "Bin Laden was a star. He had a special charisma." Reflecting upon what had befallen the jihadist movement of which he had been a key architect, Maqdisi lamented that, despite his personal respect and affection for bin Laden's successor, Zawahiri never possessed the authority or control to rebuff the threat posed to al-Qaeda by IS's insurrection. Though the solemn Egyptian's lack of charisma in comparison to bin Laden is frequently cited as a major flaw in his leadership, Maqdisi's colleague, the radical cleric Abu Qatada, provided a more pragmatic observation of al-Qaeda's problems: from the "very beginning" of his tenure, Zawahiri lacked "direct military or operational control," Qatada said. "He has become accustomed to operating in this decentralized way—he is isolated." After a decade of relentless pursuit and drone warfare, Maqdisi and Qatada admitted that the group's organizational structure had collapsed. In the eyes of its intellectual grandees, al-Qaeda—as an idea and an organization—now finds itself teetering on the verge of collapse.[59]

Al-Qaeda's loss of control over the jihadist movement has also impacted upon the levels of popular support it can count upon, as well as the levels of funding. Painting a gloomy picture of the organization he has been associated with since the days of the Afghan jihad, Dr. Munif Samara—a general practitioner who runs a free clinic treating injured Syrian fighters and refugees—lamented that "donations, which once came in waves of hundreds of thousands, have dried up," with donors either redirecting their funds to IS or withdrawing support from the self-consuming, civil-war-wracked jihadist movement altogether. Compounding the sense of organizational decline, Aimen Dean, a former al-Qaeda member who defected to become a spy for British intelligence, told journalists at the *Guardian* that one of his sources in Pakistan's tribal areas reported that the finances of al-Qaeda central in Waziristan were so desperate that in 2014 it had been reduced to "selling its laptops and cars to buy food and pay rent."[60]

While it would be unwise to dismiss al-Qaeda's core leadership, which has proven itself to be extremely survivable, the evidence would suggest that the prestige and influence of Zawahiri and his jihadi brothers sheltering in Pakistan's tribal regions has been critically diminished. What is particularly worthy of note is that while to a certain extent the decline of al-Qaeda's core has been the result of drone strikes eliminating key al-Qaeda figures, the leadership's deterioration has equally been the result of the drone's original, but frequently overlooked, defining characteristic—its endurance. The CTC's ability to sustain drone coverage over the region has had the effect of containing al-Qaeda's leadership, slowing suffocating it of exposure, support, and subsequently influence. To this effect, it is as much the psychological impact of the presence of America's unmanned aircraft upon al-Qaeda's members as the actual decapitation strategy itself that has done such significant damage to al-Qaeda's core.

Strategic Objective 2: Denial of Safe Haven

The second goal of the Obama administration's drone strategy in the AfPak region has been to prevent al-Qaeda and the Taliban's ability to reestablish a safe haven in the Pakistan-Afghanistan region.[61] The establishment of a safe haven has been a core goal of Islamic radical groups since the highly influential Egyptian radical Sayidd Qutb argued that a group's "foremost objective is to change the practices of . . . society" in order to "protect the resources and the center of the movement." In Qutb's view, a radical vanguard was needed in order to free mankind and demolish the obstacles that prevented it from attaining the freedom that submission to a *salafist* (ancestor) form of Islam would provide. This vanguard required a base from which to operate.[62] The concept of a vanguard with an established base was echoed by the influential Palestinian mujahid Abdullah Azzam, a man who is regarded by many as an early mentor to bin Laden.[63] Azzam argued that "every principle needs a vanguard to carry it forward and [to] put up with heavy tasks and enormous sacrifices." This vanguard, in Azzam's words, would need a "strong foundation," which translates in Arabic to *al-qaeda al-sulbah* (al-Qaeda is also referred to as "the Base").[64] Reflecting upon the importance of Afghanistan to al-Qaeda in providing a safe haven within which the group could develop, Ayman al-Zawahiri stated in his radical manifesto *Knights under the Prophet's Banner:* "A jihadist movement needs an arena that would act like an incubator where its seeds would grow and where it can acquire practical experience in combat, politics, and organizational matters."[65] Terrorism scholars such as Jacob N. Shapiro and Abdulkadar H. Sinno have also discussed the arguments relating to the value of havens from which terrorist groups and insurgents can safely coordinate operations and transmit information, concluding that the ability to function in locations that are out of the reach of security services significantly boosts the effectiveness of such groups.[66]

Painstaking efforts by the BIJ to map out known and suspected drone strikes across the AfPak region provide clear evidence that whole areas of territory have been put out of bounds for al-Qaeda and the Taliban. The steady decline in the number of strikes in the past five years, with seventy-five in 2011, fifty in 2012, twenty-seven in 2013, twenty-five in 2014, and thirteen in 2015, indicates the steep reduction in the number of targets from the height of the campaign in 2010, when 128 strikes were recorded, suggesting militants have evacuated the strike zones.[67] An indication of the damage done to al-Qaeda by the drone campaign, and its efficacy in making certain areas no-go territories, is provided by the organization's desperate efforts to recruit engineers to try to hack the unmanned aircraft, and by the photocopied guidance sheets recovered from jihadists in North Africa, which give tips on avoiding drones, originally written by a jihadist in Yemen but since circulated among all of the terror group's franchises.[68]

Bin Laden personally discussed precautions against drone strikes in a letter sent to Shaykh Mahmud on October 21, 2010, when the CTC's campaign was at its most intense.[69] Confirming the logic behind the Obama administration's AfPak strategy of regarding the Afghanistan-Pakistan border region as one territory, bin Laden recommended al-Qaeda fighters migrate back across the mountains from Pakistan's tribal regions to new provinces in Afghanistan, such as Kunar in the northeast, where they could utilize "rougher terrain and many mountains, rivers and trees" in order to "defend brothers from the aircraft."[70] Furthermore, demonstrating the disruption the drones' presence caused to the everyday running of the organization, bin Laden urged his fellow jihadists to "not meet on the road and move in their cars because many of them got targeted while they were meeting on the road," and cautioned that meetings should occur "no more than once or twice a week," and then only "when the clouds are heavy."[71] In a passage that revealed the impact of the CTC's informants upon al-Qaeda's security and morale, bin Laden lamented that though such measures would "defend the brothers from the aircrafts [*sic*]," they would not "defend them from the traitors."[72]

Additional communication between bin Laden and his deputy Shaykh Mahmud, released as evidence in a U.S. terrorism trial on February 15, 2015, further revealed the effectiveness of drone strikes in denying al-Qaeda safe haven, while also providing a snapshot of the sort of collateral damage caused at times by operations undertaken against targets who shelter among civilians. In one letter, Mahmud provided his leader with a detailed account of how a single drone strike had killed twenty members of the organization's affiliated Abu Bakr al Sidiq Brigade, who had gathered together—in defiance of al-Qaeda's operating procedures, which restricted meetings to five or fewer members—to celebrate Eid.[73] Another letter informed bin Laden of the loss of the organization's finance chief and Afghan commander, Mustafa abu al-Yazid, and several members of his family from "a spy plane attack" in North Waziristan on May 22, 2010. According to Mahmud's detailed account, Yazid was meeting with al-Qaeda "media brothers" when "he learned that his family . . . had come from the house, which was relatively far away, because of some repairs being made to the house, to somewhere closer, staying with one of the [al-Qaeda] supporters." The finance chief "went to visit and check on them" but, in Mahmud's words, "was not supposed to go there nor stay very long," as the house was "a well known place for the Arabs," owned by a prominent al-Qaeda supporter. Furthermore, the group was "anticipating strikes" because two days previously it had launched an assault upon the U.S.-NATO Bagram Air Base. According to the account, however, when Yazid's son arrived around 10 P.M. to relocate his father, he found him sleeping and was told by his family, "He is very tired, let him sleep." Less than an hour later, the house was hit, killing Yazid, his Egyptian wife, Umm al Shaymah, three of their daughters, and a granddaughter named Hasfah. The home's owners were also killed in the blast, along with the youngest son of another al-Qaeda member, Abu Tariq al Tunisi.[74]

Whether the CTC knew Yazid's family was with him in the supporter's house is unclear. Al-Qaeda members have kept family members close by since bin Laden's days of exile in Sudan early in the group's existence, and America's post-9/11 attitude to civilian casualties subsequently hardened after the Clinton administration's pensive approach saw strikes against bin

Laden rejected due to concern about casualties among his family members.[75] Mahmud's account suggested it was the location and timing that put Yazid's family in danger, as opposed to the ruthless nature of the CTC. "Based on our analysis," the deputy informed bin Laden, "they [the CIA] are constantly monitoring several potential, or possibly confirmed targets. But they only hit them if they discover a valuable human target inside, or a gathering, or during difficult times (like revenge attacks for example)."[76] The fact that the property was known to be an al-Qaeda safe house, and that an attack against U.S. forces had taken place days before, meant, in Mahmud's mind, that neither Yazid nor his family should have stayed in the house. Shaykh Mahmud himself was killed by a drone strike in North Waziristan on August 22, 2011, shortly after the raid that killed bin Laden, an event that was described by U.S. officials as a tremendous loss for al-Qaeda.[77]

Regardless of whether the deaths are a result of ruthless calculation or tragic bad luck, the case provides a perfect example of how civilians can become caught up in America's drone war. The attacks also illustrate the difficulty of accessing accurate information regarding the success rate of drone strikes and the number of civilian casualties. Mahmud informed bin Laden that al-Qaeda should announce Yazid's death "because he is a senior person who addressed the Ummah and the Ummah knows him," meaning "it would be impossible to keep the news secret for long," a decision that suggests other, lower-profile deaths are not admitted by the organization. Furthermore, BIJ's open-source reporting on the strike on the Abu Bakr al Sidiq Brigade initially recorded that up to nine of the twenty killed had been civilians, a figure contradicted by Mahmud's report, the later release of which confirmed that "they were all militant Mujahadeen from one of our Brigades."[78] BIJ's ongoing investigation is arguably the most thorough and accurate statistical reporting on America's drone war, but even with close attention to detail and multiple-source reporting, errors can clearly be made. Former CIA Deputy Director Michael Morell described this information gap on drone strikes as a classic dilemma. "The public hears bad things about the accuracy of these strikes," he noted. "We say, 'They are accurate, trust us.' The response is 'Show us some proof and we will believe you.' But we can't because of the sensitivities involved." Ultimately Morell's solution was that the public must trust that, through their elected representatives, "the congressional oversight process works as it should."[79] Unfortunately, public trust in Congress has been at a historic low throughout the duration of America's drone campaign, making reliance upon these representatives an unsatisfactory solution for many.[80]

The conviction of the Obama administration that its drone campaign was working can be ascertained from a speech delivered at the Oxford Union on November 30, 2012, by Jeh Johnson, then serving as the general counsel of the DoD. Entitled "The Conflict against Al Qaeda and Its Affiliates: How Will It End?" the speech reflected the administration's confidence that the CTC's drone campaign was close to destroying the AfPak-based core of al-Qaeda: "Osama bin Laden is dead. Many other leaders and terrorist operatives of al-Qaeda are dead or captured; those left in al Qaeda's core struggle to communicate, issue orders, and recruit." Though Johnson was careful to offer no prediction about when this conflict would end, he suggested, as no Obama administration official had done before, that eventually a tipping point would be reached, "a tipping point at which so many of the leaders and operatives of Al Qaeda and its affiliates have been killed or captured and the group is no longer able to attempt or launch a strategic attack against the United States, such that Al Qaeda as we know it, the organization that our Congress authorized the military to pursue in 2001, has been effectively destroyed."[81] Johnson went on to assert that America's counterterrorism policy would therefore cease to take the form of an armed conflict. Rather it would be an effort targeted at individuals, "for which the law enforcement and intelligence resources of our government are principally responsible."[82] The return to the use of law enforcement agencies such as the FBI as the primary tool against terrorism would essentially mark the end of the War on Terror, though it is unlikely any president would ever tempt fate by directly saying as much. While Johnson's language was carefully selected and provided plenty of caveats, the fact that a senior Obama administration official was giving a high-profile speech in which he discussed the likelihood of an end to the war against al-Qaeda demonstrated the confidence within the Executive that al-Qaeda's core had been decimated.

Despite this confidence, other officials provided warnings about the territorial limitations of the CTC's drone campaign. Shortly before Johnson's speech, Leon Panetta, at that point serving as secretary of defense, delivered remarks at the Center for a New American Security in Washington, D.C., in which he echoed the administration line that the United States has decimated core al-Qaeda. However, he then went on to issue a warning about the increasingly fragmented nature of the group and its successful franchise operation, cautioning that "even with these gains, the threat from al-Qaeda has not been eliminated. . . . We have slowed the primary cancer," continued the former director of the CIA, "but we know that the cancer has metastasized to other parts of the global body."[83] Panetta's warning has turned out to be extremely accurate and has perhaps highlighted one of the biggest weaknesses of the CTC's drone campaign as a primary tool for destroying al-Qaeda. While the barrage of drone attacks and constant surveillance may have succeeded in eliminating much of the leadership and denied safe haven in the AfPak region, al-Qaeda and fellow jihadist groups adapted, migrating to areas where the drones were not so easily able to follow. Though Obama authorized JSOC and the CIA to expand their operations against al-Qaeda affiliates operating in the ungoverned territories of Somalia and Yemen shortly after coming to office, the chaos and instability that followed the Arab Spring uprisings in countries such as Libya, Algeria, and in particular Syria served to create new safe havens, away from the CTC's intensive drone campaign.[84]

These new havens have become perfect incubators for radical jihadist groups, allowing the Salafist-jihadist movement not only to recover but to become stronger than it has been in decades. A report published by the Rand Corporation in June 2014 provided details of the escalating threat and of the evolution of the jihadist movement, in response both to the destruction of al-Qaeda's core in AfPak and the wider political events that shook North Africa and the Middle East. The study's author, Seth G. Jones, agreed with the Obama administration's assertion that core al-Qaeda, based in the AfPak region and targeted by the CTC's drone campaign, had sustained huge damage and was no longer a principal threat to the U.S. homeland. Despite the CTC's success in its pursuit of al-Qaeda's core leadership, the report warned that the wider jihadist threat had evolved, expanding in size rather than dissipating. In 2007 there were twenty-eight Salafi-jihadist groups such as al-Qaeda, which Rand estimated had approximately eighteen thousand to forty-four thousand active terrorists, collectively responsible for a hundred attacks. By the end of 2013 the number of jihadist groups had increased to forty-nine, with estimates of the numbers of active terrorists increasing to between forty-four thousand and 105,000. The significance of this growth is represented by the fact that Salifi-Jihadists were linked to 950 attacks in 2013 alone. The report concluded that the impact of the instability across North Africa and the Middle East since 2010 had seen a 58 percent increase in the number of jihadist groups, a doubling of jihadist fighters, and a tripling of attacks by al-Qaeda affiliates. The most significant threat to the United States, the report concluded, came from terrorist groups operating in Yemen (al-Qaeda in the Arabian Peninsula) and Syria (IS and the al-Qaeda affiliate JAN).[85]

The U.S. State Department's Bureau of Counterterrorism (BoC), in its annual assessment of global terrorism published in April 2014, also acknowledged the growth of Salafi-jihadist groups and the violence they have brought. According to the report jihadist attacks across the world increased from 6,700 in 2012 to 9,700 in 2013. Approximately eighteen thousand people perished in these attacks, and a further thirty-three thousand were injured. As with Rand's report, the DoS's analysis praised the U.S. government's success against al-Qaeda's AfPak core but also highlighted the increased threat of the group's affiliates and other independent jihadist groups. The BoC's study made specific reference to the threat emerging from Syria, which it linked back to the formation of IS. The report estimated that the number of jihadist fighters could range from seven thousand to twenty thousand.[86] Such figures show that despite the CIA's drone campaign having succeeded in denying the core of al-Qaeda safe haven in the AfPak region, like-minded jihadists have been able to escape the surveillance and strike capability of drones and establish new strongholds and incubators, beyond the current reach of the CIA.

Strategic Objective 3: Force Protection for the NATO International Security Assistance Force (ISAF) in Afghanistan

The third aim of the CIA's drone campaign is revealed by the identities of the majority of the drones' targets. As discussed above, the available data suggest that the CIA's drone campaign in the AfPak region killed sixty-five senior al-Qaeda and Taliban leaders between 2004 and June 2016.[87] During this period the same sources estimate the number of nonsenior leadership militants killed by drone strikes in the region at between

1,853 and 3,032.[88] The targeting of these militants is just as significant a part of the drone campaign as the decapitation of al-Qaeda's leadership. The CIA's mission in AfPak was not (and to a lesser degree still is not) just the targeted killing of terrorist leaders; it is also to advance the wider campaign conducted by the United States and allied NATO-led ISAF against the Taliban.[89]

During his 2008 election campaign, Obama referred to the invasion of Iraq as a "strategic mistake," making clear his view that Afghanistan and Pakistan represented the central front of the battle against al-Qaeda.[90] Despite fierce criticism from Republicans such as John McCain and the American Enterprise Institute's Frederick Kagan, Obama honored the terms of the Status of Forces Agreement established between the Bush administration and the Iraqi government, which stated that all U.S. forces would withdraw from all Iraqi territory no later than December 31, 2011.[91] While ending America's war in Iraq, Obama ordered a surge of troops into Afghanistan, using a speech on March 27, 2009, to warn the American public that "if the Afghan government falls to the Taliban or allows al-Qaeda to go unchallenged, that country will again become a base for terrorists who want to kill as many of our people as they possibly can." Indicating his administration's commitment to combating al-Qaeda and the Taliban on both sides of the AfPak border, Obama declared: "The future of Afghanistan is inextricably linked to the future of its neighbor Pakistan."[92]

Obama's new strategy revealed his intention both to support the recruitment and training of Afghanistan's security forces and to pursue al-Qaeda and the Taliban into what he described as the "remote areas of the Pakistani frontier."[93] Based upon the guidance of counterinsurgency experts, such as General David Petraeus and the general's special adviser, David Kilcullen, Obama ordered two surges of U.S. forces into Afghanistan with the aim of providing the security and stability necessary to enable the Afghan government to recruit and train members for its expanding security forces.[94] The first surge of thirty thousand troops in March 2009 almost doubled the number of American soldiers who were in Afghanistan when Obama assumed office. The second surge, ordered eight months later, saw a further thirty thousand troops deployed.[95] In a speech delivered at West Point upon the announcement of his second surge, Obama deepened his commitment to securing Afghanistan by explaining that the additional forces were being deployed to "seize the initiative" and that the United States would provide support to the Afghan government "over the long haul." At the same time, the president acknowledged the new political realities of an economically weakened America whose citizens were growing tired of the war in Afghanistan by explaining that the troops would begin to be withdrawn after eighteen months. The aim was to end the combat mission in 2014, with the caveat that this would be "dependent upon conditions on the ground."[96]

This significant deployment of troops revealed Obama's determination to secure the AfPak region and finish the fight against al-Qaeda and its Taliban hosts, something the Bush administration had failed to do. While the troop surge was the most publicly visible element of Obama's Afghan campaign, the drone surge was the second part of this strategy. A secret CIA report, published on July 7, 2009, during strategic deliberations over the Afghan surge and later leaked via WikiLeaks, reveals the Directorate of Intelligence's role in promoting the CTC's drone campaign. Entitled "Best Practices in Counterinsurgency: Making High-Value Targeting Operations an Effective Counterinsurgency Tool," the document used a series of case studies to evaluate the strategic utility of targeting operations against rebel groups. Such operations, the report notes, can achieve such objectives as "damaging an insurgent group by depriving it of effective direction and experience, deterring future guerrilla actions by demonstrating the consequences, demoralizing rank-and-file members, promoting perceptions of regime viability in providing security, and imposing punishments for past acts."[97] The DoI's analysis took care to raise potential negative effects of such targeting, warning that an overreliance upon lethal force could result in a government neglecting other important aspects of counterinsurgency operations, increase public support for insurgents, and possibly escalate the level of violence by "eroding the 'rules of the game.'" Overall, however, the report's key findings recommended that targeting programs can play a useful role when they are part of a broader counterinsurgency strategy. Such operations, the authors suggested, are "most likely to contribute to successful counterinsurgency outcomes when governments decide on a desired strategic outcome before beginning . . . operations, analyze potential effects and shaping factors, and simultaneously employ other military and non-military counterinsurgency instruments."[98]

Close inspection of the aims Obama set out to have coincide with his troop surge on November 29, 2009, further reveals that the president heeded the agency's advice, employing the CIA's drone campaign as the cornerstone of the wider Afghan mission. Of the six overall goals included in Obama's new orders, three were clearly areas in which the CIA's drones were vital tools. Denying the Taliban access to key population and production centers as well as disrupting its lines of communication have been tasks particular to the drone campaign, as U.S. soldiers were unable to operate over the border in the Waziristan population centers where Taliban fighters took sanctuary. The same is true for the goal of disrupting the Taliban in areas outside the secure areas of Afghanistan—again a reference to the AfPak borderlands NATO forces were unable to control. Obama's orders made a specific reference to the importance of preventing al-Qaeda reestablishing a sanctuary in the AfPak region, too. Finally, the high number of militant casualties reveals the CTC's role in helping meet the aim of degrading the Taliban to a level that would be manageable for the ANSF.[99]

As the U.S. military commitment to Afghanistan began to draw down, so too did that of the CIA. The Bilateral Security Agreement between the Obama administration and the newly elected government of Afghanistan, along with a status of forces agreement with NATO partners, left just ninety-eight hundred U.S. troops in the country past 2014, with the open-ended mission of "training Afghan forces and supporting counterterrorism operations against the remnants of al-Qa'ida."[100] The *Washington Post* reported that the CIA correspondingly began reducing its number of Afghan bases, from twelve to six over the course of two years. As well as reflecting the scaling down of the CIA's commitment to the Afghan war, this shift was partly due to increased pressure on the agency to respond to new challenges beyond al-Qaeda and the Taliban, such as IS and the related chaos in Syria and the wider Middle East.[101] It also reflected D/CIA Brennan's desire, signaled during his confirmation hearings in February 2013, to return the agency to its traditional espionage role and to limit its future involvement in paramilitary action. Contemplating the CIA's mission since the 9/11 attacks, Brennan described it as "a bit of an aberration," arguing that the CIA's primary role should be to deliver "the best intelligence collection [and] analytical capabilities possible," not conduct military operations.[102] While Brennan himself was a key architect of the

CIA's paramilitary role, his declaration that the agency should not be permanently engaged in this mission indicates that its campaign in AfPak was always intended to be a short term mission coupled to the U.S. military mission in the country.

It is not wholly accurate to describe the CIA's engagement in covert paramilitary activities as a departure from its original role. The agency has engaged in covert operations virtually since its founding, and though it is questionable that President Truman planned the whole panoply of 1950s dirty tricks and paramilitary operations in advance, it is likely that the executive architects of the CIA hoped from the outset that it would enable them to intervene abroad in a covert manner.[103] There is also a historical precedent for the agency being called upon to weaken an insurgent enemy on behalf of the DoD as part of a wider war aim. From 1965 to 1972 the CIA coordinated the Phoenix Program in Vietnam, a campaign designed to neutralize the infrastructure of the National Liberation Front of South Vietnam (NLF or Viet Cong). South Vietnamese forces undertook reconnaissance, intelligence gathering, interrogation (often involving torture), and targeted killings with close CIA supervision. There is some debate with regard to the effectiveness of the program, but when it began to draw significant negative publicity and became the focus of congressional hearings, it was officially shut down. Figures suggest the program neutralized 81,740 suspected NLF operatives and supporters, of whom twenty thousand to sixty thousand were killed.[104]

Did the AfPak drone campaign bring the CIA too close to the DoD at the expense of its core mission? The fact that the man who had overseen both the Bush administration's strategic surge into Iraq in 2007 and the Obama administration's surge into Afghanistan two years later, General David Petraeus, was given the role of D/CIA in September 2011, while the previous post holder, Leon Panetta, moved to the Pentagon as secretary of defense in July 2011, serves to highlight the close cooperation between these two organizations in the endeavor to achieve Obama's strategic goal in Afghanistan. Meanwhile the CIA's inability to keep the U.S. government fully informed of events on the ground in a number of crises saw the agency's intelligence efforts criticized by the Obama administration, Congress, and the press.[105] First came the outbreak of waves of protest labeled the Arab Spring, which saw the Egyptian president, and close U.S. ally, Hosni

Mubarak, removed from power. In September 2012, an attack on the U.S. consulate in Benghazi resulted in the deaths of four Americans, including the ambassador to Libya, Christopher Stevens. The United States was again caught off-guard by Putin's swift military intervention in Ukraine's Crimean Peninsula, while the Islamic extremist group IS's rapid takeover of swathes of territory in Syria and Iraq, including Iraq's second city, Mosul, left the Obama administration struggling to keep up with the pace of developments. In agreement with Brennan's assessment during his testimony, former D/CIA Michael Hayden has argued that the agency's resources were spread too thin to effectively perform its primary intelligence gathering and analysis role on a global scale. Speaking to reporters in the wake of the Crimean crisis, Hayden warned that America's spies were overfocused on the terrorist threat, and had been for some time.[106] In this context, Brennan's desire to untangle his agency from its military commitments in AfPak and focus once more upon the CIA's original core business seems all the more necessary.

Regardless of whether the drone campaign is a deviation from the agency's core business or not, one should not disregard the significant impact the drones have had in aiding the U.S. combat mission in Afghanistan. Though a full analysis of its impact is impossible to undertake due to the classified nature of most of the data on the war, one can examine casualty figures in order to ascertain the extent to which the Taliban has been disrupted and its pre-2009 momentum reversed. Following Obama's decision to deploy more troops into Afghanistan, a number of commentators began making comparisons between this decision and that of Lyndon Johnson to escalate the U.S. presence in Vietnam. *Newsweek*'s John Barry labeled the conflict "Obama's Vietnam," while the *New York Times*'s Robert Wright went further, describing the decision to send in more troops as having created a situation that was "worse than Vietnam." The same paper's Bob Herbert used a Vietnam analogy to warn of a likely Afghan quagmire.[107]

However, a study of the most vital statistics of the Vietnam War, namely, the number of U.S. casualties and the attitude of the American public toward the war, suggests that the two conflicts are dissimilar. In 1968 alone, the deadliest year of the American war in Vietnam, there were 16,899 recorded U.S. casualties. By the official end of the ISAF Afghan mission in December 2014, the total number of casualties for the whole coalition force in Afghanistan was 3,485, of which 2,356 were American.[108] While the conditions in Afghanistan were obviously very different to those in Vietnam, the U.S. casualty figures are also low when compared to those of the Soviet Union's own military campaign in Afghanistan in the 1980s. From December 1979 until its eventual withdrawal in February 1989, the Soviet military had 14,453 men killed in action, with a further 53,753 wounded.[109] With regard to the domestic response, the lower casualty levels also ensured that the American public never turned against the war in Afghanistan in the same way that it did against the Vietnam War. A decade into the fighting in Vietnam only 29 percent of Americans polled believed it had not been a mistake to send U.S. troops to the country. By the same benchmark with Afghanistan, 57 percent of the American public still believed the military invasion of Afghanistan had been the right decision, and 58 percent still believed the United States would achieve its goals in Afghanistan.[110]

Though it is impossible to measure the exact role the CIA's drone campaign played in keeping casualties in Afghanistan to comparatively low levels, it is certainly the case that the high numbers of militants killed helped reverse the Taliban's momentum, while the constant presence of drones over what had previously been their safe havens disrupted their insurgency. Furthermore, the fact that the CIA's drones allowed the United States to attack these otherwise impenetrable areas with minimum risk to American service personnel no doubt also helped to maintain public support for Obama's commitment to the war. It should be noted, however, that the drone campaign has not been totally without CIA casualties. In late 2009 a Jordanian physician named Khalil al-Balawi convinced his CIA handlers, who were operating from Camp Chapman, a small outpost at Khost airport, that he could provide them with accurate intelligence regarding the location of high-ranking al-Qaeda members. Instead, during a visit on December 30, 2009, al-Balawi detonated a suicide vest, killing seven CIA officers and two other base personnel. It was the most deadly attack upon the CIA for twenty-five years.[111]

The Impact of the Drone Campaign on Pakistan

The American public may have been sheltered from the impact of the drone campaign, but the same cannot be said for the people of Pakistan, especially those living in the FATA and North-West Frontier Province

(NWFP), where the majority of strikes have taken place. International relations theorist Wali Aslam has argued that the drone strikes are "damaging the social fabric of the society in FATA."[112] The ethicist Michael Gross agrees, arguing that "assassination subverts strongly held beliefs about integrity, trust, honor, and loyalty that hold traditional societies together."[113] These negative effects, he argues, may contribute further to "an atmosphere of lawlessness and chaos in which terrorism and militancy thrive."[114] Taking the threat of this diminishing social cohesion to its most extreme consequence, Selig Harrison, a prominent scholar on Pakistani affairs, has argued that the tensions could lead to the unification of the estimated forty-one million Pashtuns on both sides of the border, the breakup of both Pakistan and Afghanistan, and the emergence of a new national entity, "Pashtunistan," under radical Islamist leadership.[115]

Others have disagreed with the assessment that drone strikes are ripping the FATA apart. Syed Alam Mehsud, a Peshawar-based political activist from Waziristan, commented, "To those people sitting in the drawing rooms of Islamabad talking about the sovereignty of Pakistan, we say, 'What about when Arabs or Uzbeks occupy your village? What about sovereignty then?'" He continued: "Any weapon which kills these people who damaged my sovereignty is in fact helping the sovereignty of my region."[116] The attitudes of the residents in the tribal regions that have borne the brunt of the drone strikes are notoriously difficult to ascertain; however, there are two important pieces of evidence that have received little exposure but suggest that the CIA's drone campaign may have more support among locals than is often assumed. First, the results of a March 2009 poll conducted by the Aryana Institute for Regional Research and Advocacy (AIRRA), a Pakistan-based think tank staffed by researchers and political activists from the NWFP and FATA, challenged many of the preconceptions about the attitudes of those who lived in the region. When asked whether or not they saw drone strikes as causing fear and terror in the common people, 55 percent responded no; 58 percent of those polled responded that anti-American feelings in the region had not been increased by the presence of the CIA's drones, while 60 percent believed al-Qaeda and the Taliban were being damaged by the drone strikes. Given the choice of having the Pakistani military conduct targeted strikes in their region instead of the CTC's drones, 70 percent of respondents rejected the idea, favoring what they believed was the greater care and precision of the American strikes.[117] Reflecting this split in Pakistani perceptions of the drone strikes, Brian Glynn Williams explains that "the results of the poll . . . would seem to indicate that many Pashtun tribesmen welcomed the strikes even if the rest of their countrymen did not."[118] This view was further supported by a similar survey, conducted by World Public Opinion, the results of which indicated that 86 percent of residents in the predominantly Pashtun NWFP supported the government of Pakistan. Only 6 percent responded that they supported the Taliban. The same study also concluded that those in the regions occupied by the Taliban were significantly more likely to see them as critical threats to Pakistan, with a nationwide takeover regarded both as an aim of the Taliban and as a realistic possibility.[119]

Finally, on December 12 and 13 of the same year a conglomeration of FATA-based political parties, civil organizations, and influential individuals met for a two-day conference held in Peshawar. Called "Terrorism—The Way Out," the aim of the conference was to produce a united position for FATA on how to eliminate terrorism from their region and establish a sustainable peace. The document produced at the end of the conference, referred to as the "Peshawar Declaration," made a number of key statements, in particular on drone use.[120] Blaming the presence of al-Qaeda in their region on Arab expansionism disguised as jihad and claiming the growth of the Taliban to be a result of the Pakistan's failed strategic depth vision against India, the coalition demanded the initiation of immediate operations against all centers and networks of terrorism.[121] These operations, the declaration stated, were to eliminate all foreign, nonlocal, and local terrorists in FATA. The declaration recognized that drone strikes were likely to be a significant tool in these efforts: "If the people of the war-affected areas are satisfied with any counter militancy strategy, it is the Drone attacks which they support the most." Challenging Pakistani reports of high civilian casualties from drone strikes, the declaration complained that a "component of the Pakistani media, some retired generals, a few journalists/analysts and pro-Taliban political parties never tire in their baseless propaganda against Drone attacks."[122] As these very divergent perspectives upon the impact of the CTC's drone strikes reveal, the commonly repeated hypothesis that the CIA's drone campaign drives Pakistanis in the FATA and NWFP into the arms of al-Qaeda and the Taliban needs to be reevaluated.

Conclusion

As this book has shown, the Obama administration's adoption of drones as a key counterterrorism tool in the AfPak region was neither unprecedented nor a direct consequence of the 9/11 attacks. It was built upon decades of policy and technological development. Since Reagan signed NSDD 138 on April 3, 1984, the United States has regarded terrorism as a national security threat against which it is legitimate to use lethal military force to preemptively neutralize anti-American terrorist groups anywhere in the world, unilaterally if necessary. The primary goals of the CTC's drone strategy in the AfPak region—the decapitation of a terrorist group's leadership and denial of safe haven through intelligence gathering—and the timely, accurate use of lethal force accord with the goal of U.S. presidents since Reagan's unsuccessful bombing raid on Libya in 1986. Though the goals have not changed, the tools used to achieve them have. Those who try to set the conduct of the Obama administration's counterterrorism apart from its predecessors are too focused upon the technologically innovative nature of the unmanned aircraft used as opposed to the policy they enable.

In terms of the extent to which the CTC's drone campaign in AfPak has achieved its aims, a full analysis is not yet possible due to the uncertainty that remains as to the long-term outcome of the Afghanistan mission. If one is to measure success purely by examining the degree to which the campaign achieved the primary goals that were set out in Obama's orders, it would appear that the CTC's drone war was a success. Al-Qaeda's core leaders were decimated and denied safe haven in a region they had called home since late 2001. Furthermore, though not defeated, the Taliban were certainly degraded to the extent that, in the short term at least, they were prevented from overthrowing Afghanistan's U.S.-backed government, a reality that seems to have been accepted by at least some elements of the movement, with extremely tentative peace negotiations between Taliban representatives and officials from Kabul being periodically brokered by Pakistan, amid ongoing bursts of violence.[123]

The real measure of success, however—whether or not this has been enough to move the United States closer toward an end to the War on Terror—remains unclear, but it looks unlikely. To wage war against an abstraction may be inherently unachievable. It has always been tempting for U.S. policy makers to see al-Qaeda as having a hierarchal command-and-control structure like that of Western militaries and governments, something that can be tracked down, targeted, and eliminated. Indeed, for a period following its formation in 1988, al-Qaeda had both a physical infrastructure, constituting brigades of battle-hardened mujahideen, arms depots, and training camps dotted across Taliban-controlled Afghanistan, and a clear command structure, with bin Laden personally involved in the running of the organization. But this was swiftly dismantled after the overthrow of the Taliban in 2001, leaving the concept of a core al-Qaeda pulling the strings of the global jihadist movement open for debate.[124]

Following the destruction of its infrastructure, al-Qaeda became as much of an idea as an organization, its name serving as a rallying call, representing a core set of extremist Islamic beliefs centered on the desire to undertake a defensive jihad to drive the "crusader-Zionist alliance" out of the "lands of Islam," replacing political institutions and heretical man-made laws with the literal interpretation and implementation of Sharia law and Islamic customs. The Obama administration's narrow focus upon al-Qaeda's core has enabled the administration to make bold claims about the success of its counterterrorism operations. Speaking at the U.S. Military Academy at West Point on May 28, 2014, the commander in chief boasted that America had "struck huge blows against al-Qaeda core and pushed back against an insurgency" that had threatened to overrun Afghanistan, actions that had ensured al-Qaeda's centralized leadership was no longer posed a principal threat to the United States.

That the CTC's drone campaign has inflicted heavy loses upon al-Qaeda is not in doubt, as the terrorist group's own communications show. Yet, despite this success, there is a troubling disconnect between the decline of al-Qaeda's core and the staggering rise of jihadist violence. Obama acknowledged this contradiction in the same West Point address that lauded success against al-Qaeda's central leadership, warning of the emerging threat posed by decentralized al-Qaeda affiliates and extremists based in safe havens created by the collapse of governments and authoritarian rule across North Africa and the Middle East in the wake of 2010's Arab Spring uprisings. "As the Syrian civil war spills across borders," warned the president, "the capacity of battle-hardened extremist groups to come after us only increases."[125] That Obama's language shifted from specifically discussing al-Qaeda and its affiliates to the more common use of the generic term "extremists" revealed an acknowledgment that, even without al-Qaeda's core, the jihadist movement was expanding.

Does the fact that the militant jihadist movement came to control more territory following a decade of intense drone warfare than at any prior point in its history suggest that ultimately the drone campaign was a failure? The CTC was focused upon the AfPak region, where al-Qaeda's presence and influence has been significantly diminished. As Obama himself indicated, the resurgence of Islamic militancy has had much more to do with the wider instability that has resulted from the uprisings across North Africa and the Middle East. Such events were beyond the scope of the CTC to deal with. It has been argued, however, that it was precisely because so many of the CIA's resources were focused upon the drone campaign that the agency was unable to detect the wider geopolitical events. While its critics may claim that a decade of paramilitary activity and counterterrorism has left the CIA weakened in other areas of analysis, which in turn led to a sluggish response from U.S. policy makers, such a perspective only serves to make the agency a scapegoat for broader policy failings by the United States and the wider global community.

There is no doubt the upheaval that accompanied the Arab Spring protests—particularly the civil war in Syria—enabled the jihadist cause to recover, restructure, and become resurgent. Yet the notion that the CIA was spread too thin to warn about the outbreak of the Arab Spring and to predict its extraordinary impact upon the jihadist movement reveals more about unrealistic expectations of the intelligence community than it does about the impact of the CIA's focus upon the drone campaign. Documents leaked by the former NSA contractor Edward Snowden reveal that, at the time of the Arab Spring, one-quarter of the intelligence workforce and a third of the intelligence budget are devoted to counterterrorism.[126] But the same sources reveal that these resources have been funded by additional allocations rather than being diverted from existing intelligence operations. In 2013 the budget requested for the CIA was $14.7 billion, up 56 percent from 2004, surging past that of every other intelligence agency in the past decade.[127] Moreover, of the overall $52.6 billion that constituted the U.S. "black budget" for intelligence in 2013, the largest sum ($20.1 billion) was still devoted to warning U.S. leaders about critical events. Rhodri Jeffreys-Jones provides an alternative and altogether more convincing reason why the CIA was unable to warn about the next phase of America's conflict against Islamic extremism, suggesting it has more to do with the general nature of intelligence. Analysts may be able to point to certain trends and dangers, but it is unrealistic to assume that intelligence can warn of any "nasty surprises." Much like the police, the CIA can only ever solve a minority of cases, providing as much of an illusion of protection and security as the real thing. "To shower the CIA and its ilk with taxpayers' money," Jeffreys-Jones observes, "is to drop pennies into the wishing well—if you are superstitious it makes you feel better."[128]

Finally, before a conclusive judgment can be made about the extent to which the CTC's drone campaign has weakened the specific threat of al-Qaeda and wider danger posed by Islamic extremism, it remains to be seen what impact the decimation of al-Qaeda's core leadership will have on global jihadist militancy in the longer term. One of al-Qaeda's greatest strengths was that it had been able to unite Islamic extremist groups that for decades had been ineffective due to fractures and infighting. It is not yet known whether the splits that have emerged with the decline of al-Qaeda's core will once more lead to the breaking up of the jihadist cause. If this becomes the case the decapitation strategy, which was a primary aim of the CTC's campaign, will have succeeded in reducing the threat. Though multiple extremist groups will still present a threat to the United States, the lack of coordination, training, and pooling of resources that al-Qaeda's core provided will make the overall threat less critical and more manageable.

It is of course possible that in the manner of unintended consequences the role played by the CTC's drone campaign in undermining al-Qaeda's leadership of the global jihadist movement may have contributed to the unleashing of a substantially more dangerous foe in the form of the Islamic State. In the years following 9/11, the United States developed a trillion-dollar infrastructure, dubbed by the *Washington Post*'s Dana Priest and William Arkin "top secret America," to fortify its domestic defenses against terrorism. At least 263 U.S. government organizations were created or reorganized as a response to the 9/11 attacks, including the Department of Homeland Security, the National Counterterrorism Center, and the Transportation Security Administration; the budget of the NSA doubled, the thirty-five staff members of the FBI's Joint Terrorism Task Force became 106, and domestic police force budgets were swelled by Congress's generous counterterrorism funding. Fifty-one federal organizations and military commands regularly track the flow of money to and from terrorist networks, and each year this huge counterterror bureaucracy produces some fifty thousand reports on terrorism.[129] This domestic fortress, well attuned to counter the efforts of a group like al-Qaeda to penetrate its defenses with foreign jihadists in another 9/11-type attack, has helped make terrorist attacks on U.S. soil exceedingly rare. But it is not so well suited for dealing with other kinds of threats, such as that posed by the Islamic State, which presents a different sort of challenge. Despite the fact that IS has proven to be more focused upon securing its own territory as opposed to plotting covert attacks against the far enemy of the United States, the group's effective propaganda output has repeatedly called for "lone wolf" supporters within the United States to attack American citizens, essentially bypassing many of America's post-9/11 counterterrorism defenses. The threat of homegrown terror and online radicalization caused FBI Director James Comey to warn in July 2015 that the group had eclipsed the threat posed to the U.S. homeland by al-Qaeda.[130]

On a wider scale, IS has caused significant geopolitical turmoil, redefining borders, and challenging existing U.S. alliances in the region. With estimates suggesting IS's strength had reached thirty-one thousand fighters by November 2014, and financial reserves projected at approximately $2 billion, IS represents a wholly different challenge to the United States, one that its current counterterrorism capabilities are incorrectly calibrated to deal with.[131] Over the past three decades the United States has built up a counterterrorism infrastructure aimed at neutralizing specific terrorists and dismantling terror groups such as al-Qaeda and its affiliates—a system that, accelerated and bolstered by the events of 9/11, has become extremely effective at doing just that. Supported by a complex legal architecture, the United States has weaved together electronic surveillance, human intelligence, and extensive UAV coverage to drive lethal drone strikes and JSOC raids. Targeted killings and the persistent threat of overhead drones have served to undermine al-Qaeda, removing its talented field commanders and containing its leadership. Yet while armed drones have proven to be a technology well suited to the task of pursuing targets hiding in the rural areas of the FATA, sparse deserts of Yemen, and desolate camps in Somalia, where the risk of accidentally killing civilians is low, such an approach does not hold the same promise for combating the Islamic State.

First, the Islamic State's possession of sophisticated antiaircraft weaponry (with which it alleged it shot down a Jordanian F-16 in December 2014, a claim disputed by U.S. authorities who have argued mechanical failure forced the pilot to eject), presents a more hostile environment for the slow-moving drones to operate in, reducing the UAV's opportunity to loiter and hunt its targets. Second, the Islamic State's fighters and leaders cluster in the large urban areas they control, basing themselves in Iraqi and Syrian towns and cities, where, as Audrey Cronin pointed out, "they are well integrated into civilian populations and usually surrounded by buildings," making drone strikes or Special Forces raids much harder to execute. Third, rather than operating as a small, clandestine unit, the Islamic State quickly moved to establish itself as a functioning government in the territory it controlled, developing a complex administrative structure of the caliphate, military commanders, and local governors, making the group much less vulnerable to the loss of individual leaders.[132]

Operation Inherent Resolve, the American-led coalition campaign against the Islamic State waged across both Iraqi and Syrian territory, has thus seen the CIA play a secondary role to the Pentagon, with the U.S. Air Force and Navy undertaking the majority of airstrikes against the group. Drone surveillance and occasional strikes have still been an element of the Pentagon's mission, though the UAVs have played more of a reconnaissance role, similar to that their GNAT-750-45 predecessors were first deployed to undertake over Bosnia in the 1990s, utilizing the drone's ability to identify targets for the air force and navy's faster moving, more evasive jets. Evidence of this reconnaissance role was provided when the Pentagon reported that it had lost contact with one if its unarmed Predator drones over the northern Syrian port of Latakia, with Syrian forces later claiming to have shot down the unmanned aircraft.[133] The Pentagon's different approach to having to deal with the Islamic State is reflected by the sheer scale and cost of the first year of the campaign, in which the first air strikes took place on August 8, 2014. After its first year of the operation, the United States had spent $3.5 billion, with an average daily cost of $9.8 million, destroying 119 captured tanks, 340 Humvees, and more than twenty-five hundred fighting positions.[134] The immense financial commitment serves to demonstrate how seriously the United States is taking the threat of the Islamic State and reveals the comparative size and strength of the organization compared to the smaller-scale drone campaign employed to counter al-Qaeda. Even with the scale of the drone war waged by the CTC under Obama, the Pentagon has launched more air strikes in a single year of its anti-ISIS operation than the CTC has undertaken since the deployment of its armed drones in October 2001.[135]

Despite the significant scale of U.S. operations against the Islamic State, Pentagon officials have presented a bleak assessment of their impact upon the organization. In a press conference on July 17, 2015, the army's outgoing chief of staff, General Ray Odierno—a veteran of the American invasion and occupation of Iraq—warned that the conflict with the self-declared caliphate was likely to last for at least a decade, possibly two.[136] Reflecting frustration on the part of the Obama administration with the failure of these conventional strikes to degrade the group's strength and demonstrating the enduring centrality of targeted killing in U.S. counterterrorism policy, the White House subsequently initiated a new covert drone war, shifting CTC staff and resources from the waning AfPak campaign to the new theater of Syria.[137] The decision to once more rely upon high-value targeting by drone, even within the more hostile airspace above Syria, reveals the extent to which the Obama administration believed the AfPak campaign successfully degraded al-Qaeda, and the hope that this new effort, operating separately from the broader military offensive against IS, will play a similar role in undermining the proto-state's ability to organize and expand. In light of Obama's long-stated intention to eventually transition responsibility for lethal drone strikes from the CIA to a more permanent military jurisdiction, but reflective of the opposition of key lawmakers to such plans, the Syrian drone campaign has been depicted as an integration of CIA and JSOC capabilities, with the CTC's targeters continuing their role of identifying and locating the drone's targets, while JSOC personnel conduct the strikes. This hybrid model marks the latest stage in the ongoing evolution of the United States' use of targeted killing as a counterterrorism tool since its introduction in NSDD 138 in 1984.

Contrary to most assessors of the situation in Iraq and Syria, Michael Scheuer, the former head the CTC's Bin Laden Issue Station from 1996 to 1999 and special Adviser to its head from September 2001 to 2004, has contended that the rise of the Islamic State and ensuing civil war among the jihadist groups is an opportunity for the United States, not a threat. Writing candidly on his online blog, the former CIA analyst emphasized that, for the first time in the history of American counterterrorism, the key terrorist enemies of the past four decades were killing each other. In Syria, Scheuer observed, "the Assad regime, Iran, and Lebanese Hizballah are killing Sunni mujhaedin [*sic*] from all over the world, as well as their local allies and supporters. . . . In turn the Sunni Islamists in Syria are killing Assad's troops, Iranian Revolutionary Guards, and Hizballah fighters." Such bloodletting between America's enemies, Scheuer argued, was "the perfect circumstance for the United States," in which "all our enemies are killing each other and it is not costing us a cent or a life."[138]

Scheuer's detached realist argument gives no consideration to the humanitarian catastrophe that the Iraq insurgency, Syrian civil war, and jihadist infighting mean for the millions of civilians caught between the various combatants in the region, yet his raw analysis does highlight a truth that may suggest the chaos unleashed by al-Qaeda central's decline may have been partially calculated, or at least considered, prior to the event. The CIA's 2009 analysis of the likely impact of high-value targeting plainly stated a possible outcome of targeted killing to be the further "radicalizing [of] an insurgent group's remaining leaders," thus "creating a vacuum into which more radical groups can enter." Though the authors of the CIA report and the policy makers who signed off on the approach it endorsed could not have seen the full scale of the upheaval the Arab Spring revolts caused just one year after the drone campaign's conception, it is likely that to a certain extent the vicious infighting that has consumed America's primary terrorist enemies, from Hezbollah to al-Qaeda, was expected, and even encouraged through the CTC's drone campaign.

CONCLUSION: "IT SENDS ITS
BLOODHOUNDS EVERYWHERE"

The same human progress that gives us the technology to strike half a world away also demands the discipline to constrain that power—or risk abusing it.

—*President Barack Obama*

The United States' drone strategy, formulated by John Brennan, implemented by the Obama administration, and built upon two decades of innovations, has transformed the way in which the United States conducts its counterterrorism operations. Yet its overall impact goes well beyond the shadowy engagements of the War on Terror. Developed from the mind-set and entrenched in the practices of Brennan's predecessors, but equally cutting-edge in its exploitation of new technologies, the drone campaign, conceived to destroy a terrorist group and undermine an insurgency, has remodeled America's national security policy. As Gregory D. Johnson, author of a detailed profile on the quiet career bureaucrat put it, Brennan is "the man responsible for taking the raw infrastructure the Bush administration had left behind and molding it into an institution that would survive."[1] This technology has also begun to be adopted by the United States' global rivals, who have observed America's conduct of its War on Terror closely since 2001, absorbing lessons and adapting their own military forces to adopt America's successful methods, and those of its asymmetrical opponents, giving rise to a new form of hybrid warfare, within which the unmanned drone has become a key component. Furthermore, as drone technology proliferates into commercial and civilian sectors, its wider impact can be seen in the incredible opportunities it presents to transform society in positive ways, as well as presenting new possibilities to the very terrorist groups the technology was first adopted to neutralize. This concluding chapter explores these emerging trends in further detail.

Perhaps the best evidence of the extent to which drone warfare has grown beyond the pursuit of al-Qaeda and become embedded as a tool of U.S. power projection comes from the physical infrastructure that has materialized to support it. When the CTC first flew reconnaissance missions over Afghanistan in September 2000, the air force's Predator pilots operated from a temporary base in a small fenced-off section of Ramstein Air Base in Germany, consisting of no more than a satellite antenna and single Ground Control Station (GCS), a thirty-foot-square tent that served as an ops-center, and a couple of porta-potties.[2] By February 2010 the USAF had begun a $10 million upgrade of Ramstein's relay hub in order to eliminate its temporary setup and satisfy the long-term requirements for Predator, Reaper, and Global Hawk drones to conduct squadron-level, multi-theater-wide operations over Europe, Africa, and the Middle East.[3] The following year air force commanders requested an additional $15 million for the construction of a second relay station, based at the Naval Air Station in Sigonella, southern Italy, to carry half of the unmanned aerial system transmissions, as well as to act as a backup system to the Ramstein site, to avoid what the air force described as the risk of a "single point of failure."[4] By the end of his first term, the hastily improvised system Obama inherited had burgeoned into a crisscrossing network of runways, hangers, and relay stations ranging from Turkey to the Philippines to Afghanistan to the Seychelles, from which the CIA, the air force, and JSOC operate their drones, supported by data provided by the NSA and fed into the Disposition Matrix. This sprawling infrastructure, aptly dubbed by Ian Shaw a "Predator empire," may well prove to be Obama's most lasting legacy, the origins of which can be traced back through Brennan's cramped West Wing quarters to the CTC and its hardliner founders.[5]

The American historian Charles S. Maier has suggested that the United States constitutes the first post-territorial empire, a nation whose power and influence have enabled it to transcend traditional concepts of territory and frontiers to exert its influence on a global scale.[6] This Predator empire, and the capability to launch lethal strikes at a moment's notice against individuals from Pakistan to Somalia, using aircraft piloted remotely from air-conditioned trailers in the American Midwest, may at first glance suggest this to be true. The very existence of a network of drone bases, however, reveals that frontiers are far from irrelevant for U.S. policy makers. From the frequent hijackings that harassed the Reagan administration to the deadly assault on the homeland on September 11, 2001, terrorism has repeatedly demonstrated the threat posed to America by uncontrolled safe havens located at the far-flung edges of America's influence. These safe havens reveal the physical frontiers of U.S. influence—contested fault lines along which rebellion, or resistance, depending upon your perspective, takes place. "Any great Power that fails adequately to protect its frontier," warned the historian C. Collin Davies, "ceases to be great; any Empire that neglects this important duty of self-preservation is eventually over-thrown."[7] America's large-scale deployment of lethal drones is evidence of policy makers' acknowledgment that, as first articulated in NSDD 138, terrorism and the safe havens that enable it are legitimate threats to American national security, which must be neutralized.

The use of drones to enforce these frontiers has enabled the United States to maintain its intention of functioning as a fully globalized, post-territorial empire, or at least to preserve the pretense of doing so. For the past decade there has been a lethal U.S. presence in Pakistan, Somalia, and Yemen, without the need for bases in that territory. No physical demarcation has been necessary. Semipermeable frontiers patrolled by unmanned aircraft have enabled the continued flow of trade, people, and money while providing a higher degree of control and security. Furthermore, as with so many military outposts in the past, America's first drone bases in Afghanistan may have been justified through the need to defend itself from an exterior threat, in this case al-Qaeda, but the growing network of bases also serves as a useful springboard for deeper penetration into areas beyond U.S. control, such as Iranian airspace.[8]

But while the United States may be the current leader in the waging of remote warfare, history has shown that any security advantage that a monopoly of technologically advanced weaponry provides is short lived as the weapons disseminate to rival states, and even nonstate forces. The

Maxim machine gun may have enabled the British to cut down the Mahdist army in 1898, but the same technology was a significant factor in the stalemate that resulted in so many deaths in the no-man's lands of the First World War a little more than a decade later. The introduction of the tank may have helped the British to break that stalemate, but it also enabled Hitler's panzer corps to overrun the Allied defenses in the Ardennes in 1940, driving the United Kingdom's forces back over the Channel. Perhaps the best example of the fleeting security advantage technological superiority provides is that of nuclear weapons, possession of which initially granted the United States near martial immunity but quickly led to fear and instability as the technology proliferated, first leading to anxiety over mutually assured destruction during the Cold War, then over nuclear terrorism and dirty bombs as the specter of nonstate terrorism emerged.

The available evidence suggests the CTC's AfPak drone campaign is drawing down from its height in 2010–2011. While there are likely to be sporadic strikes from time to time in the region as al-Qaeda targets present themselves and the Taliban challenge the Afghan government, the days of sustained weekly attacks in the FATA seem to have passed. That is not to say that the role of drones in counterterrorism has ended; even as the first drone war draws down, a new front has opened for the CTC and JSOC in Syria. It is testimony to the perceived effectiveness of drones that the deployment of these aircraft was the primary request of former prime minister Maliki's Iraqi government as it sought to drive Islamic State fighters back over the Syrian border.[9] A warning as to the future direction of warfare beyond counterterrorism and counterinsurgency is the fact that Iran, also concerned about the spread of IS in Iraq, deployed its own drones to help defend Maliki's government.[10] Drone proliferation is a reality, and though the CIA may well be looking to scale-down its role in unmanned strikes post-Afghanistan, the United States is not, nor is the rest of the world. "It's unbelievable how much it's exploded," the U.S. Army's Manfredi Luciana, an engineer working on a counter-UAV project, told *Army Technology*. "Every country has them [drones] now, whether they are armed or not or what level of performance. This is a huge threat [that] has been coming up on everybody."[11]

For decades the United States sought a weapon that would allow it to strike at targets in places it could not otherwise reach. In drones it has this weapon, but as it pushed back the boundaries of targeted killing, the adoption of the method by other countries, and even the nonstate groups drones were developed to deal with, was inevitable. It was only a matter of time before the relatively cheap and easy Radio Shack solution Clarridge pursued in the CTC's efforts to undertake targeted killing was adopted by both America's global rivals and the very groups the CIA itself sought to target, for the same reasons the CTC uses the weapon—to strike at targets in places they could not otherwise reach.

The Pentagon's 2015 National Security Strategy specifically singled out the risks posed to U.S. forces by drones, deployed by the likes of China and Russia as part of "hybrid warfare" campaigns—a new post–War on Terror style of combat that, as Joseph Trevithick explains, "arises from the intersection of conventional military forces, insurgents and terrorists, and less traditional factors such as economic pressure and media campaign."[12] Such hybrid warfare is not purely a tool of great powers. In fact, its employment by Russia in its 2014–2015 invasion and occupation of Crimea mimics the style of warfare developed by sophisticated nonstate groups such as the Taliban, Hezbollah, and Islamic State over the past decade, as these smaller nonstate groups sought—with considerable success—to counter the conventional military advantage of their opponents.

But drones do not fly solely over the battlefields and contested territories of the world. Like so many other technological innovations that started from military investment, such as the splitting of the atom, the Global Positioning System (GPS), and the Internet, drones have begun to filter through to commercial and civilian life. And just like those technologies that preceded them, the rise of civilian drones has been met with grave concerns about how these remote aircraft may be exploited by terror organizations. There have already been early indications of the defensive challenges drones will come to pose for governments around the world: German activists belonging to the Pirate Party interrupted an election rally on September 16, 2013, crashing a sixteen-inch Parrot AR drone in front of the bemused-looking Chancellor Angela Merkel and her defense minister, Thomas de Maizière, as part of a political stunt. More ominously, between October 5 and November 2, 2014, French authorities reported at least fourteen illegal drone flights by unknown pilots over French nuclear power plants. Almost comically (except for the security gaps it exposed), in late January 2015 an inebriated off-duty employee of the U.S. government's National Geospatial-Intelligence Agency—an intelligence agency that, rather ironically, helps map territory for U.S. drone strikes—accidently lost a DJI Phantom 2 drone in the grounds of the White House, where the $300 UAV had evaded the Secret Service's radar. Three months later, in Japan authorities discovered a drone carrying radioactive sand, which had been landed on the roof of the prime minister's office by forty-year-old Yasuo Yamamoto as a protest against the government's nuclear energy policy. Finally, as evidence of the versatility of commercially available electronics and drone technology, a video was posted anonymously on YouTube in July 2015 showing a homemade drone remotely firing a pistol four times while airborne, automatically compensating for the recoil by adjusting its height and position after each shot.[13]

Much of the negativity directed toward civilian drones is misplaced. That is not to say terrorists will not seek to exploit them in violent ways, and a degree of caution and control over how drones are integrated into society is necessary, both for security and privacy purposes. Nevertheless, terrorists will always seek to exploit new technology to address the disparity between their group's limited resources and those of the states they seek to challenge. All technological innovation can be turned to nefarious purposes by those who seek to cause fear and mayhem. One thinks of dirty bombs, laced with the by-product of civilian nuclear power; al-Qaeda's use of civilian airliners as crude cruise missiles; Lashka-e-Taiba's exploitation of GPS to coordinate its 2008 attack upon Mumbai, India; the Islamic State's effective abuse of smart phones and social media to spread its propaganda; even the use of portable pressure cookers to manufacture the bombs used by the Tsarnaev brothers in the 2013 Boston bombings. Drones are no different, and their abuse by terrorists is inevitable. Yet alongside these dangers lie great possibilities. For every story raising concerns over the proliferation of drone technology, one can find another illustrating its future benefits, from enhancing disaster relief efforts to tracking illegal whalers, monitoring air quality, fighting wildfires, protecting wildlife, even providing Internet access to the two-thirds of the world otherwise lacking the necessary infrastructure.[14]

The fear of terrorist attacks upon American citizens is a virulent vision, one that has preyed on the minds of American policy makers since the

Iranian hostage crisis first demonstrated terrorism's ability to make even a superpower of the United States' stature look powerless to protect its citizens. It is the desire to prevent such events that has driven America's policy makers' aspiration to track down and eliminate the nation's enemies, eventually giving birth to a global, remotely piloted, targeted-killing campaign that has, for better or for worse, enabled the United States to do just that. Yet as the adapted technologies that preceded drones have demonstrated time and time again, innovations born from fear and created with military funding with the express purpose of delivering death can also eventually provide hope and save lives. In due time, drones, so synonymous with the controversies of America's War on Terror, will come to revolutionize not just the way we fight and die but also the way we cooperate and live.

NOTES

Prologue

1. United States v. Abid Naseer, "Letter from Shaykh Mahmud to Osama bin Laden," U.S. Department of Justice, Government Exhibit 429 10-CR-019 (S-4) (RJD), February 15, 2015, 113, accessed March 25, 2015, http://kronosadvisory.com/Abid.Naseer.Trial_Abbottabad.Documents_Exhibits.403.404.405.420thru433.pdf.

2. Joshua Partlow, "U.S. Contractor Killed, 9 Soldiers Wounded in Taliban Attack on Bagram Air Base," *Washington Post,* May 20, 2010, accessed November 11, 2015, http://www.washingtonpost.com/wp-dyn/content/article/2010/05/19/AR2010051901190.html.

3. "Letter from Osama bin Laden to Shaykh Mahmud ('Atiyya)," May 21, 2010, Letters from Abbottabad, Combating Terrorism Center, West Point, May 3, 2012, SOCOM-2012-0000019, 30, accessed July 21, 2015, https://www.ctc.usma.edu/posts/letters-from-abbottabad-bin-ladin-sidelined.

4. Jason Bamford, *The Shadow Factory: The Ultra-Secret NSA from 9/11 to the Eavesdropping of America* (New York: Anchor Books, 2008), 135.

5. Greg Miller, "Plan for Hunting Terrorists Signals U.S. Intends to Keep Adding Names to Kill Lists," *Washington Post,* October 24, 2012, accessed February 26, 2014, http://www.washingtonpost.com/world/national-security/plan-for-hunting-terrorists-signals-us-intends-to-keep-adding-names-to-kill-lists/2012/10/23/4789b2ae-18b3-11e2-a55c-39408fbe6a4b_story.html; 'IC off the Record Ant Catalog: Mobile Phones,' accessed October 26, 2015, https://nsa.gov1.info/dni/nsa-ant-catalog/mobile-phones/index.html.

6. Ryan Devereaux, "Manhunting in the Hindu Kush," "The Drone Papers," *Intercept,* October 15, 2015, accessed October 26, 2015, https://theintercept.com/drone-papers/manhunting-in-the-hindu-kush/.

7. Bamford, *The Shadow Factory,* 136.

8. Greg Miller and Julie Tate, "CIA Shifts Focus to Killing Targets," *Washington Post*, September 1, 2011, accessed October 19, 2015, https://www.washingtonpost.com/world/national-security/cia-shifts-focus-to-killing-targets/2011/08/30/gIQA7MZGvJ_story.html; Steve Coll, *Ghost Wars: The Secret History of the CIA, Afghanistan and bin Laden from the Soviet Invasion to September 10, 2001* (London: Penguin, 2005), 142.

9. Joby Warrick, "U.S. Missile Strikes in Pakistan Kill Taliban Militants," *Washington Post*, January 7, 2010, http://www.washingtonpost.com/wp-dyn/content/article/2010/01/06/AR2010010604597.html; Sheryl Gay Stolberg and Mark Mazzetti, "Suicide Bombing Puts a Rare Face on CIA Work," *New York Times*, January 6, 2010, http://www.nytimes.com/2010/01/07/world/asia/07intel.html?_r=0; "Timeline of Communications from Mustafa Abu al-Yazid," SITE Intelligence Group, https://news.siteintelgroup.com/Articles-Analysis/timeline-of-communications-from-mustafa-abu-al-yazid.html. All accessed October 19, 2015.

10. United States v. Abid Naseer, "Letter from Shaykh Mahmud to Osama bin Laden."

11. Ibid.

12. T. Mark Mccurley, "I Was a Drone Warrior for 11 Years. I Regret Nothing," *Politico*, October 16, 2015, accessed October 23, 2015, http://myemail.constantcontact.com/I-Was-a-Drone-Warrior-for-11-Years—I-Regret-Nothing-.html?soid=1114009586911&aid=0aYsIqboUug.

13. Matt J. Martin and Charles W. Sasser, *Predator: The Remote-Control Air War over Iraq and Afghanistan: A Pilot's Story* (Minneapolis: Zenith Press, 2010), 42.

14. Omer Fast, *5,000 Feet Is the Best*, 2011, accessed 19 October, 2015, http://www.gbagency.fr/en/42/Omer-Fast/#!/5-000-Feet-is-the-Best/site_video_listes/88.

15. Martin and Sasser, *Predator*, 43.

16. Andrew Cockburn, *Kill Chain: Drones and the Rise of High-Tech Assassins* (London: Verso, 2015), 4–5.

17. "Living under Drones," International Human Rights and Constitutional Resolution Clinic, Stanford Law School, Global Justice Clinic, NYU School of Law, September 2012, for a summary of U.S. statements on collateral damage from drone strikes, accessed October 14, 2014, http://livingunderdrones.org/wp-content/uploads/2012/09/Stanford_NYU_LIVING_UNDER_DRONES.pdf, 94–95.

18. Scott Swanson, "War Is No Video Game—Not Even Remotely," *Breaking Defense*, November 18, 2014, accessed October 23, 2015, http://breakingdefense.com/2014/11/war-is-no-video-game-not-even-remotely/.

19. "NWA Drone Attack Toll Mounts to 10," May 22, 2010, accessed October 26, 2015, http://www.geo.tv/5-22-2010/65372.htm.

20. "Five Civilians among 10 Killed in Drone Attack," *Dawn*, May 23, 2010, accessed October 26, 2015, http://www.dawn.com/news/947533/five-civilians-among-10-killed-in-drone-attack.

21. Haji Mujtaba, "U.S. Drone Attack Kills at Least Six in Pakistan," May 22, 2010, accessed October 26, 2015, http://www.reuters.com/article/2010/05/22/us-pakistan-violence-idUSTRE64E0UB20100522.

22. "Ten Dead in North Waziristan Drone Attacks," *Express Tribune,* May 22, 2010, accessed October 28, 2015, http://tribune.com.pk/story/15258/us-drones-fire-five-missiles-in-n-waziristan/.

23. Samson Desta, "9 Killed in Suspected U.S. Drone Strike in Pakistan," CNN, May 22, 2010, accessed October 28, 2015, http://news.blogs.cnn.com/2010/05/22/9-killed-in-suspected-u-s-drone-strike-in-pakistan/.

24. Bill Roggio, "Top al-Qaeda Leader Mustafa Abu Yazid Confirmed Killed in Airstrike in North Waziristan," *Long War Journal,* May 31, 2010, accessed October 24, 2015, http://www.longwarjournal.org/archives/2010/05/top_al_qaeda_leader_1. php; Eric Schmitt, "American Strike Is Said to Kill a Top Qaeda Leader," *New York Times,* May 31, 2010, accessed October 26, 2015, http://www.nytimes.com/2010/06/01/world/asia/01qaeda.html?_r=0.

25. Greg Miller and Craig Whitlock, "Al-Qaeda No. 3 Yazid reported killed by U.S. drone," *Washington Post,* June 1, 2010, http://www.washingtonpost.com/wp-dyn/content/article/2010/05/31/AR2010053103617.html; Frank Gardner, "Death of Mustafa Abu al-Yazid 'setback' for al-Qaeda," *BBC News,* June 1, 2015, http://www.bbc.co.uk/news/10206180; Declan Walsh, "Afghanistan head of al-Qaida 'killed in Pakistan drone strike,'" *Guardian,* June 1, 2010, http://www.theguardian.com/world/2010/jun/01/al-qaida-afghanistan-killed-pakistan-drone. All accessed October 28, 2015.

26. Peter Bergen and Katherine Tiedmann, "The Drone Wars: Killing by remote control in Pakistan," *Atlantic,* December 2010, http://www.theatlantic.com/magazine/archive/2010/12/the-drone-wars/308304/.

27. Jefferson Morley, "The Face of Collateral Damage," *Salon,* May 29, 2012, accessed October 28, 2010, http://www.salon.com/2012/05/29/the_face_of_collateral_damage/.

28. Michael S. Smith II, "Abbottabad Documents Revealed in U.S. v. Abid Naseer," Inside the Jihad, *Downrange,* March 6, 2015, accessed October 26, 2015, http://insidethejihad.com/2015/03/abbottabad-documents-revealed-in-us-v-abid-naseer/; Peter Bergen, "A gripping glimpse into bin Laden's decline and fall," CNN, March 12, 2015, http://edition.cnn.com/2015/03/10/opinions/bergen-bin-laden-al-qaeda-decline-fall/; "Letter from Shaykh Mahmud to Osama bin Laden."

Introduction

Epigraph. Daniel J. Boorstin, "History's Hidden Turning Points," *U.S. News and World Report,* April 22, 1991, 52.

1. Ibid.

2. Jeremy Scahill, *Dirty Wars: The World Is a Battlefield* (London: Serpent's Tail, 2013), 55–60; "The Drone Papers," *Intercept,* October 15, 2015, accessed November 4, 2015, https://theintercept.com/drone-papers/.

3. Greg Miller, "U.S. Launches Secret Drone Campaign to Hunt Islamic State Leaders in Syria," *Washington Post,* September 1, 2015, accessed September 2, 2015, https://www.washingtonpost.com/world/national-security/us-launches-secret-

drone-campaign-to-hunt-islamic-state-leaders-in-syria/2015/09/01/723b3e04-5033-11e5-933e-7d06c647a395_story.html.

4. Jimmy Carter, "A Cruel and Unusual Record," *New York Times,* June 24, 2012, accessed August 17, 2015, http://www.nytimes.com/2012/06/25/opinion/americas-shameful-human-rights-record.html

5. Chris Woods, "Drones: Barack Obama's Secret War," *New Statesman,* June 13, 2012, accessed June 16, 2012, http://www.newstatesman.com/politics/politics/2012/06/drones-barack-obama-war-secret.

6. Jonathan Bernstein, "At Stake in the 2012 Election, War or Peace?" *Washington Post,* August 21, 2012, http://www.washingtonpost.com/blogs/plum-line/post/at-stake-in-the-2012-election-war-or-peace/2012/08/21/f62e4564-ebd0-11e1-866f-60a00f604425_blog.html; Charles Krauthammer, "Barack Obama: Drone Warrior," *Washington Post,* June 1, 2012, http://www.washingtonpost.com/opinions/barack-obama-drone-warrior/2012/05/31/gJQAr6zQ5U_story.html; Mark Mardell, "Is Obama's Drone Doctrine Counter-productive?" *BBC News,* May 30, 2012, http://www.bbc.co.uk/news/world-us-canada-18270490; Ari Shapiro, "Are Drones Obama's Legacy in War on Terrorism?" National Public Radio, June 20, 2012, http://www.npr.org/2012/06/20/155389081/are-drones-obamas-legacy-in-war-on-terrorism; Gregory D. Johnson, "The Untouchable John Brennan," *BuzzFeed,* April 24, 2015. All accessed August 19, 2015.

7. Greg Miller, "Under Obama, an Emerging Global Apparatus for Drone Killing," *Washington Post,* December 28, 2011, accessed August 30, 2012, http://www.washingtonpost.com/national/national-security/under-obama-an-emerging-global-apparatus-for-drone-killing/2011/12/13/gIQANPdILP_story.html.

8. Ibid.

9. Ibid.

10. Scott Horton, "The Trouble with Drones," remarks delivered at New York University Law School Center on Law and Security's Seventh Annual Global Security Forum, May 1, 2010, published in *Harper's,* May 3, 2010, accessed August 28, 2012, http://harpers.org/archive/2010/05/hbc-90006980.

11. National Security Act of 1947, U.S. Senate Select Committee on Intelligence, accessed March 25, 2012, http://www.intelligence.senate.gov/pdfs/nsact1947.pdf.

12. Micah Zenko, "Transferring CIA Drone Strikes to the Pentagon," Policy Innovation Memorandum No. 31, Council on Foreign Relations, accessed August 19, 2015, http://www.cfr.org/drones/transferring-cia-drone-strikes-pentagon/p30434.

13. Horton, "The Trouble with Drones."

14. "CIA Drone Strikes in Pakistan, 2004–Present," Bureau of Investigative Journalism, accessed August 20, 2015, https://docs.google.com/spreadsheets/d/1NAfjFonM-Tn7fziqiv33HlGto9wgLZDSCP-BQaux51w/edit#gid=1000652376.

15. Spencer Ackerman, "How the CIA Became 'One Hell of a Killing Machine,'" *Wired,* September 2, 2011, accessed September 5, 2014, http://www.wired.com/dangerroom/2011/09/cia-killing-machine/.

16. Paul Harris, "Rand Paul Anti-Drone Filibuster Draws Stinging Criticism from Republicans," *Guardian,* March 7, 2013, accessed March 27, 2013, http://www.guardian.co.uk/world/2013/mar/07/rand-paul-drones-policy-filibuster.

17. Jane Mayer, "The Predator War: What Are the Risks of the C.I.A.'s Covert Drone Program?" *New Yorker,* October 26, 2009, accessed August 17, 2015, http://www.newyorker.com/reporting/2009/10/26/091026fa_fact_mayer.

18. Horton, "The Trouble with Drones."

19. David Luban, "What Would Augustine Do? The President, Drones and Just War Theory," *Boston Review,* June 6, 2012, accessed July 7 2015, http://www.bostonreview.net/david-luban-the-president-drones-augustine-just-war-theory.

20. Executive Order 11905, "United States Foreign Intelligence Activities," February 18, 1976, Gerald R. Ford Presidential Library, accessed March 29, 2012, http://www.ford.utexas.edu/library/speeches/760110e.htm; "Select Committee to Study Government Operations with Respect to Intelligence Activities (Church Committee)," U.S. Senate Select Committee on Intelligence, accessed October 27, 2014, http://www.intelligence.senate.gov/churchcommittee.html.

21. Mary Ellen O'Connell, "The Choice of Law against Terrorism," *Norte Dame Law School Legal Studies Research Paper* (2010): 10–20; Richard Murphy and Afsheen John Radson, "Due Process and Targeted Killing of Terrorists," *Cardozo Law Review* 32, no. 2 (2010): 405–450.

22. John Yoo, "Assassination or Targeted Killings after 9/11," *New York Law School Law Review* 56 (20–12): 58–79; Andrew C. Orr, "Unmanned, Unprecedented and Unresolved: The Status of American Drone Strikes in Pakistan under International Law," *Cornell International Law Journal* 44 (2011): 729–752.

23. "Remarks of John O. Brennan, Assistant to the President for Homeland Security and Counterterrorism," Harvard Law School/Brookings Conference, Cambridge, Massachusetts, September 16, 2011, accessed September 20, 2011, http://www.lawfare-blog.com/2011/09/john-brennans-remarks-at-hls-brookings-conference/; "Attorney General Eric Holder Speaks at Northwestern University School of Law," Chicago, March 5, 2012, accessed March 6, 2012, http://www.justice.gov/iso/opa/ag/speeches/2012/ag-speech-1203051.html; Jo Becker and Scott Shane, "Secret 'Kill List' Tests Obama's Principles and Will," *New York Times,* May 29, 2012, accessed May 31, 2012, http://www.nytimes.com/2012/05/29/world/obamas-leadership-in-war-on-al-qaeda.html?pagewanted=all&_r=0. See Appendix C, U.S. Statements on Civilian Casualties, 157, in "Living under Drones," International Human Rights and Constitutional Resolution Clinic, Stanford Law School, Global Justice Clinic, NYU School of Law, September 2012, for a summary of U.S. statements on collateral damage from drone strikes, accessed October 14, 2014, http://livingunderdrones.org/wp-content/uploads/2012/09/Stanford_NYU_LIVING_UNDER_DRONES.pdf; "U.S. Policy Standards and Procedures for the Use of Force in Counterterrorism Operations Outside the United States and Areas of Active Hostilities," White House, May 23, 2013, accessed August 19, 2015, https://www.whitehouse.gov/sites/default/files/uploads/2013.05.23_fact_sheet_on_ppg.pdf.

24. Andru E. Wall, "Demystifying the Title 10–Title 50 Debate: Distinguishing Military Operations, Intelligence Activities & Covert Action," *Harvard National Security Journal* 3 (2011): 85–142.

25. Victoria Parsons, "21 Calls for Transparency around U.S. Drone Strikes," Bureau of Investigative Journalism, August 19, 2014, accessed August 20, 2015, https://www.thebureauinvestigates.com/2014/08/19/interactive-twenty-calls-for-transparency-around-us-drone-strikes/.

26. "Drones," Blog of Rights, American Civil Liberties Union, August 31, 2012, http://www.aclu.org/blog/tag/drones; Center for Constitutional Rights, accessed August 31, 2012, http://ccrjustice.org/newsroom/news/families-of-us-citizens-killed-drone-strike-file-wrongful-death-lawsuit; Karen McVeigh, "Families of U.S. Citizen Killed in Drone Strike File Wrongful Death Lawsuit," *Guardian,* July 18, 2012, accessed August 28, 2012, http://www.guardian.co.uk/world/2012/jul/18/us-citizens-drone-strike-deaths; Jerome Taylor, "Pakistani Lawyer to Try and Halt Drone Strikes," *Independent,* May 9, 2011, accessed August 20, 2012, http://www.independent.co.uk/news/world/asia/pakistani-lawyer-to-try-and-halt-drone-strikes-2281538.html.

27. Vince Crawley and Amy Svitak, "Is Predator the Future of Warfare?" *Defense News,* November 11–17, 2002, 8; Lubine, "What Would Augustine Do?"

28. Philip Alston, "Report of the Special Rapporteur on Extrajudicial, Summary or Arbitrary Executions," U.N. Human Rights Council, May 28, 2010, paragraph 3, accessed June 5, 2011, http://www2.ohchr.org/english/bodies/hrcouncil/docs/14session/A.HRC.14.24.Add6.pdf.

29. Hina Shamsi, "The Legacy of 9/11: Endless War Without Oversight," *Guardian,* September 7, 2011, accessed August 30, 2012, http://www.guardian.co.uk/commentisfree/cifamerica/2011/sep/07/us-constitution-and-civil-liberties-congress.

30. "Drones and the Law: America's Attacks on Suspected Terrorists Should Be More Closely Monitored," *Economist,* October 8, 2011, accessed October 15, 2012, http://www.economist.com/node/21531477; "Yemen: Cracking Down Under Pressure," Amnesty International, August 2010, accessed August 31, 2012, http://www.amnesty.org/en/library/info/MDE31/010/2010/en.

31. "Your Interview with the President—2012," President Obama on Google+, January 21, 2013, White House YouTube channel, 26:31–30:17, accessed January 21, 2013, http://www.youtube.com/watch?v=eeTj5qMGTAI&feature=plcp.

32. "Brennan Defends U.S. Drone Attacks Despite Risks to Civilians," *Los Angeles Times,* April 29, 2012, accessed May 1, 2012, http://latimesblogs.latimes.com/world_now/2012/04/brennan-drone-attacks.html.

33. Becker and Shane, "Secret 'Kill List' Proves a Test of Obama's Principles and Will."

34. Woods, "Drones"; "Covert War on Terror," Bureau of Investigative Journalism, accessed October 21, 2012, http://www.thebureauinvestigates.com/category/projects/drones/.

35. "Year of the Drone," New American Foundation, accessed October 21, 2012, http://counterterrorism.newamerica.net/drones.

36. David Cortright, "License to Kill," in *Cato Unbound,* January 9, 2012, accessed August 30, 2012, http://www.cato-unbound.org/2012/01/09/david-cortright/license-to-kill/.

37. "Effective Counterinsurgency: The Future of the U.S. Pakistan Military Partnership," Committee on Armed Services, 111th Congress, 1st session, April 23, 2009, 21, gpo.gov, accessed January 20, 2010, http://www.gpo.gov/fdsys/pkg/CHRG-111hhrg52666/pdf/CHRG-111hhrg52666.pdf.

38. David Kilcullen and Andrew McDonald Exum, "Death from Above, Outrage Down Below," *New York Times,* May 16, 2009, accessed October 26, 2012, http://www.nytimes.com/2009/05/17/opinion/17exum.html?pagewanted=all.

39. Leila Hudson, Colin S. Owens, and Matt Flannes, "Drone Warfare: Blowback from the New American Way of War," *Middle East Policy* XVIII, no. 3 (Fall 2011): 123.

40. Noah Shachtman, "Washington Finally Feeling Drone War Backlash," *Wired,* May 10, 2010, accessed May 20, 2010, http://www.wired.com/dangerroom/2010/05/washington-finally-feeling-drone-war-backlash/.

41. Peter Bergen and Katherine Tiederman, "The Year of the Drone: An Analysis of U.S. Drone Strikes in Pakistan, 2004–2010," New America Foundation, accessed July 10, 2011, http://counterterrorism.newamerica.net/drones; Siobhan Goreman, "Suicide Bombing in Afghanistan Devastates Critical Hub for CIA Activities," *Wall Street Journal,* January 1, 2010, accessed August 25, 2012, http://online.wsj.com/article/SB126225941186711671.html; Anne E. Kornblut and Karin Brulliard, "U.S. Blames Pakistani Taliban for Times Square Bomb Plot," *Washington Post,* May 10, 2010, accessed August 31, 2012, http://www.washingtonpost.com/wp-dyn/content/article/2010/05/09/AR2010050901143.html; "Pakistanis Confront Clinton over Drones Attacks," *Washington Times,* October 30, 2009, accessed August 15, 2012, http://www.washingtontimes.com/news/2009/oct/30/clinton-rebukes-pakistan-over-al-qaeda-failure/?page=all.

42. "Mitt Romney: U.S. Can't Kill Its Way Out of Extremist Threats," October 22, 2012, accessed 26 October, 2012, http://www.huffingtonpost.com/2012/10/22/mitt-romney-arab-spring_n_2003919.html; "Full Transcript of Third Debate," *New York Times,* October 22, 2012, accessed October 26, 2012, http://www.nytimes.com/2012/10/22/us/politics/transcript-of-the-third-presidential-debate-in-boca-raton-fla.html?pagewanted=all; Mayer, "The Predator War."

43. Ibid.

44. Cortright, "License to Kill;" Alston, "Report of the Special Rapporteur on Extrajudicial, Summary or Arbitrary Executions."

45. Richard Clarke, "Strategy for Eliminating the Threat from the Jihadist Networks of al-Qida," 8, National Security Archive, accessed May 20, 2014, http://www2.gwu.edu/~nsarchiv/NSAEBB/NSAEBB147/clarke%20attachment.pdf.

1. "The Hamlet of Nations"

Epigraph. Secretary of State George P. Shultz, "Terrorism and the Modern World," Park Avenue Synagogue, New York, October 25, 1984, accessed December 6, 2012, http://www.airforce-magazine.com/MagazineArchive/Documents/2010/December%202010/1210keeperfull.pdf.

1. "Global Terrorism Database," National Consortium for the Study of Terrorism and Responses to Terrorism (START), 2012, University of Maryland, accessed December 3, 2012, http://www.start.umd.edu/gtd/.

2. Mark Bowden, "The Desert One Debacle," *Atlantic*, May 2006, accessed December 5, 2012, http://www.theatlantic.com/magazine/archive/2006/05/the-desert-one-debacle/304803/; Mark Bowden, *Guests of the Ayatollah: The Iran Hostage Crisis: The First Battle in America's War with Militant Islam* (New York: Grove Press, 2007).

3. Televised Address by Governor Ronald Reagan, "A Strategy for Peace in the '80s," October 19, 1980, Papers of Ronald Reagan, American Presidency Project, University of California, Santa Barbara, accessed December 3, 2012, http://www.presidency.ucsb.edu/ws/index.php?pid=85200.

4. President Jimmy Carter and Governor Ronald Reagan Presidential Debate, Cleveland, Ohio, October 28, 1980, Reagan Presidential Library Speeches Archive, accessed December 4, 2012, http://www.reagan.utexas.edu/archives/reference/10.28.80debate.html.

5. Theodore H. White, *America in Search of Itself: The Making of the President 1956–1980* (New York: Harper and Row, 1982), 21.

6. Richard H. Immerman, *The Hidden Hand: A Brief History of the CIA* (Chichester: Wiley, 2014), 124.

7. President Ronald Reagan, "First Inaugural Address," Washington, D.C., January 20, 1981, Miller Center, University of Virginia, accessed December 5, 2012, http://millercenter.org/president/speeches/detail/3407.

8. Steven R. Weismann, "Reagan Takes Oath as 40th President; Promises an 'Era of National Renewal'—Minutes Later, 52 U.S. Hostages in Iran Fly to Freedom after 444-Day Ordeal," *New York Times*, January 21, 1981, accessed December 5, 2012, http://www.nytimes.com/learning/general/onthisday/big/0120.html. "The Iranian Hostage Crisis," PBS American Experience series, accessed December 3, 2012, http://www.pbs.org/wgbh/americanexperience/features/general-article/carter-hostage-crisis/.

9. Neil A. Lewis, "Bani-Sadr, in U.S., Renews Charges of 1980 Deal," *New York Times*, May 7, 1991, http://www.nytimes.com/1991/05/07/world/bani-sadr-in-us-renews-charges-of-1980-deal.html; Gary Sick, "The Election Story of the Decade," *New York Times*, April 15, 1991, http://www.nytimes.com/1991/04/15/opinion/the-election-story-of-the-decade.html; Robert Parry, "Shamir's October Surprise Admission," *Consortium News*, July 3, 2012, http://consortiumnews.com/2012/07/03/shamirs-october-surprise-admission/; "Creating a Task Force to Investigate Certain Allegations Concerning the Holding of Americans as Hostages in Iran in 1980," Congressional Record, House of Representatives, February 5, 1992, Federation of American Scientists Intelligence Review Program, accessed July 9, 2014, http://fas.org/irp/congress/1992_cr/h920205-october-clips.htm.

10. For more on the Iran-Contra scandal, see chapter 3.

11. George C. Wilson, "Pentagon Scrubbed a Second Iranian Rescue Plan as Too Dangerous," *Washington Post,* January 25, 1981, accessed December 4, 2012, http://pqasb.pqarchiver.com/washingtonpost_historical/access/135680432.html?FMT=AB S&FMTS=ABS:AI&type=historic&date=Jan+25%2C+1981&author=By+George+C.+Wilson+Washington+Post+Staff+Writer&desc=Pentagon+Scrubbed+a+Second+Irani an+Rescue+Plan+as+Too+Dangerous.

12. President Ronald Reagan, "Statement on the Situation in Lebanon," February 7, 1984, Reagan Presidential Library Speeches Archive, accessed December 5, 2012, http://www.reagan.utexas.edu/archives/speeches/1984/20784d.htm.

13. William Dyess, quoted in David C. Wills, *The First War on Terrorism: Counter-Terrorism Policy During the Reagan Administration* (Lanham, Md.: Rowman and Littlefield 2003), 5.

14. President Ronald Reagan, "Remarks at the Welcoming Ceremony of the Freed American Hostages," January 27, 1981, Reagan Presidential Library Speeches Archive, accessed February 8, 2013, http://www.reagan.utexas.edu/archives/speeches/1981/12781b.htm.

15. President Ronald Reagan, "Remarks and a Question-and-Answer Session with Regional Editors and Broadcasters on the Situation in Lebanon," October 24, 1983, Public Papers of Ronald Reagan, Reagan Presidential Library, accessed February 8, 2013, http://www.reagan.utexas.edu/archives/speeches/1983/102483b.htm.

16. President Ronald Reagan, "Address to the Nation on the United States Air Strike Against Libya," April 14, 1986, Reagan Presidential Library Speeches Archive, accessed August 13, 2012, http://www.reagan.utexas.edu/archives/speeches/1986/41486g.htm.

17. President Ronald Reagan, "A Time of Remembrance for Victims of Terrorism," Proclamation 5557, October 22, 1986, Reagan Presidential Library Speeches Archive, accessed August 13, 2012, http://www.reagan.utexas.edu/archives/speeches/1986/102286b.htm.

18. George P. Shultz, *Turmoil and Triumph: My Years as Secretary of State* (New York: Macmillan, 1993), 647.

19. Haig resigned due to bureaucratic infighting on June 25, 1982. Letter Accepting the Resignation of Alexander M. Haig, Jr., as Secretary of State, June 25, 1982, Public Papers of the Presidents, Ronald Reagan, 1982, Book I, American Presidency Project, University of California, Santa Barbara, accessed January 13, 2013, http://www.presidency.ucsb.edu/ws/index.php?pid=42681.

20. Timothy Naftali, *Blind Spot: The Secret History of American Counterterrorism* (New York: Basic Books, 2005), 117–118.

21. Ibid., 122.

22. Michael Getler, "Soviets and Terrorist Activity: World of Shadows and Shading," *Washington Post,* February 7, 1981.

23. Claire Sterling, *The Terror Network: The Secret War of International Terrorism* (New York: Holt, Rinehart and Winston, 1981), 287.

24. Ibid, 287. Interview with Claire Sterling by Richard D. Heffner, *The Terror Network,* The Open Mind, 1981, The Internet Archive, accessed January 13, 2013, http://archive.org/details/openmind_ep1346.

25. Richard Halloran, "Proof of Soviet Aided Terror Is Scarce," *New York Times,* February 9, 1981, http://www.nytimes.com/1981/02/09/world/proof-of-soviet-aided-terror-is-scarce.html; Judith Miller, "U.S. Study Discounts Soviet Terror Role: A Draft CIA Report, Now Being Reviewed, Finds Insufficient Evidence for Direct Help," *New York Times,* March 29, 1981, http://www.nytimes.com/1981/03/29/world/us-study-discounts-soviet-terror-role.html; George Lardner Jr., "Assault on Terrorism: Internal Security or Witch Hunt?" *Washington Post,* April 20, 1981; Philip Taubman, "U.S. Tries to Back Up Haig on Terrorism," *New York Times,* May 3, 1981, http://www.nytimes.com/1981/05/03/world/us-tries-to-back-up-haig-on-terrorism.html; Tom Wicker, "The Great Terrorist Hunt," *New York Times,* May 5, 1981, http://www.nytimes.com/1981/05/05/opinion/in-the-nation-the-great-terrorist-hunt-by-tom-wicker.html?scp=82&sq=By%20Tom%20Wicker&st=nyt. All accessed January 13, 2013.

26. Bob Woodward, *Veil: The Secret Wars of the CIA, 1981–1987* (New York: Simon and Schuster, 1987), 103.

27. Naftali, *The Blind Spot,* 124.

28. Paul R. Pillar, *Terrorism and U.S. Foreign Policy* (Washington, D.C.: Brookings Institution Press, 2003), 43.

29. Shultz, *Turmoil and Triumph,* 643.

30. Ibid., 652.

31. Secretary of State George P. Shultz, "Power and Diplomacy in the 1980s," Statement before the Trilateral Commission, Washington, D.C., April 3, 1984, in Yonah Alexander and Michael Kraft (eds.), *Evolution of U.S. Counterterrorism Policy,* volume 1 (Westport: Praeger Security International), 93–100.

32. Edward S. Herman and Gerry O'Sullivan, *The "Terrorism" Industry: The Experts and Institutions That Shape Our View of Terror* (New York: Pantheon, 1989), 104–106; Jonathan (Yoni) Netanyahu Memorial Site, accessed June 10, 2014, http://www.yoni.org.il/.

33. Secretary of State George P. Shultz, "Terrorism: The Challenge to the Democracies," Address before the Jonathan Institute's Second International Conference on Terrorism, Washington, D.C., June 24, 1984, in Alexander and Kraft (eds.), *Evolution of U.S. Counterterrorism Policy,* volume 1, 101–104.

34. Shultz, "Terrorism and the Modern World."

35. Frank M. Benis, *U.S. Marines in Lebanon, 1982–1984* (Washington, D.C.: History and Museums Division, U.S. Marine Corps, 1987), accessed December 9, 2012, http://ia601504.us.archive.org/22/items/USMarinesInLebanon1982-1984/US%20Marines%20In%20Lebanon%201982-1984.pdf; Timothy J. Gerathy and J. Alfred M. Gray Jr., *Peacekeepers at War: Beirut 1983—The Marine Commander Tells His Story* (Washington, D.C.: Potomac Books, 2009); Lance Cpl. Paul Peterson, "29 years later: Beirut Marine shares memory of bombing," October 15, 2012, Marines.mil, accessed December 9, 2012, http://www.2ndmlg.marines.mil/News/NewsArticleDisplay/

tabid/3874/Article/128298/29-years-later-beirut-marine-shares-memory-of-bombing.aspx.

36. Shultz, *Turmoil and Triumph*, 646.

37. Brian Michael Jenkins, *New Modes of Conflict* (Santa Monica: Rand, 1983), accessed June 11, 2014, http://www.rand.org/content/dam/rand/pubs/reports/2006/R3009.pdf.

38. Osama bin Laden, *ABC Frontline* interview, May 1988, quoted in Raymond Ibrahim, *The Al-Qaeda Reader* (New York: Broadway Books, 2007), 260.

39. Ayman al-Zawahiri, *Knights under the Prophet's Banner*, translated by Laura Mansfield, in *His Own Words: A Translation of the Writings of Dr. Ayman al-Zawahiri* (New York: TLG, 2006), 219–220.

40. Shultz, *Turmoil and Triumph*, 650.

41. Shultz, "Power and Diplomacy in the 1980s," in Alexander and Kraft (eds.), *Evolution of U.S. Counterterrorism Policy*, volume 1, 93–100.

42. Shultz, *Turmoil and Triumph*, 650.

43. George Ball, "Shultz Is Wrong on Terrorism," *New York Times*, December 16, 1984, accessed December 14, 2012, http://www.nytimes.com/1984/12/16/opinion/shultz-is-wrong-on-terrorism.html.

44. Shultz, *Turmoil and Triumph*, 645.

45. Ibid., 645; David C. Martin and John Walcott, *Best Laid Plans: The Inside Story of America's War Against Terrorism* (New York: Touchstone, 1988), 156–157.

46. Ball, "Shultz Is Wrong on Terrorism."

47. American Foreign Service Association Memorial Plaque List, afsa.org, accessed January 10, 2013, http://www.afsa.org/afsa_memorial_plaque_list.aspx.

48. Woodward, *Veil*, 230.

49. Immerman, *The Hidden Hand*, 124–125.

50. John A. Bross, "William J. Casey," *Foreign Intelligence Literary Scene* 6 (March–April 1987): 2, cited in Rhodri Jeffreys-Jones, *The CIA and American Democracy* (New Haven: Yale University Press, 2003), 231.

51. William Casey quoted in Herbert E. Meyer and Mark B. Liedl (eds.), *Scouting the Future: The Public Speeches of William J. Casey* (Washington, D.C.: Regnery Gateway, 1989), 20, 32; Rhodri Jeffreys-Jones, *Cloak and Dollar: The History of American Secret Intelligence* (New Haven: Yale University Press, 2002), 236.

52. Meyer and Liedl (eds.), *Scouting the Future*, 193.

53. "Remarks of Director of Central Intelligence William J. Casey, before the Commonwealth Club of California," Sheraton Palace Hotel, San Francisco, California, May 21, 1982, CIA Electronic Reading Room, accessed January 15, 2013, http://www.foia.cia.gov/docs/DOC_0001446210/DOC_0001446210.pdf.

54. Joseph E. Persico, *Casey: The Lives and Secrets of William J. Casey from the OSS to the CIA* (New York: Penguin, 1990), 305.

55. Ibid., 286–288.

56. Director of Central Intelligence William J. Casey, "Fighting Terrorists: Identifications and Action," Address to the 14th Annual Conference of the Fletcher

School of Law and Diplomacy's National Securities Studies Program, Tufts University, Cambridge, Massachusetts, April 17, 1985, in Meyer and Liedl (eds.), *Scouting the Future,* 197–198.

57. Woodward, *Veil,* 131, Persico, *Casey,* 257–288; 306.

58. Neil C. Livingstone, *The Cult of Counterterrorism: The "Weird World" of Spooks, Counterterrorists, Adventurers, and the Not Quite Professionals* (Lexington, Mass.: Lexington Books, 1990), 232.

59. Timothy J. Geraghty and Timothy J. Grey, *Peacekeepers at War: Beirut 1983—The Marine Commander Tells His Story* (Washington, D.C.: Potomac Books, 2009), 20.

60. Martin and Walcott, *Best Laid Plans,* 153–155; Persico, *Casey,* 317; "Profile of William Francis Buckley," Arlington National Cemetery, accessed January 15, 2013, http://www.arlingtoncemetery.net/wbuckley.htm.

61. Livingstone, *Cult of Counterterrorism,* 242; Persico, *Casey,* 317.

62. Woodward, *Veil,* 231.

63. Livingstone, *Cult of Counterterrorism,* 231–232.

64. Ibid., 249–250; Wills, *First War on Terrorism,* 33.

65. Casey, April 17, 1985, in Meyer and Liedl (eds.), *Scouting the Future,* 193.

66. Ibid., 231.

67. Ibid.

68. Ibid.

69. Livingstone, *The Counterterrorism Cult,* 232.

70. Casey, "The Practice of International Terrorism Has to Be Resisted by All Legal Means," in Alexander and Kraft (eds.), *Evolution of U.S. Counterterrorism Policy,* volume 1, 229.

71. W. R. Peers, "Intelligence Operations of OSS Detachment 101" (Summer 1960), Center for the Study of Intelligence, https://www.cia.gov/library/center-for-the-study-of-intelligence/kent-csi/vol4no3/html/v04i3a11p_0001.htm; William J. Casey, "OSS: Lessons for Today" (Winter 1986), Center for the Study of Intelligence, Declassified Articles Archive. Both accessed February 10, 2013. Also see Richard H. Smith, *OSS: The Secret History of America's First Central Intelligence Agency* (Guildford: Lyons Press, 2005), and Michael Warner, *The Office of Strategic Services: America's First Intelligence Agency* (Washington, D.C.: CIA History Staff, Center for the Study of Intelligence, 2000), for an overview of the operations conducted by the OSS.

72. Livingstone, *Cult of Counterterrorism,* 231–232.

73. Oliver L. North and William Novak, *Under Fire: An American Story* (New York: HarperCollins, 1991), 197.

74. Livingstone, *Cult of Counterterrorism,* 261.

75. Ben Bradlee Jr., *Guts and Glory: The Rise and Fall of Oliver North* (New York: Donald I. Fine, 1988), 291.

76. Persico, *Casey,* 397.

77. North and Novak, *Under Fire,* 179–182; Persico, *Casey,* 397.

78. Sean Wilentz, *The Age of Reagan: A History, 1974–2008* (New York: HarperCollins, 2008), 212; Bill Summary and Status, 98th Congress (1983–1984),

H.R2968, Library of Congress, Thomas, accessed January 25, 2013, http://thomas. loc.gov/cgi-bin/bdquery/z?d098:HR02968:@@@L&summ2—&%7CTOM:/bss/ d098query.html.

79. For more on the Iran-Contra scandal, see chapter 3.

80. North and Novak, *Under Fire,* 197.

81. Ibid., 197; Oliver North, quoted in Woodward, *Veil,* 357.

82. "Report of the DoD Commission on Beirut, International Airport Terrorist Act," Long Commission, October 23, 1983, to December 20, 1983, 128–129, Federation of American Scientists, Intelligence Resource Program, accessed January 18, 2013, https://www.fas.org/irp/threat/beirut-1983.pdf.

83. Ibid., 128

84. Ibid., 132–133.

85. Naftali, *Blind Spot,* 138; Livingstone, *Cult of Counterterrorism,* 234.

86. Wills, *First War on Terrorism,* 84; "Secret Policy on Terrorism Given Airing," *Washington Post,* April 18, 1984, A1; Livingstone, *Cult of Counterterrorism,* 234.

87. Martin and Walcott, *Best Laid Plans,* 157; Wells, *The First War on Terrorism,* 87.

88. For details of why NSDD 138 was never fully implemented, see chapter 2.

89. "Combating Terrorism," National Security Decision Directive 138, April 3, 1984, Reagan Presidential Library Speeches Archive, accessed January 18, 2013, http://www. reagan.utexas.edu/archives/reference/Scanned%20NSDDS/NSDD138.pdf.

90. "Memorandum for Edwin Meese, III from Robert C. McFarlane," Background Material to Meese, August 15, 1984, National Security Archive, Counterterrorism, in Washington Decoded, accessed January 18, 2014, www.washingtondecoded.com/files/ nsdd.pdf; NSDD 138, 2; Naftali, *Blind Spot,* 138.

91. NSDD 138, 2.

92. Ibid., 3.

93. Robert M. Gates, *From the Shadows: The Ultimate Insider's Story of Five Presidents and How They Won the Cold War* (New York: Simon and Schuster, 1996), 193.

94. Interview with Noel Koch, April 3, 1987, in Livingstone, *Cult of Counterterrorism,* 235.

95. Ibid., 235.

96. Martin and Walcott, *Best Laid Plans,* 156.

97. Bradlee, *Guts and Glory,* 188.

98. NSDD 138, 4

99. Woodward, *Veil,* 358.

100. Ronald Reagan, Executive Order 12333, "United States Intelligence Activities," December 4, 1981, National Archives, accessed January 18, 2013, http://www.archives. gov/federal-register/codification/executive-order/12333.html.

101. Seymour Hersh, "Huge CIA Operation Reported in U.S. Against Anti-War Forces," *New York Times,* December 22, 1974. CIA's Family Jewels, National Security Archive, accessed July 13, 2014, http://www2.gwu.edu/~nsarchiv/NSAEBB/ NSAEBB222/.

102. "United States Foreign Intelligence Activities," Executive Order 11905, February 18, 1976, accessed July 13, 2014, http://fas.org/irp/offdocs/eo11905.htm.

103. Gerald K. Haines, "The Pike Committee Investigations and the CIA: Looking for a Rogue Elephant," *Studies in Intelligence* (Winter 1998–1999), accessed July 13, 2014, https://www.cia.gov/library/center-for-the-study-of-intelligence/csi-publications/csi-studies/studies/winter98_99/art07.html.

104. Ibid.

105. Loch K. Johnson, *A Season of Inquiry: Congress and Intelligence* (Chicago: Dorsey Press, 1988), 8–11. For more on the Church Committee hearings, see chapter 4.

106. "Alleged Assassination Plots Involving Foreign Leaders," Church Committee Interim Report, 1975, 257, Assassination Archives and Research Center, accessed January 18, 2013, http://www.aarclibrary.org/publib/church/reports/ir/html/ChurchIR_0136a.htm.

107. Katheryn S. Olmsted, *Challenging the Secret Government: The Post-Watergate Investigations of the CIA and FBI* (Chapel Hill: University of North Carolina Press, 1996), 110.

108. "History and Justification," U.S. House of Representatives Permanent Select Committee on Intelligence, accessed July 13, 2014, http://intelligence.house.gov/about/history-jurisdiction.

109. "Spook No. 2," *Time,* May 10, 1982, accessed January 18, 2013, http://www.time.com/time/magazine/article/0,9171,925363,00.html.

110. Brian Michael Jenkins, "The Lessons of Beirut: Testimony Before the Long Commission," February 1984, Rand Corporation, Reports, accessed January 23, 2013, http://www.rand.org/content/dam/rand/pubs/notes/2007/N2114.pdf, 6–8.

111. Michael C. Ruppert, *Crossing the Rubicon: The Decline of the American Empire at the End of the Age of Oil* (Vancouver: New Society, 2004), 54–55.

112. Persico, *Casey,* 240, 405.

113. Wills, *First War on Terrorism,* 83

114. Article 51, Chapter VII, Charter of the United Nations, June 26, 1945, UN.org, accessed January 18, 2013, http://www.un.org/en/documents/charter/intro.shtml; Woodward, *Veil,* 358.

115. Caspar Weinberger, *Fighting for Peace: Seven Critical Years at the Pentagon* (London: Penguin, 1990), 54.

116. William Schneider, "Conservatism, Not Interventionism: Trends in Foreign Policy Opinion, 1974–1982," in Kenneth A. Oye, Robert Lieber, and Donald Rothchild (eds.), *Eagle Defiant: United States Foreign Policy in the 1980s* (Boston: Little, Brown, 1983), 36.

117. Weinberger, *Fighting for Peace,* 28.

118. Jesús Velasco, *Neoconservatives in U.S. Foreign Policy under Ronald Reagan and George W. Bush: Voices behind the Throne* (Baltimore: Johns Hopkins University Press, 2010), 111.

119. "Weinberger Takes Tough Defense Stance," *Sarasota Herald Tribune,* January 23, 1981, accessed January 23, 2013, http://news.google.com/newspapers?nid=1755&dat=19810123&id=T-MbAAAAIBAJ&sjid=9GcEAAAAIBAJ&pg=2887,3454660.

120. Adrian R. Lewis, *The American Way of War: The History of U.S. Military Force from World War II to Operation Iraqi Freedom* (New York: Routledge, 2007), 295.

121. President Ronald Reagan, "Evil Empire," National Association of Evangelicals, Orlando, Florida, March 8, 1983, Miller Center Presidential Speech Archive, accessed January 24, 2013, http://millercenter.org/president/speeches/detail/3409.

122. Stanley Hoffman, "Carter's 'Fiasco' in Iran," *New York Times,* April 26, 1980, accessed February 10, 2013, http://query.nytimes.com/mem/archive/pdf?res=F009 14F83A5D11728DDDAF0A94DC405B8084F1D3.

123. "Weinberger Suggests Turning to the U.N. with Terrorist Woes," *New York Times,* June 26, 1984, A14; Fred Hoffman, "Weinberger: Let the U.N. Deal with International Terrorism," *Gettysburg Times,* June 25, 1984, http://news.google.com/newspapers?nid=2202&dat=19840625&id=sqkyAAAAIBAJ&sjid=UegFAAAAIBA J&pg=3440,578097; "Weinberger Wants Terrorism Handled by United Nations," *Palm Beach Post,* June 26, 1984, Google Archive, http://news.google.com/newspap ers?nid=1964&dat=19840626&id=zAAtAAAAIBAJ&sjid=6c0FAAAAIBA J&pg=3574,4026597. All accessed January 24, 2013.

124. Secretary of Defense Caspar W. Weinberger, "The Uses of Military Power," Remarks to the National Press Club, Washington, D.C., November 28, 1984, in Alexander and Kraft (eds.), *Evolution of U.S. Counterterrorism Policy,* volume 1, 117.

125. Ibid., 115.

126. Shultz, "Terrorism and the Modern World;" Weinberger, "The Uses of Military Power," 117–118.

127. Secretary of Defense Caspar W. Weinberger, "Remarks Prepared for Delivery at the Conference on Low-Intensity Warfare," Fort McNair, Washington, D.C., January 14, 1984, in Alexander and Kraft (eds.), *Evolution of U.S. Counterterrorism Policy,* volume 1, 124.

128. Wills, *The First War on Terrorism,* 30.

129. Christopher C. Joyner, "In Search of an Anti-terrorism Policy: Lessons from the Reagan Era," *Studies in Conflict and Terrorism* 11, no. 1 (1988): 29–42.

130. Weinberger, "Remarks Prepared for Delivery at the Conference on Low-Intensity Warfare," 124.

131. Chairman of the Joint Chiefs of Staff General John Vessey, "Address to the Dwight Eisenhower Society," Gettysburg, Pennsylvania, October 14, 1982, in *Selected Works of General John W. Vessey, Jr., Tenth Chairman of the Joint Chiefs of Staff,* 16, accessed March 28, 2013, http://www.dtic.mil/doctrine/doctrine/history/vessey_speeches.pdf.

132. Wills, *First War on Terror,* 42.

133. Watkins quoted in G. Davidson Smith, *Combating Terrorism* (London: Routledge, 1990), 26; James D. Watkins, Adm., USN, "Terrorism: An Already Declared War," *Wings of Gold* 9, no. 2 (Summer 1984): 19–21.

134. Robert L Pfaltzgraff Jr. and Jacquelyn K. Davis (eds.), *National Security Decisions: The Participants Speak* (Lexington, Mass.: Lexington Books, 1990), 20–24.

135. Livingstone, *Cult of Counterterrorism,* 233.

136. Michael Moran, "The Pope's Divisions," *Foreign Affairs,* February 3, 2006, accessed March 28, 2013, http://www.cfr.org/religion-and-politics/popes-divisions/p9765.

137. Jane Mayer and Doyle McManus, *Landslide: The Unmaking of the President, 1984–1988* (Boston: Houghton Mifflin, 1988), 27.

138. Ann Reilly Dowd and Barbara Hetzer, "What Managers Can Learn from Manager Reagan," *Fortune,* September 15, 1986, accessed July 28, 2015, http://archive.fortune.com/magazines/fortune/fortune_archive/1986/09/15/68052/index.htm.

139. Ronald Reagan, *Ronald Reagan: An American Life* (New York: Pocket Books, 1990), 161.

140. Wilentz, *The Age of Reagan,* 151.

141. Robert Timberg, *The Nightingale's Song* (New York: Touchstone, 1995), 448.

142. Haynes Johnson, *Sleepwalking Through History: America in the Reagan Years* (New York: W. W. Norton, 2003), 50.

143. Garry Wills, *Reagan's America: Innocents at Home* (New York: Penguin, 2000), 423–425.

144. James A. Nathan, "Decisions in the Land of the Pretend: U.S. Foreign Policy in the Reagan Years," *Virginia Quarterly Review* 65 (Winter 1989), accessed July 28, 2015, http://www.vqronline.org/essay/decisions-land-pretend-us-foreign-policy-reagan-years.

145. Wills, *Reagan's America,* 423.

146. Francis Fitzgerald, *Way Out There in the Blue: Reagan, Star Wars and the End of the Cold War* (New York: Touchstone, 2000), 15.

147. Bob Schieffer and Gary Paul Gates, *The Acting President: Ronald Reagan and the Supporting Players Who Helped Him Create the Illusion That Held America Spellbound* (New York: E. P. Dutton, 1992), 89.

148. Nathan, "Decisions in the Land of the Pretend."

149. Fitzgerald, *Way Out There in the Blue,* 219.

150. Colin Powell with Joseph Persico, *My American Journey* (New York: Ballantine Books, 1996), 334.

151. Mayer and McManus, *Landslide,* 27.

152. Ibid., 27.

153. Lou Cannon, *President Reagan: The Role of a Lifetime* (New York: Simon and Schuster, 1991), 41.

154. Fitzgerald, *Way Out There in the Blue,* 219.

155. Malcolm Byrne, *Iran-Contra: Reagan's Scandal and the Unchecked Abuse of Presidential Power* (Lawrence: University of Kansas Press, 2014), 12.

156. Alexander M. Haig, *Caveat: Realism, Reagan and Foreign Policy* (New York: Macmillan, 1984), 85. See 67, 84–86, 93, 115 for commentary on obstacles within the Reagan administration to policy making.

157. Persico, *Casey,* 305–306.

158. Johnson, *Sleepwalking Through History,* 303–304.

159. "Reagan and International Arms Agreements," interview with Caspar Weinberger, August 21, 1990, in Kenneth W. Thompson (ed.), *Foreign Policy in the Reagan Presidency: Nine Intimate Perspectives* (Lanham, Md.: University Press of America, 1993), 44.

160. Weinberger, *Fighting for Peace,* 23–24.

161. Timberg, *The Nightingale's Song,* 359; Robert McFarlane and Zofia Smardz, *Special Trust* (New York: Cadell and Davies, 1994), 268.

162. Ibid., 359.

163. Lou Cannon, *President Reagan,* 78.

164. Byrne, *Iran-Contra,* 12.

165. Naftali, *Blind Spot,* 131–132.

166. "Responding to the Lebanon Crisis," National Security Decision Directive 109, October 23, 1983, Federation of American Scientists Intelligence Resource Program, accessed January 31, 2013, http://www.fas.org/irp/offdocs/nsdd/nsdd-109.htm.

167. McFarlane and Smardz, *Special Trust,* 270.

168. Weinberger, *Fighting for Peace,* 112.

169. Ibid., 112; McFarlane, *Special Trust,* 270–271.

170. Reagan, *Ronald Reagan,* 463–464.

171. Martin and Walcott, *Best Laid Plans,* 292–293.

172. Public testimony of Richard Armitage, National Commission on the Terrorist Attacks upon the United States, March 24, 2004, 175, accessed August 17, 2012, http://govinfo.library.unt.edu/911/archive/hearing8/9-11Commission_Hearing_2004-03-24.pdf.

173. Ibid, 176.

174. Naftali, *Blind Spot,* 133.

175. Byrne, *Iran-Contra,* 159.

176. Ron Reagan, *My Father at 100: A Memoir* (New York: Viking, 2011), 205, 218.

177. Interview with John Poindexter, January 14, 2013, in Byrne, *Iran-Contra,* 287.

178. Visar Berisha, Shuai Wang, Amy LaCross, and Julie Liss, "Tracking Discourse Complexity Preceding Alzheimer's Disease Diagnosis: A Case Study Comparing the Press Conferences of Presidents Ronald Reagan and George Herbert Walker Bush," *Journal of Alzheimer's Disease* 45, no. 3 (2015); Lawrence K. Altman, "Parsing Ronald Reagan's Words for Early Signs of Alzheimer's," *New York Times,* March 30, 2015, accessed July 29, 2015, http://www.nytimes.com/2015/03/31/health/parsing-ronald-reagans-words-for-early-signs-of-alzheimers.html?_r=0.

179. Nathan, "Decisions in the Land of the Pretend."

180. *The Tower Commission Report: Full Text of the President's Special Review Board* (New York: Bantam Books, 1987), 79–80.

181. William Kristol and Robert Kagan, "Toward a Neo-Reaganite Foreign Policy," *Foreign Affairs* (July/August 1996): 18–32.

182. Senate Joint Resolution 23, 107th Congress, September 18, 2001, gpo.gov, accessed February 11, 2013, http://frwebgate.access.gpo.gov/cgi-bin/getdoc.cgi?dbname=107_cong_public_laws&docid=f:publ040.107.

183. "United States National Strategy for Counterterrorism," June 2011, 2, white-house.gov, accessed February 11, 2013, http://www.whitehouse.gov/sites/default/files/counterterrorism_strategy.pdf.

184. President George W. Bush, "Graduation Speech at West Point," United States Military Academy, West Point, New York, June 1, 2002, G. W. Bush White House Archive, http://georgewbush-whitehouse.archives.gov/news/releases/2002/06/20020601-3.html; "National Security Strategy of the United States, 2002," 15, George W. Bush White House Archive, http://georgewbush-whitehouse.archives. gov/nsc/nss/2002/nss.pdf. Both accessed February 11, 2013.

185. "Lawfulness of Lethal Operation Directed against a U.S. Citizen Who Is a Senior Level Operational Leader of Al-Qa'ida or an Associated Force," Department of Justice White Paper, 7–8, *MSNBC News,* accessed February 11, 2013, http://msnbcmedia.msn.com/i/msnbc/sections/news/020413_DOJ_White_Paper.pdf.

186. Osama bin Laden, "World Islamic Front Declaration of Jihad Against Jews and Crusaders," February 23, 1998, accessed February 17, 2013, http://www.fas.org/irp/world/para/docs/980223-fatwa.htm; Zawahiri, *Knights under the Prophet's Banner,* translated by Mansfield, in *His Own Words,* 223.

187. "Suspect Reveals 9/11 Planning," *BBC News,* September 22, 2003, accessed February 17, 2013, http://news.bbc.co.uk/1/hi/world/south_asia/3128802.stm.

188. John Yoo, "Assassination or Targeted Killings after 9/11," *New York Law School Law Review* 56 (2011–2012): 58–79, 73. See also Michael N. Schmitt, "State-Sponsored Assassination in International and Domestic Law," *Yale Journal of International Law* (1992): 609–648; Gregory M. Travalio, "Terrorism, International Law, and the Use of Military Force," *Wisconsin International Law Journal* 18 (2000): 145–173.

189. "Lawfulness of Lethal Operation Directed against a U.S. Citizen Who Is a Senior Level Operational Leader of Al-Qa'ida or an Associated Force," DoJ.

190. Assistant to the President for Homeland Security and Counterterrorism John O. Brennan, "Ensuring al-Qaeda's demise," Paul H. Nitze School of Advanced International Studies, Washington DC, June 29, 2011, accessed February 11, 2013, http://www.sais-jhu.edu/sebin/a/h/2011-06-29-john-brennan-remarks-sais.pdf.

191. "The Year of the Drone: Leaders Killed," Counterterrorism Strategy Initiative, New American Foundation Drones Database, accessed February 17, 2013, http://counterterrorism.newamerica.net/about/militants. For more on Brennan and the goals of the drone campaign, see Chapter 6.

192. Assistant to the President for Homeland Security and Counterterrorism John O. Brennan, "The Efficacy and Ethics of The President's Counterterrorism Strategy," Wilson Center, April 30, 2012, Wilson Center International Security Studies, accessed February 17, 2013, http://www.wilsoncenter.org/event/the-efficacy-and-ethics-us-counterterrorism-strategy. See also "Additional Prehearing Questions for Mr. John O. Brennan on his nomination to be Director of the Central Intelligence Agency," 24–28, *Wall Street Journal* Public Resources, accessed February 15, 2013, http://online.wsj. com/public/resources/documents/brennanquestionnaire2013-add.pdf.

193. "Lawfulness of Lethal Operation Directed Against a US Citizen Who Is a Senior Level Operational Leader of Al-Qa'ida or an Associated Force," DoJ.

194. NSDD 138.

195. Vice President Dick Cheney, interview with Tim Russert, *NBC News,* September 14, 2003, MSNBC.com, accessed February 12, 2013, http://www.msnbc.msn.com/id/3080244/.

2. "Let's Find a Way to Go after Them"

Epigraph. Robert Timberg, *The Nightingale's Song* (New York: Simon and Schuster, 1995), 337.

1. John Tirman, *The Deaths of Others: The Fate of Civilians in America's Wars* (New York: Oxford University Press, 2011), 182.

2. "U.S. Relations with the USSR," National Security Decision Directive 75, January 17, 1983, Federation of American Scientists, Intelligence Resource Program, accessed March 13, 2013, http://www.fas.org/irp/offdocs/nsdd/nsdd-075.htm.

3. Steve Coll, *Ghost Wars: The Secret History of the CIA, Afghanistan and bin Laden from the Soviet Invasion to September 10, 2001* (London: Penguin, 2005), 65, 121. For more on the role Congress played in securing this funding, see George Crile, *Charlie Wilson's War* (New York: Grove Press, 2003); Richard Sobel, "Contra Aid Fundamentals: Exploring the Intricacies and the Issues," *Political Science Quarterly* 110, no. 2 (Summer 1995): 287–306.

4. President Ronald Reagan, State of the Union Address, Washington, D.C., February 6, 1985, American Presidency Project, University of California, Santa Barbara, accessed March 13, 2013, http://www.presidency.ucsb.edu/ws/index.php?pid=38069.

5. Tirman, *The Deaths of Others,* 185.

6. Thanassis Cambanis, "Grand Ayatollah Fadlallah, Shiite Cleric, Dies at 75," *New York Times,* July 4, 2010, accessed March 15, 2013, http://www.nytimes.com/2010/07/05/world/middleeast/05fadlallah.html?_r=0.

7. Bob Woodward, *Veil: The Secret Wars of the CIA, 1981–1987* (New York: Simon and Schuster, 1987), 397.

8. Ibid., 397.

9. "United States Intelligence Activities," Ronald Reagan, Executive Order 12333, National Archives, accessed January 18, 2013, http://www.archives.gov/federal-register/codification/executive-order/12333.html.

10. Woodward, *Veil,* 394.

11. Woodward, *Veil,* 393–399; "Target America: Interview: Bob Woodward," *PBS Frontline,* September, 2001, http://www.pbs.org/wgbh/pages/frontline/shows/target/interviews/woodward.html; Richard Zoglin, Jay Peterzell, and Bruce van Voorst, "Did a Dead Man Tell No Tales?" *Time,* October 12, 1987, accessed July 17, 2014, http://content.time.com/time/printout/0,8816,965712,00.html.

12. Joseph E. Persico, *Casey: The Lives and Secrets of William J. Casey: From the OSS to the CIA* (New York: Penguin Books, 1990), 428–440, 580; "Target America: Interview: Robert McFarlane," *PBS Frontline,* September, 2001, accessed July 15, 2014,

http://www.pbs.org/wgbh/pages/frontline/shows/target/interviews/mcfarlane.
html.

13. Persico, *Casey,* 443.

14. Article 51, Chapter VII, Charter of the United Nations, June 26, 1945, UN.org, accessed March 17, 2013, http://www.un.org/en/documents/charter/intro.shtml.

15. "War and International Humanitarian Law," International Committee of the Red Cross, accessed March 17, 2013, http://www.icrc.org/eng/war-and-law/index. jsp.

16. "Respecting the Laws and Customs of War on Land, with Annex of Regulations," Convention (No. IV), United States Department of State, January 26, 1910, Avalon Project, Documents in Law, History and Diplomacy, Yale Law School, Lillian Goldman Law Library, accessed March 17, 2013, http://avalon.law.yale.edu/20th_century/ hague04.asp.

17. Interview with Ambassador Robert Oakley, June 25, 2008, in Richard J. Chasdi, *An Analysis of Counterterror Practice Failure: The Case of the Fadlallah Assassination Attempt,* 11, Center for Peace and Conflict Studies and the Center of Academic Excellence in National Security Intelligence Studies, Wayne State University, Detroit Michigan, accessed March 15, 2013, http://www.roa.org/site/DocServer/Chasdi_-_ An_Analysis_of_Counterterror_Practice_Failure_-.pdf?docID=29561.

18. Executive Order 12958, February 7, 2002, Torturing Democracy, National Security Achive, http://www.torturingdemocracy.org/documents/20020207-2.pdf; Erin Chlopak, "Dealing with the Detainees at Guantanamo Bay: Humanitarian and Human Rights Obligations under the Geneva Conventions," *Human Rights Brief* 9, no. 3 (2002): 1–13, Washington College of Law, accessed March 18, 2013, http://www. wcl.american.edu/hrbrief/09/3guantanamo.cfm.

19. President Barack Obama, "A Just and Lasting Peace," Nobel Peace Prize Address, Oslo, Norway, December 10, 2009, accessed March 17, 2013, http://www.nobelprize. org/nobel_prizes/peace/laureates/2009/obama-lecture.html.

20. Bob Woodward and Charles R. Babcock, "Anti-terrorist Plan Rescinded after Unauthorized Bombing," *Washington Post,* May 12, 1985; Stuart Taylor, "Lebanese Group Linked to CIA Is Tied to Car Bombing Fatal to 80," *New York Times,* May 13, 1985, accessed March 17, 2013, http://www.nytimes.com/1985/05/13/world/ lebanese-group-linked-to-cia-is-tied-to-car-bombing-fatal-to-80.html.

21. "Gilani to Codel Snowe: Help Us Hit Targets," U.S. State Department cable, Islamabad Embassy, February 18, 2008, accessed December 6, 2010, http://wikileaks. sbr.im/cable/2008/11/08ISLAMABAD3586.html.

22. Director of Central Intelligence George J. Tenet, "Remarks before the National Commission on Terrorist Attacks upon the United States," April 14, 2004, CIA Speeches and Testimony Archive, https://www.cia.gov/news-information/speeches-testimony/2004/tenet_testimony_04142004.html; Spencer Ackerman, "How the CIA Became 'One Hell of a Killing Machine,'" *Wired,* September 2, 2011, http://www. wired.com/dangerroom/2011/09/cia-killing-machine/. Both accessed March 18, 2013. For more on the CIA's use of foreign agents to guide drone strikes, see chapter 5.

23. Craig Whitlock and Greg Miller, "U.S. Covert Paramilitary Presence in Afghanistan Much Larger Than Thought," *Washington Post,* September 22, 2010, http://www.washingtonpost.com/wp-dyn/content/article/2010/09/22/ AR2010092206241.html?sid=ST2010092106707; Spencer Ackerman, "CIA Snitches Are Pakistan Drone-Spotters," *Wired,* September 23, 2010, http://www.wired.com/ dangerroom/2010/09/cia-snitches-are-pakistan-drone-spotters/. Both accessed March 18, 2013.

24. Abu Yahya al-Libi, "Guide to the Law Regarding Muslim Spies," June 2009, http://xa.yimg.com/kq/groups/19377022/1660762922/name/%5Bthe+teacher+i n+the+government+spy+Muslim+-+Sheikh+Abu+Yahya+al-Libi%5D+06.30.09+ Archangel.pdf, accessed March 18, 2013, translated in Report 2438 (July 9, 2009), Middle East Research Institute, MEMRI, Jihad and Terrorism Threat Monitor.

25. "Al-Qaeda Confirms Death of Bin Laden Confidant Libi," *Reuters,* September 11, 2012, http://www.reuters.com/article/2012/09/11/us-security-qaeda- idUSBRE88A04L20120911; Robert Mackey, "Qaeda Leader Confirms Death of His Deputy," *New York Times,* September 11, 2012, http://thelede.blogs.nytimes. com/2012/09/11/qaeda-leader-confirms-death-of-his-deputy/. Both accessed March 18, 2013. See chapter 5 for further details on the use of Afghan agents.

26. Coll, *Ghost Wars,* 126–127; Robert Gates, *From the Shadows: The Ultimate Insider's Story of Five Presidents and How They Won the Cold War* (New York: Simon and Schuster, 1996), 349; Kirsten Lundberg, "Politics of a Covert Action: The U.S., the *Mujahideen,* and the Stinger Missile," Harvard Case Study 25 (1999): 1–25.

27. Ayman al-Zawahiri, *Knights under the Prophet's Banner,* translated by Laura Mansfield, in *His Own Words: A Translation of the Writings of Dr Ayman al Zawahiri* (New York: TLG, 2006), 38.

28. Charles G. Cogan, "Shawl of Lead," *Conflict* 10, no. 3 (1990): 189–204.

29. Interview with Hamid Gailani, former chairman of the Mujahideen Group, May 14, 2002, Kabul, Afghanistan, in Coll, *Ghost Wars,* 183.

30. Coll, *Ghost Wars,* 290; William Maley, *The Afghanistan Wars* (New York: Palgrave Macmillan, 2009), 218–250.

31. Chidanand Rajghatta, "Rana, Headley implicate Pak ISI in Mumbai attack during ISI chief's visit to US," *Times of India,* April 12, 2011, http://articles.timesofindia. indiatimes.com/2011-04-12/us/29409412_1_rana-and-headley-isi-tahawwur-hussain- rana; Bill Roggio, "Even More Shocking: ISI Behind Mumbai Massacre," Long War Journal, April 12, 2011, http://www.longwarjournal.org/threat-matrix/ archives/2011/04/even_more_shocking_isi_behind.php. Both accessed March 18, 2013.

32. Graham T. Allison, *Essence of Decision: Explaining the Cuban Missile Crisis* (New York: Harper Collins 1971), 83, 162–181; Robert J. Art, "Bureaucratic Politics and American Foreign Policy: A Critique," *Policy Sciences* 4 (1973): 467–487.

33. Robert Kupperman and Jeff Kamen, *Final Warning: Averting Disaster in the New Age of Terrorism* (New York: Doubleday, 1989), 121–122; Paul R. Pillar, *Terrorism and U.S. Foreign Policy* (Washington, D.C.: Brookings Institution Press, 2001), 28;

Martha Crenshaw, "Terrorism, Strategies and Grand Strategies," in Russell D. Howard, Reid L. Sawyer, and Natasha E. Bajema (eds.), *Terrorism and Counterterrorism: Understanding the New Security Environment, Readings and Interpretations* (New York: McGraw Hill, 2004), 444–459; Boaz Ganor, *The Counter-Terrorism Puzzle* (New Brunswick: Transaction Books, 2005), 41; Roger N. McDermott, *Countering Global Terrorism: Developing The Anti-Terrorist Capabilities of Central Asian Militaries* (Carlisle: Strategic Studies Institute, U.S. Army War College, 2004), 7–8; Ian O. Lesser, "Countering the New Terrorism: Implications for Strategy," in *Countering the New Terrorism* (Santa Monica: Rand Corporation, 1999), 126; J. Paul de Taillon, *Hijacking and Hostages: Government Response to Terrorism* (Westport: Praeger, 2002), 78.

34. "Centers in the CIA," cia.gov, accessed May 28, 2012, https://www.cia.gov/library/publications/additional-publications/the-work-of-a-nation/cia-director-and-principles/centers-in-the-cia.html. For further details of the founding of the CTC, see Chapter 3.

35. Robert M. Chesney, "Beyond the Battlefield, Beyond al-Qaeda: The Destabilizing Legal Architecture of Counterterrorism," *Michigan Law Review* 11, no. 163 (November 2013): 163.

36. Chasdi, *An Analysis of Counterterror Practice Failure*, 40.

37. William E. Smith, "Terror Aboard Flight 847," *Time*, June 24, 2001, accessed March 18, 2013, http://www.time.com/time/magazine/article/0,9171,142099,00.html; Wills, *First War on Terrorism*, 91.

38. Faisal Shahzad, quoted in Lorraine Adams with Ayesha Nasir, "Inside the Mind of the Times Square Bomber," *Observer*, September 19, 2010, accessed March 19, 2013, http://www.guardian.co.uk/world/2010/sep/19/times-square-bomber.

39. Ibid.

40. Interview with Hafiz Hanif (alias), in Sami Yousafzai and Ron Moreau, "Inside al-Qaeda," *Newsweek*, September 4, 2010, accessed March 19, 2013, http://www.thedailybeast.com/newsweek/2010/09/04/inside-al-qaeda.html.

41. Interview with Hafiz Hanif (alias), in Sami Yousafzai and Ron Moreau, "Al-Qaeda on the Ropes: One Fighter's Inside Story," *Newsweek*, January 2, 2012, accessed March 19, 2013, http://www.thedailybeast.com/newsweek/2012/01/01/al-qaeda-on-the-ropes-one-fighter-s-inside-story.html.

42. Michael K. Bohn, *The* Achille Lauro *Hijacking: Lessons in the Politics and Prejudice of Terrorism* (Washington, D.C.: Potomac Books, 2004), 1–19.

43. "United Nations Security Council Resolution 573," October 4, 1985, UN Security Council Achieve, http://www.un.org/ga/search/view_doc.asp?symbol=S/RES/573(1985); "Operation 'Wooden Leg,'" Israeli Air Force, http://www.iaf.org.il/4694-33087-en/IAF.aspx; William E. Smith, "Israel's 1,500 Mile Raid," *Time*, October 14, 1985, accessed February 24, 2013, http://www.time.com/time/magazine/article/0,9171,960108-1,00.html.

44. "Gunmen Kill Three Israelis Aboard Yacht: Palestinian Trio Surrender After Shootings in Cyprus," *Los Angeles Times*, September 25, 1985, accessed February 24, 2013, http://articles.latimes.com/1985-09-25/news/mn-19794_1_israelis-aboard-yacht.

45. Simon Jenkins, "Drones Are a Fool's Gold: They Prolong Wars We Cannot Win," *Guardian,* January 10, 2013, http://www.guardian.co.uk/commentisfree/2013/jan/10/drones-fools-gold-prolong-wars; Shamim-uh-Rahman, "Drones Fuel Terrorism, Says Imran," *Dawn,* May 22, 2011, http://dawn.com/2011/05/22/drones-fuel-terrorism-says-imran/. All accessed February 24, 2013.

46. "America Reacts with Rejoicing: Capture Proves U.S. Still Strong," *Washington Post,* October 13, 1985, accessed September 24, 2013, http://news.google.com/newspapers?nid=1948&dat=19851013&id=-6lJAAAAIBAJ&sjid=8BwNAAAAIBAJ&pg=1780,734973.

47. Ronald Reagan, *An American Life* (New York: Pocket Books, 1990), 509.

48. Timothy Naftali, *Blind Spot: The Secret History of American Counterterrorism* (New York: Basic Books, 2005), 171–174.

49. Ibid., 173–174.

50. Deborah H. Strober and Gerald S. Strober, *Reagan: The Man and His Presidency, The Oral History of an Era* (New York: Houghton Mifflin, 1998), 376–379; Howard Teicher and Gayle Radley Teicher, *Twin Pillars to Desert Storm: America's Flawed Vision in the Middle East from Nixon to Bush* (New York: William Morrow, 1993), 339; David C. Wills, *The First War on Terrorism: Counterterrorism Policy during the Reagan Administration* (Lanham, Md.: Rowman and Littlefield, 2003), 151–155.

51. Wills, *First War on Terrorism,* 153.

52. John F. Lehman Jr., *Command of the Seas: Building the 600 Ship Navy* (New York: Charles Scribner's Sons, 1988), 365.

53. Bernard Gwertzman, "U.S. Intercepts Jet Carrying Hijackers; Fighters Divert It to NATO Base in Italy; Gunmen Face Trial for Slaying of Hostage," *New York Times,* October 11, 1985, accessed March 3, 2013, http://www.nytimes.com/1985/10/11/world/us-intercepts-jet-carrying-hijackers-fighters-divert-it-nato-base-italy-gunmen.html.

54. Wills, *First War on Terrorism,* 161; David Ensor, "U.S. Captures Mastermind of Achille Lauro Hijacking," CNN, April 16, 2003, http://edition.cnn.com/2003/WORLD/meast/04/15/sprj.irq.abbas.arrested/; "Terrorist Abu Abbas Dies in Iraq," *Fox News,* March 9, 2004, http://www.foxnews.com/story/0,2933,113719,00.html. Both accessed March 3, 2013.

55. Lawrence Joffe, "Abu Abbas Obituary," *Guardian,* March 11, 2004, accessed March 3, 2013, http://www.theguardian.com/news/2004/mar/11/guardianobituaries.israel.

56. Andrew L. Liput, "An Analysis of the *Achille Lauro* Affair: Towards an Effective and Legal Method of Bringing International Terrorists to Justice," *Fordham International Law Journal* 9, no. 2 (1985): 328–372. Also see Antonio Cassese, *Terrorism, Politics and Law: The* Achille Lauro *Affair* (Princeton: Princeton University Press, 1989); John M. Rogers, "Prosecuting Terrorists: When Does Apprehension in Violation of International Law Preclude Trial?" *University of Miami Law Review* (1987–1988): 447–465.

57. Liput, "An Analysis of the *Achille Lauro* Affair," 365.

58. "Bilateral Extradition Treaty, Egypt and the United States," August 11, 1874, International Extradition Lawyers, http://internationalextraditionblog.files.wordpress.com/2011/03/egypt.pdf; International Convention against the Taking of Hostages, United Nations, New York, December 17, 1979, http://www.unodc.org/documents/treaties/Special/1979%20International%20Convention%20against%20the%20Taking%20of%20Hostages.pdf. Both accessed March 3, 2013.

59. Emmanuel Sivan, *Radical Islam: Medieval Theology and Modern Politics* (New York: Yale University Press, 1990), 101–102; Devin R. Springer, James L. Regens, and David N. Edgar, *Islamic Radicalism and Global Jihad* (Washington, D.C.: Georgetown University Press, 2009), 31, "Muhammad Rashid Rida," in *Encyclopaedia of the Middle East,* accessed March 3, 2013, http://www.mideastweb.org/Middle-East-Encyclopedia/muhammad_rashid_rida.htm; Albert J. Bergesen (ed.), *The Sayyid Qutb Reader: Selected Writing on Politics, Religion and Society* (London: Routledge, 2007), 3–31; Springer, Regens, and Edgar, *Islamic Radicalism and Global Jihad,* 33–37; Lawrence Wright, *The Looming Tower: Al-Qaeda's Road to 9/11* (New York: Penguin, 2006), 7–31; Giles Kepel, *The Prophet and the Pharaoh: Muslim Extremism in Egypt* (California: University of California Press, 2003); Bruce Riedel, *The Search for Al-Qaeda: Its Leadership, Ideology and Future* (Washington, D.C.: Brookings Institution Press, 2008), 14–36.

60. Liput, *An Analysis of the* Achille Lauro *Affair,* 362; Naftali, *Blind Spot,* 173.

61. "United Nations Security Council Resolution 1368," September 12, 2001, UN.org, accessed November 15, 2012, http://daccess-dds-ny.un.org/doc/UNDOC/GEN/N01/533/82/PDF/N0153382.pdf?OpenElement.

62. "The Escalating Casualties in Pakistan, 2005–2010," Costs of War.org, Watson Institute for International Studies, Brown University, http://costsofwar.org/article/pakistani-civilians; Neta C. Crawford, "War Related Death and Injury in Pakistan, 2004–2011," Boston University, September 9, 2011, http://costsofwar.org/sites/default/files/articles/16/attachments/Crawford%20Pakistan%20Casualties.pdf. Both accessed March 6, 2013.

63. Syed Saleem Shahzad, "Al-Qaeda Claims Bhutto killing," December 28, 2007, World Security Network, accessed March 6, 2013, http://www.worldsecuritynetwork.com/Terrorism/Shahzad-Syed-Saleem/Al-Qaeda-claims-Bhutto-killing.

64. "Zadari Urges U.S. to Stop Drone Attacks," *Dawn,* March 25, 2011, http://www.dawn.com/2011/03/24/us-should-stop-drone-attacks-says-zardari.html; Nick Paton Walsh and Nasir Habib, "Pakistani Leaders Condemn Suspected U.S. Drone Strike," CNN, http://articles.cnn.com/2011-03-18/world/pakistan.drone.strike_1_drone-strike-intelligence-officials-volatile-tribal-region?_s=PM:WORLD; Qasim Nauman, "Pakistan Condemns U.S. Drone Strikes," Reuters, June 4, 2012, http://www.reuters.com/article/2012/06/04/us-pakistan-usa-drones-idUSBRE8530MS20120604. All accessed March 11, 2013.

65. Ben Emmerson, "Pakistan: Statement by the U.N. Special Rapporteur on Human Rights and Counterterrorism," March 15, 2013, United Nations Human Rights, accessed March 17, 2013, http://www.ohchr.org/EN/NewsEvents/Pages/DisplayNews.aspx?NewsID=13148&LangID=E.

66. Jeremy Page, "Google Earth Reveals Secret History of US Base in Pakistan," *Times,* February 19, 2009, http://www.thetimes.co.uk/tto/news/world/asia/article2609737.ece; "Admiral Fallon Discusses Security Cooperation with General Kayani," U.S. State Department cable, Islamabad Embassy, February 11, 2008, accessed March 11, 2013, http://wikileaks.org/cable/2008/02/08ISLAMABAD609.html.

67. Steve Kroft, interview with former prime minister Pervez Musharraf, "In the Line of Fire," 60 Minutes, CBS, September 25, 2006, accessed March 11, 2013, http://www.cbsnews.com/video/watch/?id=2036261n&tag—ncol;lst;1.

68. U.N. Security Council Resolution 748, March 31, 1992, accessed March 11, 2013, http://www.un.org/ga/search/view_doc.asp?symbol=S/RES/748 (1992).

69. Bruce Riedel, *Deadly Embrace: Pakistan, America, and the Future of the Global Jihad* (New York: Brookings Institution Press, 2012), 1–16; "Statement of Admiral Mullen, Chairman of Joint Chiefs of Staff before Senate Armed Services Committee," September 22, 2011, http://armed-services.senate.gov/statemnt/2011/09%20September/Mullen%2009-22-11.pdf; Elisabeth Bumiller and Jane Perlez, "Pakistan's Spy Agency Is Tied to Attack on U.S. Embassy," *New York Times,* September 22, 2011, http://www.nytimes.com/2011/09/23/world/asia/mullen-asserts-pakistani-role-in-attack-on-us-embassy.html?pagewanted=all. All accessed March 6, 2013.

70. Philip Alston, "Report of the Special Rapporteur on Extrajudicial, Summary or Arbitrary Executions," United Nations, May 28, 2010, 11–12, accessed March 11, 2013, http://www2.ohchr.org/english/bodies/hrcouncil/docs/14session/A.HRC.14.24.Add6.pdf.

71. See Peter Bergen, *Manhunt: From 9/11 to Abbottabad—The Ten-Year Search for Osama bin Laden* (London: Bodley Head, 2012), and Mark Bowden, *The Finish: The Killing of Osama bin Laden* (New York: Grove Atlantic, 2012), for details on how the CIA located bin Laden.

72. Wills, *First War on Terrorism,* 155.

73. Ibid., 159; Michael Ross, "Mubarak Accuses U.S. of 'Act of Piracy': Will Strain Relations for 'Long Time to Come,' Egypt Leader Says," *Los Angeles Times,* October 13, 1985, accessed March 11, 2013, http://articles.latimes.com/1985-10-13/news/mn-15798_1_united-states.

74. "Terrorism Review," Directorate of Intelligence, CIA, October 21, 1985, CIA Library Electronic Reading Room.

75. Woodward, *Veil,* 418; "Near East and South Asian Review," Directorate of Intelligence, CIA, October 25, 1985, quoted in Wills, *First War on Terrorism,* 255.

76. David C. Martin and John Walcott, *Best Laid Plans: The Inside Story of America's War against Terrorism* (New York: Simon and Schuster, 1989), 253.

77. "Hostages and Hijackers: An Affront to Egypt's Pride; Text of U.S. Statement," *New York Times,* October 14, 1985, http://www.nytimes.com/1985/10/14/world/hostages-and-hijackers-an-affront-to-egypt-s-pride-text-of-us-statement.html; presidential spokesman Larry Speakes, quoted in "Yugoslavia refuses to hold suspect, U.S. says," *Pittsburgh Post-Gazette,* October 14, 1985, http://news.google.com/newspap

ers?nid=1144&dat=19851014&id=1-wdAAAAIBAJ&sjid=jmIEAAAAIBA
J&pg=4865,9442390 Both accessed March 12, 2013.

78. Martin and Walcott, *Best Laid Plans,* 253; Wills, *First War on Terrorism,* 160–161.

79. Naftali, *Blind Spot,* 174; Woodward, *Veil,* 417; Deborah Hart Strober and Gerald S. Strober, *The Reagan Presidency: An Oral History of the Era,* Revised Edition (Washington, D.C.: Brassey's, 2003), 380.

80. *Washington Post/ABC News* Poll, February 1–4, 2012, *Washington Post,* accessed March 12, 2013, http://www.washingtonpost.com/wp-srv/politics/polls/postabc-poll_020412.html.

81. "Global Opinion of Obama Slips, International Policies Faulted," Pew Research, Global Attitudes Project, June 13, 2012, accessed March 12, 2013, http://www.pewglobal.org/2012/06/13/global-opinion-of-obama-slips-international-policies-faulted/.

82. Jamie Fuller, "Americans Are Fine with Drone Strikes. Everyone Else in the World? Not so Much," *Washington Post,* July 15, 2014, accessed November 11, 2015, https://www.washingtonpost.com/news/the-fix/wp/2014/07/15/americans-are-fine-with-drone-strikes-everyone-else-in-the-world-not-so-much/.

83. "Pakistani Public Opinion on the Swat Conflict, Afghanistan, and the U.S.," World Public Opinion.org, University of Maryland, July 1, 2009, accessed March 12, 2013, http://www.worldpublicopinion.org/pipa/articles/brasiapacificra/619.php.

84. "Pakistani Public Opinion Ever More Critical of U.S.," Pew Research Global Attitudes Project, June 27, 2012, accessed March 12, 2013, http://www.pewglobal.org/2012/06/27/pakistani-public-opinion-ever-more-critical-of-u-s/.

85. Reza Sayah, "Anti-Drone Peace March Halted in Pakistan," CNN, October 8, 2012, http://edition.cnn.com/2012/10/07/world/asia/pakistan-us-drone-protest; Murtaza Ali Shah, "Drones Radicalising Western Muslims, says Imran Khan," *The News,* November 22, 2012, accessed March 12, 2013, http://www.thenews.com.pk/Todays-News-2-144356-Drones-radicalising-western-Muslims,-says-Imran-Khan.

86. Pew Research, Global Attitudes Project, June 13, 2012. See chapter 6 for more details on Pakistani attitudes toward drones.

87. Ibid.

88. Brendon O'Conner, "The Anti-American Tradition: A History in Four Phases," in Brendon O' Conner and Martin Griffiths (eds.), *The Rise of Anti-Americanism* (Abingdon: Routledge, 2006), 11–24.

89. Chris Cillizza, "Barack Obama 2012 = Ronald Reagan 1984?" *Washington Post,* October 5, 2012, http://www.washingtonpost.com/blogs/the-fix/wp/2012/10/05/barack-obama-2012-ronald-reagan-1984/; Joshua Greenman, "Election Day 2012: President Obama Is the New Ronald Reagan," *New York Daily News,* November 7, 2012. Both accessed April 2, 2013.

90. Michael Duffy, "The Role Model: What Obama Sees in Reagan," *Time,* January 27, 2011, accessed April 2, 2013, http://www.time.com/time/magazine/article/0,9171,2044712,00.html.

91. Senator Barack Obama interview with *Reno-Gazette-Journal*'s Editorial Board, January 16, 2008, in Ben Smith, "Transformation Like Reagan," *Politico,* accessed April 2, 2013, http://www.politico.com/blogs/bensmith/0108/Transformation_like_Reagan.html.

92. Wills, *First War on Terrorism,* 155.

3. "We Have to Find a Better Way to Send a Message"

Epigraph. Duane R. Clarridge with Digby Diehl, *A Spy for All Seasons: My Life in the CIA* (New York: Charles Scribner's Sons, 1997), 339.

1. David C. Wills, *The First War on Terrorism: Counter-Terrorism Policy during the Reagan Administration* (Lanham, Md.: Rowman and Littlefield, 2003), 87.

2. Robert Timberg, *The Nightingale's Song* (New York: Simon and Schuster, 1995), 337.

3. Clarridge with Diehl, *A Spy for All Seasons: My Life in the CIA* (New York: Charles Scribner's Sons, 1997), 320; Steve Coll, *Ghost Wars: The Secret History of the CIA, Afghanistan and bin Laden, from the Soviet Invasion to September 10, 2001* (New York: Penguin Books, 2005), 139.

4. Malcolm Byrne, *Iran-Contra: Reagan's Scandal and the Unchecked Abuse of Presidential Power* (Kansas: University Press of Kansas, 2014), 40.

5. Timothy Naftali, *Blind Spot: The Secret History of American Counterterrorism* (New York: Basic Books, 2005), 181.

6. Ronald Kessler, *The CIA at War: Inside the Secret Campaign against Terror* (New York: St. Martin's Griffin, 2003), 119–120; Joseph E. Persico, *Casey: The Lives and Secrets of William J. Casey: From the OSS to the CIA* (New York: Penguin Books, 1990), 265–266.

7. Rhodri Jeffreys-Jones, *The CIA and American Democracy* (New Haven: Yale University Press, 2003), 239.

8. John Prados, *Presidents' Secret Wars: CIA and Pentagon Covert Operations from World War II Through the Persian Gulf* (Chicago: Elephant Paperbacks, 1996), 418.

9. "Psychological Operations in Guerilla Warfare," Central Intelligence Agency, Federation of American Scientists, Intelligence Review Program, accessed March 19, 2013, http://www.fas.org/irp/cia/guerilla.htm.

10. See chapter 1 for more detail on the counterterrorism hardliners and the introduction of NSDD 138.

11. Loch K. Johnson, *National Security Intelligence* (Malden, Mass.: Polity Press, 2012), 96.

12. Persico, *Casey,* 290.

13. Loch K. Johnson, *A Season of Inquiry: Congress and Intelligence* (Belmont: Dorsey Press, 1988), 263.

14. Ibid., 428–429.

15. Clarridge with Diehl, *A Spy for All Seasons,* 320; Coll, *Ghost Wars,* 139.

16. Coll, *Ghost Wars,* 140.

17. Clarridge with Diehl, *A Spy for All Seasons,* 321.

18. Ibid., 321; Naftali, *Blind Spot,* 181.

19. Coll, *Ghost Wars,* 40.

20. "The National Program for Combatting Terrorism," National Security Decision Directive 207, National Security Achieve, accessed May 21, 2012, http://www.gwu.edu/~nsarchiv/NSAEBB/NSAEBB55/nsdd207.pdf.

21. NSDD 207, 1–2.

22. Coll, *Ghost Wars,* 139–140.

23. Clarridge with Diehl, 321.

24. Jason Burke, *Al-Qaeda: The True Story of Radical Islam* (London: Penguin Books, 2003), 1–22; Christina Hellmich, *Al-Qaeda: From Global Network to Local Franchise* (Black Point, Nova Scotia: Fernwood, 2011), 21–60.

25. John Prados, *Safe for Democracy: The Secret Wars of the CIA* (Chicago: Ivan R. Dee, 2006), 576.

26. Clarridge with Diehl, *A Spy for All Seasons,* 323–329; "Centers in the CIA," cia. gov, accessed May 28, 2012, https://www.cia.gov/library/publications/additional-publications/the-work-of-a-nation/cia-director-and-principles/centers-in-the-cia.html.

27. Robert Gates, *From the Shadows: The Ultimate Insider's Story of Five Presidents and How They Won the Cold War* (New York: Simon and Schuster, 1996), 293–294.

28. Naftali, *Blind Spot,* 182.

29. Gary Berntsen, *Jawbreaker: The Attack on bin Laden and al-Qaeda* (New York: Three Rivers Press, 2005), 75; Gary C. Schroen, *First In: An Insider's Account of How the CIA Spearheaded the War on Terror in Afghanistan* (New York: Ballantine Books, 2007), 21.

30. Greg Miller and Julie Tate, "CIA Shifts Focus to Killing Targets," *Washington Post,* September 1, 2011, accessed October 28, 2014, http://www.washingtonpost.com/world/national-security/cia-shifts-focus-to-killing-targets/2011/08/30/gIQA7MZGvJ_story.html.

31. Nick Turse and Tom Engelhardt, *Terminator Planet: The First History of Drone Warfare, 2001–2050* (London: Dispatch Books, 2013), 71–74.

32. Robert F. Worth, Mark Mazzetti and Scott Shane, " 'Drone Strikes' Risks to Get Rare Moment in the Public Eye," *New York Times,* February 5, 2013, accessed March 20, 2013, http://www.nytimes.com/2013/02/06/world/middleeast/with-brennan-pick-a-light-on-drone-strikes-hazards.html?_r=0.

33. NSDD 207, 2–3.

34. Clarridge with Diehl, *A Spy for All Seasons,* 322.

35. Ibid., 322.

36. Miller and Tate, "CIA Shifts Focus to Killing Targets.".

37. Ibid.

38. See Dino A. Brugioni, *Eyes in the Sky: Eisenhower, the CIA and Cold War Aerial Espionage* (Annapolis: Naval Institute Press, 2010), 90–171, for background on the development of the U-2; 361–386 for details on the Corona satellite program. See Richard H. Graham, *SR-71 Revealed: The Inside Story* (Minneapolis: Zenith Press, 1996), 41–44, for details on the development of the SR-71.

39. Clarridge with Diehl, *A Spy for All Seasons,* 323.

40. Jeffery T. Richelson, "The Wizards of Langley: The CIA's Directorate of Science and Technology," in Rhodri Jeffreys-Jones and Christopher Andrew (eds.), *Eternal Vigilance? 50 Years of the CIA* (London: Frank Cass, 1997), 82.

41. "Libya: Reviewing Terrorist Capabilities," CIA, April 1989, CIA Electronic Reading Room, accessed May 28, 2012, http://www.foia.cia.gov/docs/DOC_0000259675/DOC_0000259675.pdf.

42. President Ronald Reagan, "Remarks at the Annual Convention of the American Bar Association," July 8, 1985, Reagan Presidential Library Speeches Archive, accessed March 21, 2013, http://www.reagan.utexas.edu/archives/speeches/1985/70885a.htm.

43. John Prados, *Keepers of the Keys: A History of the National Security Council from Truman to Bush* (New York: William Morrow, 1991), 503–504; Timberg, *The Nightingale's Song,* 375; Wills, *First War on Terrorism,* 172–173.

44. Prados, *Keepers of the Keys,* 505–506.

45. Christopher Andrew, *For the President's Eyes Only: Secret Intelligence and the American Presidency from Washington to Bush* (London: HarperCollins, 1995), 483; Caspar Weinberger, *Fighting for Peace: Seven Critical Years at the Pentagon* (London: Michael Joseph, 1990), 132–133.

46. Ronald Reagan, *An American Life* (New York: Pocket Books, 1990), 518.

47. George P. Shultz, *Turmoil and Triumph: My Years as Secretary of State* (New York: Charles Scribner's Sons, 1993) 685–687; Howard Teicher and Gayle Radley Teicher, *Twin Pillars to Desert Storm: America's Flawed Vision in the Middle East from Nixon to Bush* (New York: William Morrow, 1993), 349–350.

48. President Ronald Reagan, "Address to the Nation on the United States Air Strike against Libya," Oval Office, Washington, D.C., April 14, 1986, Reagan Presidential Library Speeches Archive, accessed March 21, 2013, http://www.reagan.utexas.edu/archives/speeches/1986/41486g.htm.

49. Commander Mark E. Kosnik, "An Analysis of the United States Use of Force against Terrorism," Weatherhead Center for International Affairs, Harvard University, May 1999, accessed March 21, 2013, http://programs.wcfia.harvard.edu/files/fellows/files/kosnik.pdf, 10–26.

50. Brian L. Davis, *Qaddafi, Terrorism, and the Origins of the U.S. Attack on Libya* (New York: Praeger, 1990), 169.

51. Reagan, *An American Life,* 520; Seymour Hersh, "Target Qaddafi," *New York Times,* February 22, 1987, accessed 21 March, 2013, http://www.nytimes.com/1987/02/22/magazine/target-qaddafi.html?pagewanted=all&src=pm.

52. Teicher and Teicher, *Twin Pillars to Desert Storm,* 350.

53. CIA, "Libya: Reviewing Terrorist Capabilities."

54. Hersh, "Target Qaddafi."

55. Reagan, *An American Life,* 518; Weinberger, *Fighting for Peace,* 135.

56. Hersh, "Target Qaddafi;" Weinberger, *Fighting for Peace,* 139.

57. Damien McElroy, "Col Muammar Gaddafi's Daughter Hana 'Still Alive,'" *Telegraph,* August 10, 2011, accessed July 30, 2015, http://www.telegraph.co.uk/news/

worldnews/africaandindianocean/libya/8693628/Col-Muammar-Gaddafis-daughter-Hana-still-alive.html.

58. Andrew, *For the President's Eyes Only,* 484; Prados, *Keepers of the Keys,* 506; Teicher and Teicher, *Twin Pillars to Desert Storm,* 350; Weinberger, *Fighting for Peace,* 138; "Hana Gaddafi, Libyan Leader's Presumed Dead Daughter, May Be Still Alive," *Huffington Post,* August 9, 2011, accessed March 21, 2013, http://www.huffingtonpost.com/2011/08/09/hana-gaddafi-libyan-leader-daughter-alive-_n_922043.html.

59. Kosnik, "An Analysis of the United States Use of Force against Terrorism," 10–26.

60. Ramsey Clark, quoted in Neil C. Livingstone, "The Raid on Libya and the Use of Force in Combating Terrorism," in Neil C. Livingstone and Terrell E. Arnold (eds.), *Beyond the Iran-Contra Crisis: The Shape of U.S. Anti-terrorism Policy in the Post-Reagan Era* (Washington, D.C.: Lexington Books, 1988), 74.

61. Marion Dean, "Barack Obama Must Be Impeached over Drone Strike Issue, Demands Code Pink," *Global News Desk,* February 12, 2013, accessed March 21, 2013, http://www.globalnewsdesk.co.uk/north-america/barack-obama-impeached-us-drone-strikes/03305/.

62. Ben Bradlee Jr., *Guts and Glory: The Rise and Fall of Oliver North* (New York: Donald I. Fine, 1988), 353–354.

63. Ibid., 354; Wills, *First War on Terrorism,* 200.

64. See chapter 5 for details on Clinton's use of cruise missiles.

65. Clarridge with Diehl, *A Spy for All Seasons,* 339.

66. Ibid., 339.

67. Steve Coll interview with Duane Clarridge, December 28, 2001, in *Ghost Wars,* 143–144.

68. "Initial DCI-Hostage Location Task Force Report," CIA, January 3, 1986, CIA Electronic Reading Room, accessed May 29, 2012, http://www.foia.cia.gov/docs/DOC_0000139471/DOC_0000139471.pdf.

69. See chapter 1 for details on Operation Eagle Claw.

70. Coll interview with Duane Clarridge, December 28, 2001, in *Ghost Wars,* 143.

71. Director of Central Intelligence William Casey, "Address before the Jewish American Committee," Thursday May 15, 1986, Marriott Hotel, Washington, D.C., CIA Electronic Reading Room, accessed May 23, 2012, http://www.foia.cia.gov/docs/DOC_0000247983/DOC_0000247983.pdf.

72. Coll, *Ghost Wars,* 143–144; Clarridge with Diehl, *A Spy for All Seasons,* 339–340.

73. Richard A. Clarke, *Against All Enemies: Inside America's War on Terror* (London: Free Press, 2004), 181–184, 220; "1998 Embassy Bombings," Global Security.org, accessed March 22, 2013, http://www.globalsecurity.org/security/ops/98emb.htm.

74. *The 9/11 Commission Report: Final Report of the National Commission on Terrorist Attacks upon the United States* (New York: W. W. Norton, 2004), 190. Also available online, accessed March 22, 2013, http://www.9-11commission.gov/report/911Report_Ch6.pdf.

75. *9/11 Commission Report,* 190; Clarke, *Against All Enemies,* 221; George Tenet, *At the Center of the Storm: My Years at the CIA* (New York: Harper Luxe, 2007), 194.

76. Clarke, *Against All Enemies,* 220.

77. Clarridge with Diehl, *A Spy for All Seasons,* 339.

78. "U.S. Directory of U.S. Rockets and Missiles, Appendix 4, Undesignated Vehicles—Amber," accessed May 25, 2012, http://www.designation-systems.net/dusrm/app4/amber.html; Curtiss Peebles, *Dark Eagles: A History of Top Secret U.S. Aircraft Programs* (New York: Simon and Schuster, 1995), 207–209.

79. Peebles, *Dark Eagles,* 207; Richard Whittle, *Predator: The Secret Origins of the Drone Revolution* (New York: Henry Holt, 2014), 50.

80. "About DARPA," DARPA.mil, accessed July 9, 2015. http://www.darpa.mil/about-us/about-darpa.

81. Whittle, *Predator,* 51–53.

82. In response to a FOIA request regarding the Eagle Program, the CIA stated: "The fact of the existence or nonexistence of records is . . . protected from disclosure by Section 6 of the CIA Act of 1949 . . . and section 102(A)(i)(1) of the National Security Act of 1947." Michele Meeks, CIA Information and Privacy Coordinator, May 3, 2013; Response to FOIA Request 13-F-0623, Don Nichelson, Office of Freedom of Information, Department of Defense, Pentagon, Washington, D.C., February 1, 2014.

83. "Aquila Remotely Piloted Vehicle: Its Potential Battlefield Contribution Still in Doubt," United States General Accounting Office, October 1987, GAO.gov, accessed July 9, 2015, http://www.gao.gov/assets/210/209657.pdf.

84. John H. Cushman Jr., "Flaws Force New Delay in Army Drone," *New York Times,* October 6, 1987, accessed July 9, 2015, http://www.nytimes.com/1987/10/06/us/flaws-force-new-delay-in-army-drone.html.

85. Peebles, *Dark Eagles,* 208; Whittle, *Predator,* 58.

86. Whittle, *Predator,* 66; Chris Woods, *Sudden Justice: America's Secret Drone Wars* (London: Hurst, 2015), 31.

87. E-mail interview with Kimberly A. Kasitz, public relations and communications manager, General Atomics Aeronautical Systems, Inc., June 7, 2012.

88. Wood, *Sudden Justice,* 29.

89. Charles Duhigg, "The Pilotless Plane That Only Looks Like Child's Play," *New York Times,* April 15, 2007, accessed June 15, 2014, http://www.nytimes.com/2007/04/15/business/yourmoney/15atomics.html?pagewanted=all&_r=0.

90. Ibid.

91. Barney Gimbel, "The Predator: A Profile of Neal Blue," *Fortune,* October 31, 2008, accessed June 15, 2014, http://archive.fortune.com/2008/10/28/magazines/fortune/predator_gimbel.fortune/index.htm.

92. Woods, *Sudden Justice,* 30.

93. Sidney Blumenthal, *The Clinton Wars* (New York: Penguin, 2003), 61–62.

94. Secretary of State Baker, quoted in ibid., 62.

95. Tony Smith, *America's Mission: The United States and the Worldwide Struggle for Democracy* (New Jersey: Princeton University Press, 2012), 349.

96. Zbigniew Brzezinski, *Second Chance: Three Presidents and the Crisis of American Superpower* (New York: Basic Books, 2007), 116–117. The triumphalism of the time is perhaps best represented by Francis Fukuyama's best-selling *End of History and the Last Man* (New York: Penguin, 1993).

97. Richard Holbrooke, *To End a War* (New York: Modern Library, 1999), 102.

98. Whittle, *Predator,* 71.

99. Francis X. Clines, "Conflict in the Balkans: The Rescue; Downed U.S. Pilot Rescued in Bosnia in Daring Raid," *New York Times,* June 9, 1995, accessed July 31, 2015, http://www.nytimes.com/1995/06/09/world/conflict-balkans-rescue-downed-us-pilot-rescued-bosnia-daring-raid.html?pagewanted=all&src=pm.

100. "Final Report of the United Nations Commission of Experts on the Former Yugoslavia," May 27, 1994, 44–46, accessed July 31, 2015, http://www.icty.org/x/file/About/OTP/un_commission_of_experts_report1994_en.pdf.

101. Garry R. Mace, "Dynamic Targeting and the Mobile Missile Threat," Department of Joint Military Operations, U.S. Naval War College, 1999, 4, cited in William Rosenau, *Special Operations Forces and Elusive Enemy Ground Targets: Lessons from Vietnam and the Persian Gulf War* (Santa Monica: Rand, 2001), 33, accessed 20 July, 2015, http://www.rand.org/pubs/monograph_reports/MR1408.html.

102. "Iraq's Scud Ballistic Missiles," Office of the Special Assistant for Gulf War Illnesses, Department of Defense, July 25, 2000, accessed 20 July, 2015, http://www.defense.gov/Releases/Release.aspx?ReleaseID=2604; George N. Lewis, Steve Fetter, and Lisbeth Gronlund, "Casualties and Damage from Scud Attacks in the 1991 Gulf War," Defense and Arms Control Studies Program, MIT Center for International Studies, March 1993, accessed 21 July, 2015, http://web.mit.edu/ssp/publications/working_papers/wp93-2.pdf.

103. Whittle, *Predator,* 69–70.

104. Lou Cannon et al., "Airborne MX Reported Eyed by Weinberger," *Washington Post,* July 16, 1981, accessed July 9, 2015, http://www.washingtonpost.com/archive/politics/1981/07/16/airborne-mx-reported-eyed-by-weinberger/1d368f91-d405-4318-9568-7f39c7148ec0/.

105. Whittle, *Predator,* 70.

106. "General Atomic GNAT-750 Lofty View," Federation of American Scientists Intelligence Resource Program, accessed April 1, 2013, https://www.fas.org/irp/program/collect/gnat-750.htm.

107. Frank Strickland, "The Early Evolution of the Predator Drone," *Studies in Intelligence* 57, no. 1 (March 2013): 1–6, CIA Library, accessed July 14, 2015. https://www.cia.gov/library/center-for-the-study-of-intelligence/csi-publications/csi-studies/studies/vol.-57-no.-1-a/vol.-57-no.-1-a-pdfs/Strickland-Evolution%20of%20the%20Predator.pdf.

108. Woods, *Sudden Justice,* 33.

109. Ian Shaw, "The Rise of the Predator Empire: Tracing the History of U.S. Drones," in *Understanding Empire* (2014), WordPress blog, accessed July 31, 2015, https://understandingempire.wordpress.com/2-0-a-brief-history-of-u-s-drones/.

110. "General Atomics Predator," *Spyflight,* accessed July 31, 2015, http://spyflight.co.uk/Predator.htm.

111. Mark Mazzetti, *The Way of the Knife: The CIA, a Secret Army, and a War at the Ends of the Earth* (New York: Penguin Press, 2013), 90–91; Shaw, "The Rise of the Predator Empire"; Whittle, *Predator,* 81–82.

112. Yenne, *Birds of Prey: Predators, Reapers and America's Newest UAVs in Combat* (North Branch, Minn.: Specialty Press, 2010), 39.

113. "Appendix 4: Undesignated Vehicles—GNAT," *Directory of U.S. Military Rockets and Missiles,* accessed July 13, 2015, http://www.designation-systems.net/dusrm/app4/gnat.html; "Unmanned Aircraft Systems Roadmap 2005–2013," Office of the Secretary of Defense, August 4, 2005, Federation of American Scientists, Intelligence Review Program, accessed July 13, 2015, https://www.fas.org/irp/program/collect/uav_roadmap2005.pdf.

114. Whittle, *Predator,* 78, 85.

115. Shaw, "The Rise of the Predator Empire."

116. Joakim Kasper Oestergaard Balle, "Defense: MQ-1 Predator/MQ-9 Reaper," BGA-Aeroweb, accessed July 31, 2015, http://www.bga-aeroweb.com/Defense/MQ-1-Predator-MQ-9-Reaper.html; "RQ-1 Predator MAE UAV Technical Specifications," in Federation of American Scientists Intelligence Resource Program, accessed July 31, 2015, http://fas.org/irp/program/collect/predator.htm.

117. Linda D. Kozaryn, "Predators Bound for Bosnia," American Forces Press Service, Department of Defense, February 8, 1996, accessed May 13, 2014, http://www.defense.gov/news/newsarticle.aspx?id=40516; Woods, *Sudden Justice,* 34.

118. Kenneth Israel, "UAV Annual Report FY 1996," Defense Airborne Reconnaissance Office, November 6, 1996, declassified in Federation of American Scientists, accessed July 14, 2015, http://fas.org/irp/agency/daro/uav96/page1%262.html.

119. Kozaryn, "Predators Bound for Bosnia."

120. Shaw, "The Rise of the Predator Empire."

121. Kenneth Israel, "UAV Annual Report FY 1996," Defense Airborne Reconnaissance Office, November 6, 1996, declassified in Federation of American Scientists, accessed July 14, 2015, http://fas.org/irp/agency/daro/uav96/page1%262.html.

122. "RQ-1 Predator MAE UAV," Global Security.org, accessed March 22, 2013, http://www.globalsecurity.org/intell/systems/predator.htm; "Unmanned Aircraft Systems Roadmap 2005–2013."

123. "HR 1775 Intelligence Authorization Act for Fiscal Year 1998," House Permanent Select Committee on Intelligence, Title VI, Sec. 603, Congress.gov, accessed July 13, 2015, https://www.congress.gov/bill/105th-congress/house-bill/1775.

124. Bill Grimes, *The History of Big Safari* (Richmond: Archway, 2014), 329.

125. Scott Swanson, "War Is No Video Game—Not Even Remotely," *Breaking Defense,* November 18, 2014, accessed December 20, 2014, http://breakingdefense.com/2014/11/war-is-no-video-game-not-even-remotely/.

126. Whittle, *Predator,* 131.

127. David Cenciotti, " 'Vega 31': The First and Only F-117 Stealth Fighter Jet Shot Down in Combat," *Aviationist,* March 27, 2014, accessed 13 July, 2015, http://theaviationist.com/2014/03/27/vega-31-shot-down/.

128. Whittle, *Predator,* 130–131.

129. Grimes, *The History of Big Safari,* 330–331.

130. Whittle, *Predator,* 141–142; Woods, *Sudden Justice,* 36.

131. Jumper quoted in interview with Whittle, *Predator,* 163–170.

132. Grimes, *The History of Big Safari,* 331–332; Whittle, *Predator,* 172.

133. Coll, *Ghost Wars,* 526–527.

134. Clarridge with Diehl, *A Spy for All Seasons,* 340; David E. Kaplan, Kevin Whitelaw, and Monica M. Ekman, "Mission Impossible: The Inside Story of How a Band of Reformers Tried—and Failed—to Change America's Spy Agencies," *U.S. News and World Report* 137, no. 3 (August 2004).

135. "Initial DCI-Hostage Location Task Force Report," Memorandum for Vice Admiral John M. Poindexter from Charles E. Allen, January 3, 1986, CIA FOIA, accessed August 3, 2015, http://www.foia.cia.gov/sites/default/files/document_conversions/89801/DOC_0000139471.pdf.

136. Clarridge with Diehl, *A Spy for All Seasons,* 340.

137. Whittle, *Predator,* 145.

138. Ibid., 148–149.

139. Coll, *Ghost Wars,* 530.

140. Ibid., 530; see chapter 4 for details on Clinton's series of MONs.

141. Whittle, *Predator,* 149.

142. Whit Peters, quoted in Coll, *Ghost Wars,* 534.

143. "Department of Defense FY 2001 Budget Amendment—Counterterrorism," "Department of Defense, FY 2001 Request for Other Procurement, Air Force," and "Central Intelligence Agency FY 2001 Amendment Request for Counterterrorism Funding," May 2000, in National Security Council, Millennium Threat Budget After Action Review, Transnational Threats, Records of the 2000 Millennium Attack Plots, Box 1, Clinton Presidential Records, Little Rock, Arkansas; *The 9/11 Commission Report,* 506, note 113.

144. Woods, *Sudden Justice,* 39.

145. Ibid., 39; Micha Zenko, "Ten Things You Didn't Know about Drones," *Foreign Policy,* February 27, 2012, accessed August 4, 2015, http://foreignpolicy.com/2012/02/27/10-things-you-didnt-know-about-drones/.

146. Micah Zenko and Emma Welch, "Where the Drones Are: Mapping the Launch Pads for Obama's Secret Wars," *Foreign Policy,* May 29, 2012, accessed August 4, 2015, http://foreignpolicy.com/2012/05/29/where-the-drones-are/.

147. Grimes, *The History of Big Safari,* 332–333; Jeremy Scahill, "Germany is the Tell-Tale Heart of America's Drone War," *Intercept,* April 17, 2015, accessed April 17, 2015, https://firstlook.org/theintercept/2015/04/17/ramstein/; Spiegel Staff, "A War Waged from German Soil: U.S. Ramstein Base Key in Drone Attacks," *Spiegel,*

April 22, 2015, accessed April 23, 2015, http://www.spiegel.de/international/germany/ramstein-base-in-germany-a-key-center-in-us-drone-war-a-1029279.html.

148. Coll, *Ghost Wars,* 531.

149. Richard Clarke, "Strategy for Eliminating the Threat from the Jihadist Networks of al-Qida," 8, National Security Archive, accessed May 20, 2014, http://www2.gwu.edu/~nsarchiv/NSAEBB/NSAEBB147/clarke%20attachment.pdf.

150. Barton Gellman, "A Strategy's Cautious Evolution," *Washington Post,* January 20, 2002, accessed March, 22, 2013, http://www.washingtonpost.com/wp-dyn/content/article/2006/06/09/AR2006060900885.html; Yeene, *Birds of Prey,* 13–16; "MQ-1B Predator Factsheet," US Air Force, accessed April 1, 2014, http://www.af.mil/information/factsheets/factsheet.asp?fsID=122.

151. *The 9/11 Commission Report,* 189.

152. Clarke, "Strategy for Eliminating the Threat from the Jihadist Networks of al Qida"; Gellman, "A Strategy's Cautious Evolution."

153. Benjamin and Simon, *The Age of Sacred Terror: Radical Islam's War against America* (New York: Random House, 2003), 345–346; Shaw, "The Rise of the Predator Empire"; Robert Windrem, "How the Predator Went from Eye in the Sky to War on Terror's Weapon of Choice," *NBC News,* June 5, 2013, accessed December 9, 2013, http://investigations.nbcnews.com/_news/2013/06/05/18780716-how-the-predator-went-from-eye-in-the-sky-to-war-on-terrors-weapon-of-choice.

154. David Gow, "Bush Gives Green Light to CIA for Assassination of Named Terrorists," *Guardian,* October 28, 2001, accessed December 16, 2013, http://www.theguardian.com/world/2001/oct/29/afghanistan.terrorism3; Brian Glyn Williams, *Predators: The CIA's Drone War on Al-Qaeda* (Washington, D.C.: Potomac Books 2013), 28; John Rizzo, interview with *PBS Frontline,* September 6, 2011, accessed August 4, 2015, http://www.pbs.org/wgbh/pages/frontline/iraq-war-on-terror/topsecretamerica/john-rizzo-the-lawyer-who-approved-cias-most-controversial-programs/.

155. Woods, *Sudden Justice,* 26.

156. "CIA 'Killed al-Qaeda Suspects' in Yemen," *BBC News,* November 5, 2002, accessed April 1, 2013, http://news.bbc.co.uk/1/hi/2402479.stm.

157. Robert Baer, *See No Evil: The True Story of a Ground Soldier in the CIA's War against Terrorism* (London: Arrow Books, 2002), 126.

158. Ibid., 84–85.

159. See chapter 4 for a more detailed discussion of Iran-Contra.

160. Johnson, *National Security Intelligence,* 97.

161. Church quoted in Johnson, *A Season of Inquiry,* 143. An internal White House poll taken in 1983 revealed that only 37 percent of the public had heard anything about Nicaragua, and of those only 18 percent were in favor of the Reagan administration's support to the Contra rebels. Jane Mayer and Doyle McManus, *Landslide: The Unmaking of the President, 1984–1988* (Boston: Houghton Mifflin, 1988), 15.

162. Johnson, *A Season of Inquiry,* 255–256.

163. Coll, interview with Vincent Cannistraro, Rosslyn, Virginia, January 8, 2002, quoted in *Ghost Wars,* 142.

164. Persico, *Casey*, 572.

165. "Understanding the Iran-Contra Affairs: The Legal Aftermath," National Security Achieve, accessed April 1, 2013, http://www.gwu.edu/~nsarchiv/coldwar/interviews/episode-18/clarridge2.html; Clarridge with Diehl, *A Spy For All Seasons*, 363–386.

166. "Understanding the Iran-Contra Affairs: The Legal Aftermath," Brown University Good Government Project/National Security Archive, accessed April 1, 2013, http://www.brown.edu/Research/Understanding_the_Iran_Contra_Affair/h--on-gallery.php.

167. Rhodri Jeffreys-Jones, *Cloak and Dollar: The History of American Secret Intelligence* (New Haven: Yale University Press, 2002), 244.

168. Coll, interview with Cannistraro, quoted in *Ghost Wars*, 142.

169. "Hostage Recovery Activities," President Obama's Presidential Policy Directive 30, White House, June 24, 2015, accessed August 5, 2015, https://www.whitehouse.gov/the-press-office/2015/06/24/presidential-policy-directive-hostage-recovery-activities.

170. David Axe, "8,000 Miles, 96 Hours, 3 Dead Pirates: Inside a Navy SEAL Rescue," *Wired*, October 17, 2012, accessed August 5, 2015, http://www.wired.com/2012/10/navy-seals-pirates/.

171. Karen DeYoung and Greg Jaffe, "Navy SEALs Rescue Kidnapped Aid Workers Jessica Buchanan and Poul Hagen Thisted in Somalia," *Washington Post*, January 25, 2012, accessed August 5, 2015, https://www.washingtonpost.com/world/national-security/us-forces-rescue-kidnapped-aid-workers-jessica-buchanan-and-poul-hagen-thisted-in-somalia/2012/01/25/gIQA7WopPQ_story.html; "Statement by the President on Successful Hostage Rescue," January 25, 2012, accessed August 5, 2015, https://www.whitehouse.gov/the-press-office/2012/01/25/statement-president-successful-hostage-rescue.

172. Sean Rayment, "How the British Hostages Were Rescued in Afghanistan," *Telegraph*, June 3, 2012, accessed August 5, 2015, http://www.telegraph.co.uk/news/worldnews/asia/afghanistan/9307833/How-the-British-hostages-were-rescued-in-Afghanistan.html.

173. "Aid Worker Linda Norgrove Was Killed by US grenade," BBC, December 2, 2010, accessed August 5, 2015, http://www.bbc.co.uk/news/uk-scotland-highlands-islands-11900709.

174. Kareem Fahim and Eric Schmitt, "2 Hostages Killed in Yemen as U.S. Rescue Effort Fails," *New York Times*, December 6, 2014, accessed August 5, 2015, http://www.nytimes.com/2014/12/07/world/middleeast/hostage-luke-somers-is-killed-in-yemen-during-rescue-attempt-american-official-says.html.

175. Peter Baker, "Obama Apologizes after Drone Kills American and Italian Held by Al Qaeda," *New York Times*, accessed August 5, 2015, http://www.nytimes.com/2015/04/24/world/asia/2-qaeda-hostages-were-accidentally-killed-in-us-raid-white-house-says.html.

176. Lawrence Wright, "Five Hostages," *New Yorker*, July 6, 2015, accessed August 5, 2015, http://www.newyorker.com/magazine/2015/07/06/five-hostages.

177. "Political Divide: 'Don't Negotiate with Terrorists' vs. 'Leave No Man Behind,'" CNN, May 31, 2014, http://politicalticker.blogs.cnn.com/2014/05/31/political-divide-dont-negotiate-with-terrorists-vs-leave-no-man-behind/.

178. Brendan McGarry, "Drones Most Accident-Prone U.S. Air Force Craft: BGOV Barometer," *Bloomberg News,* June 18, 2012, accessed March 22, 2013, http://www.bloomberg.com/news/2012-06-18/drones-most-accident-prone-u-s-air-force-craft-bgov-barometer.html.

179. "RC South: (non-combat event) Equipment Failure RPT JDOC: 0 INJ/DAM," November 20, 2008, Wikileaks Afghanistan War Log, accessed April 2, 2013, http://wikileaks.org/afg/sort/date/2008_11_21.html.

180. "RC East: (non-combat event) Equipment Failure RPT TF Currahee: 0 INJ/DAM," December 27, 2008, Wikileaks Afghanistan War Log, accessed April 2, 2013, http://wikileaks.org/afg/event/2008/12/AFG20081227n1607.html.

181. "RC Capital: (non-combat event) Equipment Failure RPT Predator UAV: 0 INJ/DAM," September 4, 2009, Wikileaks Afghanistan War Log, accessed April 2, 2013, http://wikileaks.org/afg/event/2009/09/AFG20090904n2089.html.

182. "'Reaper' Moniker Given to MQ-9 Unmanned Aerial Vehicle," Department of Defense, September 14, 2006, accessed May 29, 2015, http://www.defense.gov/transformation/articles/2006-09/ta091406a.html.

183. David S. Cloud, "U.S. Begins Using Predator Drones in Libya," *Los Angeles Times,* April 22, 2011, accessed April 2, 2013, http://articles.latimes.com/2011/apr/22/world/la-fg-gates-libya-20110422.

184. Chris Woods and Alice K. Ross, "Revealed: U.S. and Britain Launched 1,200 Drone Strikes in Recent Wars," December 4, 2012, Covert Drone War, Bureau of Investigative Journalism, accessed April 2, 2013, http://www.thebureauinvestigates.com/2012/12/04/revealed-us-and-britain-launched-1200-drone-strikes-in-recent-wars/.

185. Kareem Fahim, Anthony Shadid, and Rick Gladstone," Violent End to an Era as Qaddafi Dies in Libya," *New York Times,* October 20, 2011, accessed June 5, 2012, http://www.nytimes.com/2011/10/21/world/africa/qaddafi-is-killed-as-libyan-forces-take-surt.html?_r=2&pagewanted=all.

4. "Talking about Capturing bin Laden"

Epigraph. John Rizzo, *Company Man: Thirty Years of Controversy and Crisis in the CIA* (New York: Charles Scribner's Sons, 2014), 161.

1. Richard A. Clarke, *Against All Enemies: Inside America's War on Terror* (London: Free Press, 2004), 89.

2. *The 9/11 Commission Report: Final Report of the National Commission on Terrorist Attacks upon the United States* (New York: W. W. Norton, 2004), accessed February 19, 2014, http://www.9-11commission.gov/report/. Also see Daniel Benjamin and Steven Simon, *The Age of Sacred Terror: Radical Islam's War against America* (New York: Random House, 2003) for an NSC insider's perspective, and Timothy Naftali, *Blind Spot: The Secret History of American Counterterrorism* (New York: Basic Books, 2005) for a contextual overview.

3. Robert M. Chesney, "Beyond the Battlefield, Beyond al-Qaeda: The Destabilizing Legal Architecture of Counterterrorism," *Michigan Law Review* 112 (November 2013): 163–224, at 163.

4. Ibid., 171.

5. For details on CIA activities during the Afghan anti-Soviet jihad and funding for the Nicaraguan Contras see John Prados, *Safe for Democracy: The Secret Wars of the CIA* (Chicago: Ivan R. Dee, 2006), 507–538; Milton Bearden and James Risen, *The Main Enemy: The CIA's Battle with the Soviet Union* (London: Random House, 2003), 205–367. For an assessment of the CIA's policy in Afghanistan, see Richard Immerman, *The Hidden Hand: A Brief History of the CIA* (Chichester: Wiley, 2014), 153–160.

6. For example, see the Select Committee's questioning of John O. Brennan on U.S. drone policy during his nomination hearing for the D/CIA post: "Open Hearing on the Nomination of John O. Brennan to Be Director of the Central Intelligence," U.S. Senate Select Committee on Intelligence, Washington, D.C., February 7, 2013, http://www.intelligence.senate.gov/130207/transcript.pdf; Alice K. Ross, "Is Congressional Oversight Tough Enough on Drones?" *Bureau of Investigative Journalism*, August 1, 2013, accessed February 19, 2014, http://www.thebureauinvestigates.com/2013/08/01/is-congressional-oversight-tough-enough-on-drones/.

7. Remarks of President Barack Obama, National Defense University, May 23, 2013, whitehouse.gov, accessed February 18, 2014, http://www.whitehouse.gov/the-press-office/2013/05/23/remarks-president-barack-obama.

8. Dana Priest and William M. Arkin, " 'Top Secret America': A Look at the Military's Joint Special Operations Command," *Washington Post*, September 2, 2011, https://www.washingtonpost.com/world/national-security/top-secret-america-a-look-at-the-militarys-joint-special-operations-command/2011/08/30/gIQAvYuAxJ_story.html; Iona Craig, "What Really Happened When a U.S. Drone Hit a Yemeni Wedding Convoy?" Al Jazeera America, January 20, 2014, http://america.aljazeera.com/watch/shows/america-tonight/america-tonight-blog/2014/1/17/what-really-happenedwhenausdronehitayemeniweddingconvoy.html; Greg Miller, "Obama's New Drone Policy Leaves Room for CIA Role," *Washington Post*, May 25, 2013, https://www.washingtonpost.com/world/national-security/obamas-new-drone-policy-has-cause-for-concern/2013/05/25/0daad8be-c480-11e2-914f-a7aba60512a7_story.html. All accessed September 7, 2015.

9. Greg Miller, "Lawmakers Seek to Stymie Plan to Shift Control of Drone Campaign from CIA to Pentagon," *Washington Post*, January 15, 2014, accessed September 7, 2014, https://www.washingtonpost.com/world/national-security/lawmakers-seek-to-stymie-plan-to-shift-control-of-drone-campaign-from-cia-to-pentagon/2014/01/15/c0096b18-7e0e-11e3-9556-4a4bf7bcbd84_story.html.

10. Richard W. Stevenson and Ashley Parker, "A Senator's Stand on Drones Scrambles Partisan Lines," *New York Times*, March 7, 2013, http://www.nytimes.com/2013/03/08/us/politics/mccain-and-graham-assail-paul-filibuster-over-drones.html?pagewanted=all&_r=0; Ken Dilanian, "Sen. Levin's Bid to Boost Drone Oversight Falters in Congress," *Los Angeles Times*, http://www.latimes.com/world/worldnow/

la-fg-wn-levin-drone-oversight-20140212,0,2114536.story?utm_source=Sailthru&utm_
medium=email&utm_term=%2ASituation%20Report&utm_campaign=SITREP%20
FEB%2013#ixzz2tC6flahu, both accessed February 20, 2014; President Barack Obama,
"Executive Order—United States Policy on Pre- and Post-Strike Measures to Address
Civilian Casualties in U.S. Operations Involving the Use of Force," White House, July
1, 2016, https://www.whitehouse.gov/the-press-office/2016/07/01/executive-order-
united-states-policy-pre-and-post-strike-measures; Robert Chesney, "President
Obama's Executive Order on Pre/Post Strike Airstrike Policies and Practices," *Lawfare*,
July 1, 2016, https://www.lawfareblog.com/president-obamas-executive-order-prepost-
airstrike-policies-and-practices.

11. "The NSA Files," *Guardian*, http://www.theguardian.com/world/the-nsa-files;
"Nicaragua Downs Plane and Survivor Implicates CIA," *New York Times*, October 12,
1986, http://www.nytimes.com/1986/10/12/weekinreview/in-summary-nicaragua-
downs-plane-and-survivor-implicates-cia.html; Thomas Blanton (ed.), "The CIA's
Family Jewels: Agency Violated Charter for 25 Years," National Security Archive,
Electronic Briefing Book No. 222, http://www2.gwu.edu/~nsarchiv/NSAEBB/
NSAEBB222/; "Church Committee Reports," Assassination Archives and Research
Center Public Library, http://www.aarclibrary.org/publib/contents/church/
contents_church_reports.htm, all accessed February 20, 2014.

12. For a detailed account of the Church Committee hearings, see Loch K. Johnson,
A Season of Inquiry: Congress and Intelligence (Chicago: Dorsey Press, 1988).

13. Richard Immerman, "A Brief History of the CIA," in Athan Theoharis, Richard
Immerman, and Kathryn Olmsted (eds.), *The Central Intelligence Agency: Security
under Scrutiny* (Westport, Conn.: Greenwood Press, 2006), 50.

14. L. Britt Snider, *The Agency and the Hill: CIA's Relationship with Congress,
1946–2004* (Washington, D.C.: Center for the Study of Intelligence, 2008), 275. Also
available in digital format from the CIA Library, accessed March 12, 2014, https://
www.cia.gov/library/center-for-the-study-of-intelligence/csi-publications/books-and-
monographs/agency-and-the-hill/The%20Agency%20and%20the%20Hill_
Book_1May2008.pdf.

15. Kathryn S. Olmsted, *Challenging the Secret Government: The Post-Watergate
Investigations of the CIA and FBI* (Chapel Hill: University of North Carolina Press,
1996), 81.

16. Mark Mazzetti, *The Way of the Knife: The CIA, a Secret Army, and a War at the
Ends of the Earth* (New York: Penguin Press, 2013), 46.

17. "Final Report of the Select Committee to Study Governmental Operations with
Respect to Intelligence Activities," United States Senate, April 26, 1976, in U.S. Select
Committee on Intelligence, Hearings, accessed March 12, 2014 http://www.intelli-
gence.senate.gov/churchcommittee.html.

18. Johnson, *A Season of Inquiry*, 253, 262.

19. Executive Order 11905, "United States Foreign Intelligence Activities," February
18, 1976, Federation of American Scientists Intelligence Resource Program, accessed
March 12, 2014, http://www.fas.org/irp/offdocs/eo11905.htm. Executive Order

12333, "United States Intelligence Activities," December 4, 1981, National Archives. gov, accessed July 18, 2010, http://www.archives.gov/federal-register/codification/executive-order/12333.html#2.11.

20. Mazzetti, *Way of the Knife*, 46.

21. Joseph E. Persico, *Casey: The Lives and Secrets of William J. Casey from the OSS to the CIA* (New York: Penguin, 1990), 216.

22. Johnson, *A Season of Inquiry*, 263.

23. See chapter 2 for details of Casey's efforts to reform the CIA.

24. For detailed accounts of the Iran-Contra affair, see: Malcolm Byrne, *Iran-Contra: Reagan's Scandal and the Unchecked Abuse of Presidential Power* (Lawrence: Kansas University Press, 2014); Theodore Draper, *A Very Thin Line: The Iran-Contra Affairs* (New York: Touchstone Books, 1992); Jane Mayer and Doyle McManus, *Landslide: The Unmaking of the President, 1984–1988* (Boston: Houghton Mifflin, 1988); Lawrence E. Walsh, *Firewall: The Iran-Contra Conspiracy and Cover-Up* (New York: W. W. Norton, 1999).

25. Byrne, *Iran-Contra*, 260; Robert M. Gates, *From the Shadows: The Ultimate Insider's Story of Five Presidents and How They Won the Cold War* (New York: Simon and Schuster, 1996), 418.

26. Rizzo, *Company Man*, 130.

27. Douglas Martin, "Clair George, Spy and Iran-Contra Figure, Dies at 81," *New York Times*, August 20, 2011, accessed February 17, 2014 http://www.nytimes.com/2011/08/21/us/21george.html?_r=0.

28. "Clair E. George, CIA Officer Who Figured in Iran-Contra Scandal, Dies at 81," *Washington Post*, August 13, 2011, accessed March 12, 2014, http://www.washington-post.com/local/obituaries/clair-e-george-cia-officer-who-figured-in-iran-contra-scandal-dies-at-81/2011/08/12/gIQAADpzBJ_story.html.

29. Rizzo, *Company Man*, 132.

30. "Sporkin, Stanley," in *Biographical Directory of Federal Judges*, Federal Judicial Center, accessed February 17, 2014, http://www.fjc.gov/public/home.nsf/hisj.

31. Byrne, *Iran-Contra*, 325–326.

32. David Johnston, "Bush Pardons 6 in Iran Affair, Aborting a Weinberger Trial; Prosecutor Assails 'Cover-Up,'" *New York Times*, December 25, 1992, accessed August 6, 2015, https://www.nytimes.com/books/97/06/29/reviews/iran-pardon.html.

33. Duane R. Clarridge with Digby Diehl, *A Spy for All Seasons: My Life in the CIA* (New York: Charles Scribner's Sons, 1997), 347–348.

34. Ibid., 359–360; "Jordanian Man Sentenced in 1982 Bombing of Pan Am Flight from Tokyo to Honolulu," U.S. Department of Justice, accessed February 21, 2014, http://www.justice.gov/opa/pr/2006/March/06_crm_172.html.

35. Clarridge with Diehl, *A Spy for All Seasons*, 371.

36. Ibid., 371.

37. Byrne, *Iran-Contra*, 32; Malcolm Byrne, Peter Kornbluh, and Thomas Blanton, "The Iran-Contra Affair 20 Years On," National Security Archive, accessed March 12, 2014, http://www2.gwu.edu/~nsarchiv/NSAEBB/NSAEBB210/.

38. Loch K. Johnson, *Secret Agencies: U.S. Intelligence in a Hostile World* (New Haven: Yale University Press, 1996), 116.

39. Johnson interview with Webster, May 2, 1991, in Johnson, *Secret Agencies,* 116.

40. Gates, *From the Shadows,* 419.

41. Rhodri Jeffreys-Jones, *The CIA and American Democracy* (New Haven: Yale University Press, 2003), 229; Immerman, "A Brief History of the CIA," 57.

42. Paul R. Pillar, *Terrorism and U.S. Foreign Policy* (Washington, D.C.: Brookings Institution Press, 2003), 43; Clarke, *Against All Enemies,* 90.

43. Thomas L. Friedman, "Clinton Keeping Foreign Policy on a Back Burner," *New York Times,* February 8, 1993, accessed February 17, 2014, http://www.nytimes.com/1993/02/08/world/clinton-keeping-foreign-policy-on-a-back-burner.html?src=pm.

44. Will Marshall and Martin Schram, *Mandate for Change* (New York: Berkley Books, 1993); Charles F. Allen and Jonathan Portis, *The Life and Career of Bill Clinton: The Comeback Kid* (New York: Birch Lane Press, 1992), 153–157; Bob Woodward, *The Agenda: Inside the Clinton White House* (New York: Simon and Schuster, 1995), xi.

45. Douglas Jehl, "CIA Nominee Wary of Budget Cuts," *New York Times,* February 3, 1993, http://www.nytimes.com/1993/02/03/us/cia-nominee-wary-of-budget-cuts.html; "Budget Cuts Force CIA to Recruit Youths," *Baltimore Sun,* June 11, 1993, http://articles.baltimoresun.com/1993-06-11/news/1993162250_1_cia-students-recruit. All accessed February 20, 2014.

46. "Hearing before the Senate Select Committee on Intelligence on the Nomination of R. James Woolsey, to Be Director of the Central Intelligence Agency," February 2–3, 1993, U.S. Senate Select Committee on Intelligence, accessed February 17, 2014, http://www.intelligence.senate.gov/pdfs103rd/103296.pdf, 76; Douglas F. Garthoff, *Directors of Central Intelligence as Leaders of the U.S. Intelligence Community 1946–2005* (Pittsburgh: Government Printing Office, 2005), 221.

47. "Ramzi Yousef: A New Generation of Sunni Islamic Terrorists," FBI, cited in Steve Coll, *Ghost Wars: The Secret History of the CIA, Afghanistan and bin Laden from the Soviet Invasion to September 10, 2001* (London: Penguin, 2005), 278–279; Pillar, *Terrorism and U.S. Foreign Policy,* 49.

48. "Preparing for the 21st Century: An Appraisal of U.S. Intelligence," March 1, 1996, Federation of American Scientists Intelligence Review Program, accessed February 20, 2014, http://www.fas.org/irp/offdocs/report.html; "Congressional Action on Terrorism," in Richard Clarke Terrorism Legislation, Transnational Threats, Box 9, Clinton Presidential Archive, Little Rock, Arkansas; Richard Sale, *Clinton's Secret Wars: The Evolution of a Commander in Chief* (New York: Thomas Dunne Books, 2009), 290.

49. Rizzo, *Company Man,* 135.

50. Jeffreys-Jones, *The CIA and American Democracy,* 10.

51. Loch K. Johnson, *National Security Intelligence* (Malden, Mass.: Polity Press, 2012), 90, 98.

52. Walter Pincus, "CIA: Ames Betrayed 55 Operations; Inspector General's Draft Report Blames Supervisors for Failure to Plug Leak," *Washington Post,* September 24,

1994, http://www.highbeam.com/doc/1P2-911040.html; "Aldrich Hazen Ames," Famous Cases and Criminals, Federal Bureau of Investigation, http://www.fbi.gov/about-us/history/famous-cases/aldrich-hazen-ames. Both accessed February 17, 2014.

53. Loch K. Johnson, *The Threat on the Horizon: An Inside Account of America's Search for Security after the Cold War* (New York: Oxford University Press, 2011), 99; "Hearing before the Senate Select Committee on Intelligence on the Nomination of John M. Deutch to Be Director of the Central Intelligence Agency," April 26 and May 3, 1995, U.S. Archive.org, accessed February 17, 2014, https://ia600304.us.archive.org/2/items/nominationofjohn1995unit/nominationofjohn1995unit.pdf, 14–34.

54. John Deutch, quoted in Ronald Kessler, *The CIA at War: Inside the Secret Campaign against Terror* (New York: St. Martin's Griffin, 2003), 21–22.

55. Johnson, *The Threat on the Horizon,* 157.

56. Ibid., 107, 159, 198.

57. New York Times Opinion, "Mr. Deutch's Message," *New York Times,* April 28, 1995, accessed February 17, 2014, http://www.nytimes.com/1995/04/28/opinion/mr-deutch-s-message.html.

58. "Report on the Guatemala Review," President's Intelligence Oversight Board, June 28, 1996, University of Texas iSchool, accessed February 17, 2014, https://www.ischool.utexas.edu/~gpasch/tesis/pages/guatemala/otro2/iob.tag.

59. Rizzo, *Company Man,* 147.

60. Johnson, *The Threat on the Horizon,* 249.

61. James Risen, "CIA to Issue Guidelines on Hiring Foreign Operatives," *Los Angeles Times,* June 20, 1995, http://articles.latimes.com/1995-06-20/news/mn-15022_1_cia-director; Sandra Sobieraj, "After Guatemala, CIA Developing New Spy-Recruitment Guidelines," Associated Press, September 12, 1995, http://www.apnewsarchive.com/1995/After-Guatemala-CIA-Developing-New-Spy-Recruitment-Guidelines/id-078e4d9040fd397c6893a4325d8b7599; New York Times Opinion, "Making the CIA Accountable," *New York Times,* August 18, 1996, http://www.nytimes.com/1996/08/18/opinion/making-the-cia-accountable.html. All accessed February 17, 2014.

62. William Lofgren, quoted in Kessler, *The CIA at War,* 28.

63. Thomas Powers, "Computer Security; The Whiz Kid Vs. the Old Boys," *New York Times Magazine,* December 3, 2000, accessed August 6, 2015, http://www.nytimes.com/2000/12/03/magazine/computer-security-the-whiz-kid-vs-the-old-boys.html.

64. George Tenet, *At the Center of the Storm: My Years at the CIA* (New York: HarperCollins, 2007), 20.

65. Coll, *Ghost Wars,* 359.

66. Melissa Boyle Mahle, *Denial and Deception: An Insider's View of the CIA from Iran-Contra to 9/11* (New York: Nation Books, 2006), 249.

67. George Tenet, "Does America Need the CIA?" November 19, 1997, CIA Speeches and Testimony Archive, cia.gov, accessed April 12, 2011, https://www.cia.gov/news-information/speeches-testimony/1997/dci_speech_111997.html.

68. Tenet, *At the Center of the Storm,* 23.

69. "Hearing before the Senate Select Committee on Intelligence, First Session on the Nomination of George J. Tenet to Be Director of the Central Intelligence Agency," May 6, 1997, U.S. Senate Select Committee on Intelligence, accessed 3 December, 2012, www.intelligence.senate.gov/pdfs105th/105314.pdf.; "Excerpts of Prepared Statement by George J. Tenet to Senate Select Committee on Intelligence," May 6, 1997, Federation of American Scientists Intelligence Resource Program, accessed December 3, 2012 http://www.fas.org/irp/congress/1997_hr/s970506t.htm.

70. Kessler, *The CIA at War,* 27.

71. Immerman, *The Hidden Hand,* 161.

72. Ibid., 162.

73. See chapter 1 for a detailed discussion of the content of NSDD 138 and its introduction.

74. Rizzo, *Company Man,* 168.

75. Barbara Lee quoted in Gregory D. Johnsen, "60 Words and a War without End: The Untold Story of the Most Dangerous Sentence in U.S. History," *Buzzfeed,* January 16, 2014, accessed April 14, 2014, http://www.buzzfeed.com/gregorydjohnsen/60-words-and-a-war-without-end-the-untold-story-of-the-most.

76. "Senate Joint Resolution 23," 107th Congress, September 18, 2001, Congressional Record, vol. 147 (2001), accessed April 14, 2014, http://frwebgate.access.gpo.gov/cgi-bin/getdoc.cgi?dbname=107_cong_public_laws&docid=f:publ040.107.

77. Johnsen, "60 Words and a War without End."

78. "UN Security Council Resolution 1368," September 12, 2001, UN.org, accessed April 14, 2014, http://daccess-dds-ny.un.org/doc/UNDOC/GEN/N01/533/82/PDF/N0153382.pdf?OpenElement.

79. "Article 51, Chapter VII, Charter of the United Nations," June 26, 1945, UN.org, accessed April 14, 2014, http://www.un.org/en/documents/charter/intro.shtml.

80. Rizzo, *Company Man,* 168.

81. David Gow, "Bush Gives Green Light to CIA for Assassination of Named Terrorists," *Guardian,* October 28, 2001, accessed December 16, 2013; http://www.theguardian.com/world/2001/oct/29/afghanistan.terrorism3; Brian Glyn Williams, *Predators: The CIA's Drone War on Al-Qaeda* (Washington, D.C.: Potomac Books, 2013), 28; Rizzo, *Company Man,* 161.

82. "Tim Russert Interview with Vice-President Dick Cheney," Meet the Press, *NBC News,* September, 14, 2003, accessed April 14, 2014, http://www.nbcnews.com/id/3080244/ns/meet_the_press/t/transcript-sept/#.UovHzbFwa7o.

83. "Bring Me the Head of bin Laden," *BBC News,* May 4, 2005, accessed April 14, 2014, http://news.bbc.co.uk/1/hi/world/americas/4511943.stm.

84. Barton Gellman, "CIA Weighs 'Targeted Killing' Missions: Administration Believes Restraints Do Not Bar Singling Out Individual Terrorists," *Washington Post,* October 28, 2001, A01.

85. George W. Bush, "Detention, Treatment and Trial of Certain Non-Citizens in the War Against Terrorism," November 13, 2001, George W. Bush White House Archive,

accessed December 20, 2013, http://georgewbush-whitehouse.archives.gov/news/releases/2001/11/print/20011113-27.html.

86. Claire Finkelstein, "Targeted Killing as Preemptive Action," in Claire Finkelstein, Jens David Ohlin, and Andrew Altman (eds.), *Targeted Killings: Law and Morality in an Asymmetrical World* (Oxford: Oxford University Press, 2012), 159.

87. Executive Order 13355, "Strengthened Management of Intelligence Community," August 27, 2004, Federal Register, vol. 69, no. 169, September 1, 2004, accessed December 20, 2013, http://www.gpo.gov/fdsys/pkg/FR-2004-09-01/pdf/04-20051.pdf.

88. "Yemen: Reported U.S. Covert Actions 2001–2011," Bureau of Investigative Journalism, March 29, 2012, http://www.thebureauinvestigates.com/2012/03/29/yemen-reported-us-covert-actions-since-2001/; "CIA 'Killed al-Qaeda Suspects' in Yemen," *BBC News,* November 5, 2002, http://news.bbc.co.uk/1/hi/2402479.stm. Both accessed February 20, 2014.

89. See chapter 3 for details on the origin of the Eagle Program.

90. Philip Alston, "Report of the Special Rapporteur on Extrajudicial, Summary or Arbitrary Executions," United Nations Human Rights Council, May 28, 2010, 21, http://www2.ohchr.org/english/bodies/hrcouncil/docs/14session/A.HRC.14.24.Add6.pdf; "Targeted Killing," American Civil Liberties Union, accessed August 6, 2015, https://www.aclu.org/issues/national-security/targeted-killing. Both accessed August 6, 2015.

91. Gary Solis, "CIA Drone Attacks Produce America's Own Unlawful Combatants," *Washington Post,* March 12, 2010, accessed March 20, 2010, http://www.washingtonpost.com/wp-dyn/content/article/2010/03/11/AR2010031103653.html.

92. Nathan Hodge, "Drone Pilots Could Be Tried for 'War Crimes,' Law Prof Says," *Wired,* April 28, 2010, accessed July 27, 2015, http://www.wired.com/2010/04/drone-pilots-could-be-tried-for-war-crimes-law-prof-says/.

93. Solis, "CIA Drone Attacks Produce America's Own Unlawful Combatants."

94. Charles G. Kels, "Why There's Nothing Illegal about CIA Drone Pilots," *Small Wars Journal,* August 3, 2012, accessed December 9, 2013, http://smallwarsjournal.com/blog/why-theres-nothing-illegal-about-cia-drone-pilots. For further details on the act of perfidy, see Article 37—Prohibition of Perfidy, Protocol Additional to the Geneva Conventions of 12 August 1949, and Relating to the Protection of Victims of International Armed Conflicts (Protocol I), in Adam Roberts and Richard Guelff, *Documents on the Laws of War* (New York: Oxford University Press, 2000), 442.

95. See chapter 3 for details on the partnership between the CIA, the U.S. Air Force, and GA-ASI.

96. Mazzetti, *Way of the Knife,* 94.

97. "Headquarters Air Combat Command, Special Orders GB-52," September 18, 2001 and "GB-73," May 29, 2002, National Security Archive, accessed February 5, 2015, http://www2.gwu.edu/~nsarchiv/NSAEBB/NSAEBB484/docs/Predator-Whittle%20Document%2013%20-%20Predator%20squadron%20formed%20to%20fly%20for%20CIA.pdf.

98. "National Security Act of 1947," July 26, 1947, Archives.gov, http://research. archives.gov/description/299856; "NSC 10/2, National Security Council Directive on Office of Special Projects," June 18, 1948, Foreign Relations of the United States, 1948, Office of the Historian, U.S. Department of State, https://history.state.gov/ historicaldocuments/frus1945-50Intel/d292; "Title 50, 3093, U.S. Code, Presidential Approval and Reporting of Covert Actions," Cornell University Law School, http:// www.law.cornell.edu/uscode/text/50/3093. All accessed February 12, 2015.

99. Chesney, "Beyond the Battlefield," 174.

100. President Barack Obama, "A Just and Lasting Peace," Nobel Lecture, Oslo City Hall, December 10, 2009, accessed April 14, 2014, http://www.nobelprize.org/ nobel_prizes/peace/laureates/2009/obama-lecture.html.

101. Chesney, "Beyond the Battlefield," 176–177. For examples of the Obama administration's public speeches on the legality of its counterterrorism actions, see Kenneth Anderson, "Readings: The Canonical National Security Law Speeches of the Obama Administration Senior Officials and General Councils," *Lawfare*, August 28, 2012, accessed April 14, 2014, http://www.lawfareblog.com/2012/08/readings-the-canonical-national-security-law-speeches-of-obama-administration-senior-officials-and-general-counsels/.

102. Dave Boyer, "Bush Policies Still Alive in Obama White House," *Washington Times*, April 24, 2013, http://www.washingtontimes.com/news/2013/apr/24/bush-policies-still-alive-in-obama-white-house/?page=all; Tom Curry, "Obama Continues, Extends Some Bush Terrorism Policies," *NBC News Politics*, June 6, 2013, http:// nbcpolitics.nbcnews.com/_news/2013/06/06/18804146-obama-continues-extends-some-bush-terrorism-policies?lite; Emily Ekins, "Reason-Rupe September 2013 National Survey," September 10, 2013, Reason-Rupe Poll, http://reason.com/poll/2013/09/10/ reason-rupe-september-2013-national-surv. All accessed February 21, 2014.

103. Trevor McCrisken, "Obama's War on Terrorism in Rhetoric and Practice," in Michelle Bentley and Jack Holland (eds.), *Obama's Foreign Policy: Ending the War on Terror* (London: Routledge, 2014), 17.

104. Transcript of First Presidential Debate between Senators John McCain and Barack Obama, Oxford, Mississippi, September 26, 2008, *New York Times*, Election 2008, accessed February 21, 2014, http://elections.nytimes.com/2008/president/ debates/transcripts/first-presidential-debate.html.

105. Ibid.

106. Jack Holland, "Why Is Change so Hard? Understanding Continuity in Barack Obama's Foreign Policy," in Bentley and Holland (eds.), *Obama's Foreign Policy*, 1–16.

107. "Lawfulness of Lethal Operation Directed against a U.S. Citizen Who Is a Senior Level Operational Leader of Al-Qa'ida or an Associated Force," Department of Justice White Paper, *MSNBC News*, http://msnbcmedia.msn.com/i/msnbc/ sections/news/020413_DOJ_White_Paper.pdf; "Anwar al-Awlaki," in Times Topics, *New York Times*, http://topics.nytimes.com/top/reference/timestopics/people/a/ anwar_al_awlaki/. Both accessed February 18, 2014.

108. Obama, National Defense University, May 23, 2013.

109. "Letter from Attorney General Eric H. Holder to Chairman of Senate Committee on the Judiciary Patrick J. Leahy," May 22, 2013, Justice.gov, accessed February 18, 2014, http://www.justice.gov/ag/AG-letter-5-22-13.pdf.

110. Dana Priest, "U.S. Military Teams, Intelligence Deeply Involved in Aiding Yemen on Strikes," *Washington Post*, January 27, 2010, accessed April 13, 2014, http://www.washingtonpost.com/wp-dyn/content/article/2010/01/26/AR2010012604239.html.

111. Williams, *Predators*, 42–43; Michael Powell and Dana Priest, "U.S. Citizen Killed by CIA Linked to N.Y. Terror Case," *Washington Post*, November 9, 2002, https://groups.google.com/forum/#!topic/misc.legal/qhTUoBpOtrI; Roya Aziz and Monica Lam, "Profiles: The Lackawanna Cell," *PBS Frontline*, December, 2003, http://www.pbs.org/wgbh/pages/frontline/shows/sleeper/inside/profiles.html. Both accessed April 14, 2014.

112. "Assassination by Remote Control," *Economist*, November 5, 2002, accessed April 14, 2014, http://www.economist.com/node/1427862.

113. "Remarks by President Bush at Arkansas Welcome," Northwest Arkansas Regional Airport, Bentonville, Arkansas, November 4, 2002, George W. Bush White House Archive, accessed April 14, 2014, http://georgewbush-whitehouse.archives.gov/news/releases/2002/11/20021104-7.html.

114. Benjamin and Simon, *The Age of Sacred Terror*, 7–19; Naftali, *Blind Spot*, 232–234; Lawrence Wright, *The Looming Tower: Al-Qaeda's Road to 9/11* (New York: Penguin, 2007), 76–179; Rizzo, *Company Man*, 159.

115. Rizzo, *Company Man*, 159. See my chapter 1 for details on Shultz's position on preempting terrorist attacks.

116. Naftali, *Blind Spot*, 232–234.

117. Emad Salem, quoted in Benjamin and Simon, *The Age of Sacred Terror*, 18.

118. Naftali, *Blind Spot*, 234.

119. "1994/1995 Patterns of Global Terrorism," U.S. Department of State, Intelligence Review Program, Federation of American Scientists, accessed February 18, 2014, http://www.fas.org/irp/threat/terror.htm; Coll, *Ghost Wars*, 261.

120. "Usama bin Laden Islamic Extremist Financier," CIA Bio Sketch, Bin Laden File, National Security Archive, accessed February 18, 2014, http://www.gwu.edu/~nsarchiv/NSAEBB/NSAEBB343/osama_bin_laden_file01.pdf.

121. For details of bin Laden's call to jihad and plans to attack the "far enemy" of the United States, see: Osama bin Laden, "The Invasion of Arabia," c. 1995/1996, in Bruce Lawrence, *Messages to the World: The Statements of Osama bin Laden* (New York: Verso, 2005), 15–19; Osama bin Laden, "Declaration of *Jihad* against the United States," August 23, 1996 in Global Security.org, http://www.globalsecurity.org/security/profiles/osama_bin_laden_declares_jihad_text.htm; Osama bin Laden, "World Islamic Front Declaration of Jihad against Jews and Crusaders," February 23, 1998, Intelligence Review Program, Federation of American Scientists, http://www.fas.org/irp/world/para/docs/980223-fatwa.htm. All accessed April 15, 2014.

122. George Tenet, "Written Statement before the 9/11 Commission," March 24, 2004, 9/11 Commission Hearings, accessed February 18, 2014, http://www.9-11commission.gov/hearings/hearing8/tenet_statement.pdf.

123. Tenet, *Center of the Storm*, 102.

124. "Indictment of Osama bin Laden," United States District Court, Southern District of New York, November 4, 1998, Findlaw.com, http://fl1.findlaw.com/news.findlaw.com/hdocs/docs/binladen/usbinladen1.pdf; "Attacks on US Embassies in Kenya and Tanzania," Global Security.org, http://www.globalsecurity.org/security/ops/98emb.htm. Both accessed February 23, 2014. Also see Clarke, *Against All Enemies*, 181–184; Wright, *The Looming Tower*, 270–272.

125. Bruce Riedel, *The Search for Al-Qaeda: Its Leadership, Ideology and Future* (Washington, D.C.: Brookings Institution Press, 2008), 49; Wright, *Looming Tower*, 163–212.

126. "Chapter 5: Terrorist Safe Havens," Country Reports on Terrorism 2012, Office of the Coordinator for Counterterrorism, U.S. State Department, May 30, 2013, accessed April 15, 2014, http://www.state.gov/j/ct/rls/crt/2012/209987.htm.

127. "UN Security Council Resolution 1104," January 1996, "UN Security Council Resolution 1054," April 1996, UN.org, accessed February 21, 2014, http://www.un.org/Docs/scres/1996/scres96.htm.

128. *The 9/11 Commission Report*, 64–65.

129. "Afghanistan: Taliban Agree to Visits of Militant Training Camps, Admit bin Laden Is Their Guest," U.S. State Department cable, January 9, 1997, Bin Laden File, National Security Archive, accessed September 10, 2011, http://www.gwu.edu/~nsarchiv/NSAEBB/NSAEBB343/osama_bin_laden_file10.pdf; Riedel, *The Search for Al-Qaeda*, 61–84.

130. Clarke, *Against All Enemies*, 208–209. Also see Sajit Gandhi (ed.), "Volume VII: The Taliban File," September 11th Sourcebooks, National Security Archive, accessed February 21, 2014, http://www2.gwu.edu/~nsarchiv/NSAEBB/NSAEBB97/.

131. Coll, *Ghost Wars*, 371–371, 377–378. For more on the hunt for Kasi, see my chapter 5.

132. *The 9/11 Commission Report*, 113.

133. Rizzo, *Company Man*, 161.

134. Lawrence Wright, interviews with Scheuer, Dale Watson, Mark Rossini, Daniel Coleman, and Richard A. Clarke, in *The Looming Tower*, 265–266

135. George Tenet, *At the Center of the Storm: My Years at the CIA* (New York: Harper Collins, 2007), 171–173.

136. See chapter 2 for details on the Fadlallah assassination attempt and the reported links to the CIA.

137. Tenet, *At the Center of the Storm*, 172.

138. Rizzo, *Company Man*, 162.

139. Ibid., 162; Sale, *Clinton's Secret Wars*, 299.

140. Coll, *Ghost Wars*, 499–501, 506–507.

141. Ibid., 499.

142. *The 9/11 Commission Report,* 139.

143. Dan Eggen and John Mintz, "9/11 Panel Critical of Clinton, Bush," *Washington Post,* March 24, 2004, accessed February 24, 2014, http://www.washingtonpost.com/wp-dyn/articles/A18972-2004Mar23.html; *The 9/11 Commission Report,* 139.

144. Bill Clinton, *My Life* (New York: Knopf, 2004), 804.

145. Clarke, *Against All Enemies,* 204.

146. Gary Berntsen, *Jawbreaker: The Attack on Bin Laden and Al-Qaeda: A Personal Account by the CIA's Key Field Commander* (New York: Three Rivers Press, 2005), 32.

147. Wright, *Looming Tower,* 291–292.

148. Michael Scheuer, *Marching toward Hell: America and Islam after Iraq* (New York: Free Press, 2008).

149. Michael Scheuer, *Imperial Hubris: Why the West Is Losing the War on Terror* (Washington, D.C.: Potomac Books, 2004), 241–242.

150. Ibid., 242.

151. Robert Baer, *See No Evil: The True Story of a Ground Soldier in the CIA's War on Terrorism* (London: Arrow Books, 2002), 126, 398–399.

152. Ibid., 399.

153. Berntsen, *Jawbreaker,* 32.

154. Ibid., 33–36.

155. Paul Pillar, interview with Chris Woods, *Sudden Justice: America's Secret Drone Wars* (London: Hurst, 2015), 48.

156. *The 9/11 Commission Report,* 133.

157. Mazzetti, *Way of the Knife,* 81.

158. Jeremy Scahill and Glenn Greenwald, "The NSA's Secret Role in the U.S. Assassination Program," *Intercept,* February 10, 2014, https://firstlook.org/theintercept/article/2014/02/10/the-nsas-secret-role/; Erik Wemple, "Glenn Greenwald and the US 'assassination' program," *Washington Post,* February 10, 2014, http://www.washingtonpost.com/blogs/erik-wemple/wp/2014/02/10/glenn-greenwald-and-the-u-s-assassination-program/. Both accessed April 15, 2014.

159. Thomas B. Hunter, "Targeted Killing: Self-Defense, Preemption, and the War on Terrorism," *Journal of Strategic Studies* 2, no. 2 (May 2009): 1–52, 5.

160. James DeShaw Rae, *Analysing the Drone Debates: Targeted Killing, Remote Warfare, and Military Technology* (New York: Palgrave Macmillan, 2014), 52.

161. "CIA: Maker of Policy or Tool?" *New York Times,* April 25, 1966, Scribd digital library, accessed April 15, 2014, http://www.scribd.com/doc/186175212/CIA-Maker-of-Policy-or-Tool-Wicker-Finney-Frankel-Kenworthy-NY-Times-25-April-1966.

162. Immerman, *The Hidden Hand,* 122–124.

163. E. J. Dionne Jr., "Poll Shows Reagan Approval Rating at 4-Year Low," *New York Times,* March 3, 1987, http://www.nytimes.com/1987/03/03/us/poll-shows-reagan-approval-rating-at-4-year-low.html; Frank Newport, Jeffrey M. Jones, and Lydia Saad, "Ronald Reagan From the People's Perspective: A Gallup Poll Review," Gallup,

June 7, 2004, accessed April 15, 2014, http://www.gallup.com/poll/11887/ronald-reagan-from-peoples-perspective-gallup-poll-review.aspx.

164. Robert Busby, "The Scandal That Almost Destroyed Ronald Reagan," *Salon*, February 4, 2011, accessed April 15, 2014, http://www.salon.com/2011/02/04/busby_iran_contra/; Robert Busby, *Reagan and the Iran-Contra Affair: The Politics of Presidential Recovery* (New York: Palgrave MacMillan, 1999).

165. Clarke, *Against All Enemies*, 186.

166. Tenet, *At the Center of the Storm*, 175–176.

167. Ibid., 176.

168. Madeleine Albright, *Madame Secretary: A Memoir* (New York: Miramax Books, 2003), 468–469. For more on the operation, see my chapter 4 and "Infinite Reach, 20 August 1998," Global Security.org, accessed April 16, 2014, http://www.globalsecurity.org/military/ops/infinite-reach_news.htm.

169. "Flashback: Conservative Lawmakers Decried Clinton's Attacks against Osama as 'Wag the Dog,'" *ThinkProgress*, September 25, 2006, accessed April 16, 2014, http://thinkprogress.org/politics/2006/09/25/7679/wag-the-dog/.

170. President Harry S. Truman, "Address before a Joint Session of Congress," March 12, 1947, Avalon Project, accessed April 16, 2014, http://avalon.law.yale.edu/20th_century/trudoc.asp.

171. President William J. Clinton, "Address to the Nation on Military Action against Terrorist Sites in Afghanistan and Sudan," White House, Washington, D.C., August 20, 1998, Yonah Alexander and Michael B. Kraft, *Evolution of U.S. Counterterrorism Policy*, volume 1 (Westport, Conn.: Praeger Security International, 2008), 79–81. For a digital copy of the speech, see the American Presidency Project, accessed April 16, 2014, http://www.presidency.ucsb.edu/ws/print.php?pid=54799.

172. "U.S. Government Statements and Transcripts," Infinite Reach, August 20, 1998, Global Security.org, accessed April 16, 2014, http://www.globalsecurity.org/military/ops/infinite-reach_news.htm#usg.

173. Rizzo, *Company Man*, 164.

174. *The 9/11 Commission Report*, 132, Rizzo, *Company Man*, 164–165.

175. Rizzo, *Company Man*, 165; Tenet, *At the Center of the Storm*, 166.

176. "Transcript of Former Attorney General Janet Reno's Testimony to the 9/11 Commission," April 13, 2004, *NBC News*, accessed February 24, 2014, http://www.nbcnews.com/id/4732236/ns/us_news-security/t/janet-reno-commission-statement/#.UwtdU7FFC7o.

177. For details on the Munich massacre and subsequent Israeli retaliation, see Simon Reeve, *One Day in September: The Full Story of the 1972 Munich Olympics Massacre and the Israeli Revenge Operation "Wrath of God"* (New York: Arcade, 2011).

178. Hunter, *Targeted Killing*, 67–70; Doug Mellgren, "Norway Solves Riddle of Mossad Killing," *Guardian*, March 2, 2000, accessed April 16, 2014, http://www.theguardian.com/world/2000/mar/02/israel.

179. Sale, *Clinton's Secret Wars*, 289.

180. Ibid., 294.

181. Eliot A. Cohen, "History and the Hyperpower," *Foreign Affairs* (July/August 2004), accessed April 19, 2014, http://www.foreignaffairs.com/articles/59919/eliot-a-cohen/history-and-the-hyperpower.

182. John W. Dietrich (ed.), *The George W. Bush Foreign Policy Reader* (New York: M. E. Sharpe, 2005), 17; Secretary of Defense Dick Cheney, Under-Secretary of Defense Policy Paul Wolfowitz, and Deputy Under-Secretary for Strategy and Resources I. Lewis Libby, "1994–99 Defense Planning Guidance," February 18, 1992, National Security Archive, http://www.gwu.edu/~nsarchiv/nukevault/ebb245/doc03_extract_nytedit.pdf; Patrick E. Tyler, "U.S. Strategy Plan Calls for Ensuring No Rivals Develop," *New York Times,* March 8, 1992, accessed April 19, 2014, http://www.nytimes.com/1992/03/08/world/us-strategy-plan-calls-for-insuring-no-rivals-develop.html?scp=1&sq=us+strategy+plan+calls+for+ensuring 0+rivals+develop+March+8%2C+1992&st=cse.

183. Secretary of Defense Dick Cheney, "Defense Strategy for the 1990s: The Regional Defense Strategy," accessed April 19, 2014, http://www.gwu.edu/~nsarchiv/nukevault/ebb245/doc15.pdf.

184. Alvin Z. Rubinstein, Albina Shayevich, and Boris Zlotnikov (eds.), *The Clinton Foreign Policy Reader* (New York: M. E. Sharpe, 2000), 11–12.

185. President William J. Clinton, "Inaugural Address," January 20, 1993, Miller Center for Public Affairs, accessed April 19, 2014, http://millercenter.org/president/clinton.

186. Clinton, "Strategic Outlook for a New World Order," in Rubinstein, Shayevich, and Zlotnikov (eds.), *The Clinton Foreign Policy Reader,* 7; Dumbrell, *Clinton's Foreign Policy,* 130–131; Tudor A. Onea, *U.S. Foreign Policy in the Post–Cold War Era: Restraint versus Assertiveness from George H. W. Bush to Barack Obama* (New York: Palgrave Macmillan, 2013), 99–101; Patrick Hagopian, *American Immunity: War Crimes and the Limits of International Law* (Boston: University of Massachusetts Press, 2013), 152–164.

187. Sale, *Clinton's Secret Wars,* 291.

188. Onea, *U.S. Foreign Policy in the Post–Cold War Era,* 93.

189. "Article II.2, Intermediate-Range Nuclear Forces Treaty," December 8, 1987, U.S. Department of State, accessed July 7, 2015. http://www.state.gov/t/avc/trty/102360.htm.

190. "Predator Weaponization and INF Treaty," E-mail from Colonel James Clarke to Lt. Col. Kenneth Johns, September 8, 2000, National Security Archive Electronic Briefing Book No. 484, accessed July 8, 2015, http://nsarchive.gwu.edu/NSAEBB/NSAEBB484/.

191. Richard Whittle, *Predator: The Secret Origins of the Drone Revolution* (New York: Henry Holt, 2014), 171–183.

192. Governor George W. Bush, "A Distinctly American Internationalism," Ronald Reagan Library, Simi Valley, California, November 19, 1999, Federation of American Scientists, accessed April 19, 2014, http://www.fas.org/news/usa/1999/11/991119-bush-foreignpolicy.htm; Condoleezza Rice, "Promoting National Interest," *Foreign Affairs* (January/February 2000), 2.

193. "National Security Strategy 2010," May 2010, Whitehouse.gov, accessed June 22, 2014, http://www.whitehouse.gov/sites/default/files/rss_viewer/national_security_strategy.pdf.

194. Remarks by President Barack Obama at the Acceptance of the Nobel Peace Prize, Oslo City Hall, Oslo, Norway, December 10, 2009, whitehouse.gov, accessed May 19, 2014, http://www.whitehouse.gov/the-press-office/remarks-president-acceptance-nobel-peace-prize.

195. Sale, *Clinton's Secret Wars,* 290.

196. Barton Gellman and Greg Miller, "'Black Budget' Summary Details U.S. Spy Network's Successes, Failures and Objectives," *Washington Post,* August 29, 2013, accessed June 30, 2014, http://www.washingtonpost.com/world/national-security/black-budget-summary-details-us-spy-networks-successes-failures-and-objectives/2013/08/29/7e57bb78-10ab-11e3-8cdd-bcdc09410972_story.html.

197. Andrew C. Orr, "Unmanned, Unprecedented and Unresolved: The Status of American Drone Strikes in Pakistan Under International Law," *Cornell International Law Journal* 44 (2011): 729–752.

198. J. M. Berger, "War on Error," *Foreign Policy,* February 5, 2014, accessed June 22, 2014, http://www.foreignpolicy.com/articles/2014/02/04/war_on_error_al_qaeda_terrorism.

199. Christopher Greenwood, "War, Terrorism and International Law," *Current Legal Problems* 56 (2004): 505–529.

200. Berger, "War on Error."

201. "Senate Joint Resolution 23," 107th Congress, September 18, 2001, gpo.gov, http://frwebgate.access.gpo.gov/cgi-bin/getdoc.cgi?dbname=107_cong_public_laws&docid=f:publ040.107; United States National Strategy for Counterterrorism, June 2011, 2, whitehouse.gov, http://www.whitehouse.gov/sites/default/files/counterterrorism_strategy.pdf. Both accessed February 26, 2014.

202. Greg Miller, "Plan for Hunting Terrorists Signals U.S. Intends to Keep Adding Names to Kill Lists," *Washington Post,* October 24, 2012, accessed February 26, 2014, http://www.washingtonpost.com/world/national-security/plan-for-hunting-terrorists-signals-us-intends-to-keep-adding-names-to-kill-lists/2012/10/23/4789b2ae-18b3-11e2-a55c-39408fbe6a4b_story.html.

203. Glenn Greenwald, "Obama Moves to Make the War on Terror Permanent," *Guardian,* October 24, 2012, accessed February 26, 2014, http://www.theguardian.com/commentisfree/2012/oct/24/obama-terrorism-kill-list.

204. Greg Miller, "Under Obama, an Emerging Global Apparatus for Drone Killing," *Washington Post,* December 28, 2011, accessed August 30, 2012, http://www.washingtonpost.com/national/national-security/under-obama-an-emerging-global-apparatus-for-drone-killing/2011/12/13/gIQANPdILP_story.html.

205. For more on the Phoenix Program, see: Mark Moyar, *Phoenix and the Birds of Prey: Counterinsurgency and Counterterrorism in Vietnam* (Winnipeg: Bison Books, 1997); Colonel Andrew R. Finlayson, USMC (Ret.), "A Retrospective on Counterinsurgency Operations: The Tay Ninh Provincial Reconnaissance Unit and Its

Role in the Phoenix Program, 1969–70," *Studies in Intelligence* 51, no. 2 (12 June, 2007), accessed February 26, 2014, https://www.cia.gov/library/center-for-the-study-of-intelligence/csi-publications/csi-studies/studies/vol51no2/a-retrospective-on-counterinsurgency-operations.html. For more on efforts to assassinate Castro, see: Don Bohning, *The Castro Obsession: U.S. Covert Operations against Cuba, 1959–1965* (Washington, D.C.: Potomac Books, 2005).

206. "Alleged Assassination Plots Involving Foreign Leaders," Church Committee Interim Report, 1975, 257, Assassination Archives and Research Center, accessed June 4, 2012, http://www.aarclibrary.org/publib/church/reports/ir/html/ChurchIR_0136a.htm.

207. Executive Order 11905, "United States Foreign Intelligence Activities," February 18, 1976, Gerald R. Ford Presidential Library, accessed May 29, 2012, http://www.ford.utexas.edu/library/speeches/760110e.htm#assassination.

208. Reagan quoted in Morton H. Halperin, "The CIA's Distemper," *New Republic* (9 February, 1980), 21–22, cited in Rhodri Jeffreys-Jones, *Cloak and Dollar: The History of American Secret Intelligence* (New Haven: Yale University Press, 2002), 237–238.

209. Jeffreys-Jones, *The CIA and American Democracy,* xiii, 248.

210. "The Torture Archive," National Security Archive, accessed February 26, 2014, http://www2.gwu.edu/~nsarchiv/torture_archive/.

211. Scott Shane, "U.S. Engaged in Torture after 9/11, Review Concludes," *New York Times,* April 16, 2013, http://www.nytimes.com/2013/04/16/world/us-practiced-torture-after-9-11-nonpartisan-review-concludes.html?_r=1&; Patrick Goodenough, "U.N. Torture Expert Says U.S. Should Probe Bush-Era Torture Claims with Intention to Prosecute," *CNS News,* November 18, 2010; "Report of the Special Rapporteur on the Promotion and Protection of Human Rights and Fundamental Freedoms While Countering Terrorism," March 1, 2013, U.N. Human Rights Council, http://www.ohchr.org/Documents/HRBodies/HRCouncil/RegularSession/Session22/A-HRC-22-52_en.pdf. All accessed February 26, 2014.

212. Dschabner, "President Obama Discusses Possible Prosecutions of Bush Administration Officials," April 21, 2009, *ABC News,* http://abcnews.go.com/blogs/politics/2009/04/president-ob-25/; "Obama Names Intel Picks, Vows No Torture," Associated Press, January 9, 2009, http://www.nbcnews.com/id/28574408/ns/politics-white_house/#.Uw4eLrFFDIU; David Stout, "Holder Tells Senators Waterboarding Is Torture," *New York Times,* January 15, 2009, http://www.nytimes.com/2009/01/16/us/politics/16holdercnd.html; "Transcript of Interview between Brian Williams and CIA Director Leon Panetta," *NBC News,* May 3, 2011, http://www.nbcnews.com/id/42887700/ns/world_news-death_of_bin_laden/t/transcript-interview-cia-director-panetta/#.Uw4dk7FFDIU. All accessed February 26, 2014.

213. "Living Under Drones: Death, Injury and Trauma to Civilians from U.S. Drone Practices in Pakistan," International Human Rights and Conflict Resolution Clinic, Stanford Law School and Global Justice Clinic, NYU School of Law, September, 2012, 103–124, accessed August 10, 2015, https://www.law.stanford.edu/publications/living-under-drones-death-injury-and-trauma-to-civilians-from-us-drone-practices-in-

pakistan; C. Christine Fair, "Drones, Spies, Terrorists, and Second-Class Citizenship in Pakistan," *Small Wars and Insurgencies* 25, no. 1 (2014): 205–235, 210.

5. "Ninja Guys in Black Suits"

Epigraph. Richard Clarke, *Against All Enemies: Inside America's War on Terror* (London: Free Press, 2004), 160.

1. Amrit Singh, *Globalizing Torture: CIA Secret Detention and Extraordinary Rendition* (New York: Open Society Foundation, 2013), 5. Digital copy available from Open Society Foundation.org, accessed April 25, 2014, http://www.opensocietyfoundations.org/sites/default/files/globalizing-torture-20120205.pdf.

2. John Rizzo, *Company Man: Thirty Years of Controversy and Crisis in the CIA* (New York: Charles Scribner's Sons, 2014), 178; Jose A. Rodriguez Jr. with Bill Harlow, *Hard Measures: How Aggressive CIA Actions after 9/11 Saved American Lives* (New York: Threshold Editions, 2012), 40; Michael Kirk (dir.), "Top Secret America: From 9/11 to the Boston Bombing," *Frontline PBS*, September 6, 2011, accessed May 19, 2014, http://www.pbs.org/wgbh/pages/frontline/iraq-war-on-terror/topsecretamerica/transcript-6/.

3. Singh, *Globalizing Torture*, 29–61.

4. Kathleen T. Rhem, "DoD Releases Names of 759 Current, Former Guantanamo Detainees," May 15, 2006, http://www.defense.gov/news/newsarticle.aspx?id=15754; "List of Individuals Detained by the Department of Defense at Guantanamo Bay, Cuba from January 2002 through May 15, 2006," accessed April 25, 2014, http://www.defense.gov/news/May2006/d20060515%20List.pdf.

5. "Comprehensive Crime Control Act of 1984," U.S. House of Representatives, S.1762, February 9, 1984, National Criminal Justice Reference System, accessed April 22, 2014, https://www.ncjrs.gov/pdffiles1/Digitization/123365NCJRS.pdf.

6. "The Case of the Yachted Terrorist," FBI Famous Cases and Criminals Archive, accessed April 22, 2014, http://www.fbi.gov/about-us/history/famous-cases/fawaz-younis/fawaz-younis.

7. "HR 4151, Omnibus Diplomatic Security and Antiterrorism Act of 1986," August 27, 1986, govtrack.us, accessed April 22, 2014, https://www.govtrack.us/congress/bills/99/hr4151/text.

8. President Ronald Reagan, "Statement on Signing the Omnibus Diplomatic Security and Antiterrorism Act of 1986," August 27, 1986, Reagan Presidential Library Speeches Archive, accessed April 22, 2014, http://www.reagan.utexas.edu/archives/speeches/1986/082786a.htm.

9. Ibid.

10. See chapter 2 for further details on the *Achille Lauro* hijacking and the U.S. response.

11. Kenneth B. Noble, "Lebanese Suspect in '85 Hijacking Arrested by the FBI While at Sea," *New York Times*, September 18, 1987, accessed April 22, 2014, http://www.nytimes.com/1987/09/18/world/lebanese-suspect-in-85-hijacking-arrested-by-the-fbi-while-at-sea.html.

12. Tim Naftali, "Milan Snatch: Extraordinary Rendition Comes Back to Bite the Bush Administration," *Slate,* June 30, 2005, accessed April 22, 2014, http://www.slate.com/articles/news_and_politics/war_stories/2005/06/milan_snatch.html.

13. David C. Wills, *The First War on Terrorism: Counter-Terrorism Policy during the Reagan Administration* (Lanham, Md.: Rowman and Littlefield, 2003), 137; Robin Wright, "U.S. Sentences Arab Hijacker to 30 Years," *Los Angeles Times,* October 5, 1989, accessed April 22, 2014, http://articles.latimes.com/1989-10-05/news/mn-911_1_united-states; "The Case of the Yachted Terrorist," FBI.

14. Attorney General Edwin Meese, quoted in Noble, "Lebanese Suspect in '85 Hijacking Arrested by the FBI While at Sea." For more on the Reagan administration's counterterrorism hardliners, see chapter 1.

15. For details on the attacks carried out by Kasi and Yousef, see chapter 4.

16. Steve Coll, *Ghost Wars: The Secret History of the CIA, Afghanistan and bin Laden from the Soviet Invasion to September 10, 2001* (London: Penguin, 2005), 272; "The Long Arm of the Law: FBI International Presence Key to Bringing Terrorists to Justice," February 6, 2004, FBI.gov, accessed April 23, 2014, http://www.fbi.gov/news/stories/2004/february/law020604.

17. Peg Tyre, " 'Proud Terrorist' Gets Life for Trade Center Bombing," CNN, January 8, 1998, accessed April 23, 2014, http://edition.cnn.com/US/9801/08/yousef.update/.

18. "Presidential Decision Directive 39, U.S. Policy on Counterterrorism," June 21, 1995, Federation of American Scientists Intelligence Review Program, accessed April 23, 2014, http://www.fas.org/irp/offdocs/pdd39.htm; Nicholas D. Kristof, "Terror in Tokyo: The Overview," *New York Times,* March 21, 1995, http://www.nytimes.com/learning/general/onthisday/big/0320.html#article; James M. Lindsay, "Lessons Learned: Tokyo Sarin Gas Attack," Council on Foreign Relations, March 20, 2012, http://www.cfr.org/japan/lessons-learned-tokyo-sarin-gas-attack/p27685. Both accessed April 23, 2014.

19. Presidential Decision Directive 39.

20. Steve LaMontagne, "India-Pakistan Sanctions Legislation Fact Sheet," Center for Arms Control and Non-Proliferation, accessed April 23, 2014, http://armscontrolcenter.org/issues/nonproliferation/articles/india_pakistan_sanctions/.

21. Coll, *Ghost Wars,* 371–376. See my chapter 4 for details on how the bin Laden unit planned to use the TRODPINT assets.

22. Ibid., 371–372.

23. Tim Weiner, "Killer of Two at CIA Draws Death Sentence," *New York Times,* January 24, 1998, http://www.nytimes.com/1998/01/24/us/killer-of-two-at-cia-draws-death-sentence.html?ref—iramalkansi&version—eter+at+7&region=FixedCenter&pgtype=Article&priority=true&module=RegiWall-Regi&action=click; "Arrest and Trial of Mir Aimal Kasi," *Dawn,* November 14, 2002, accessed April 23, 2014, http://www.dawn.com/news/66694/arrest-and-trial-of-mir-aimal-kasi.

24. Cyril Almeida, "Lying to Ourselves," *Dawn,* October 27, 2013, http://www.dawn.com/news/1052086; Craig Whitlock and Greg Miller, "U.S. Covert Military Presence in Afghanistan Much Larger Than Thought," *Washington Post,* September

22, 2010, accessed April 23, 2014, http://www.washingtonpost.com/wp-dyn/content/article/2010/09/22/AR2010092206241.html.

25. Robin Wright and Joby Warrick, "U.S. Steps Up Unilateral Strikes in Pakistan," *Washington Post,* March 27, 2008, http://www.washingtonpost.com/wp-dyn/content/article/2008/03/27/AR2008032700007.html?sid=ST2008032700935. Accessed April 23, 2014.

26. Ibid.

27. Bob Woodward, *Obama's Wars: The Inside Story* (New York: Simon and Schuster, 2010), 6.

28. Director of Central Intelligence George J. Tenet, "Remarks before the National Commission on Terrorist Attacks upon the United States," April 14, 2004, CIA Speeches and Testimony Archive, https://www.cia.gov/news-information/speeches-testimony/2004/tenet_testimony_04142004.html; Spencer Ackerman, "How the CIA Became 'One Hell of a Killing Machine,'" *Wired,* September 2, 2011, http://www.wired.com/dangerroom/2011/09/cia-killing-machine/. Both accessed March 18, 2013. For more on the CIA's use of proxy agents, see chapter 2.

29. "Drone Wars Pakistan: Analysis," New American Foundation, accessed April 24, 2014, http://natsec.newamerica.net/drones/pakistan/analysis.

30. Executive Order 13491, "Ensuring Lawful Interrogations," January 22, 2009, whitehouse.gov, http://www.whitehouse.gov/the_press_office/EnsuringLawfulInterrogations; "Secret Prisons: Obama's Order to Close 'Black Sites,'" *Guardian,* January 23, 2009, http://www.theguardian.com/world/2009/jan/23/secret-prisons-closure-obama-cia. Both accessed April 25, 2014.

31. Craig Whitlock, "Renditions Continue under Obama, Despite Due-Process Concerns," *Washington Post,* January 2, 2013, accessed 25 April, 2014, http://www.washingtonpost.com/world/national-security/renditions-continue-under-obama-despite-due-process-concerns/2013/01/01/4e593aa0-5102-11e2-984e-f1de82a7c98a_story.html.

32. David Kirkpatrick, Nicholas Kulish, and Eric Schmitt, "U.S. Raids in Libya and Somalia Strike Terror Targets," *New York Times,* October 5, 2013, http://www.nytimes.com/2013/10/06/world/africa/Al-Qaeda-Suspect-Wanted-in-US-Said-to-Be-Taken-in-Libya.html; "Embassy Bombings Figure Nabbed by U.S. Forces in Libya," *CBS News,* October 5, 2013, http://www.cbsnews.com/news/embassy-bombings-figure-nabbed-by-us-forces-in-libya/; Benjamin Weiser and Eric Schmitt, "U.S. Said to Hold Qaeda Suspect on Navy Ship," *New York Times,* October 6, 2013, http://www.nytimes.com/2013/10/07/world/africa/a-terrorism-suspect-long-known-to-prosecutors.html?emc=edit_na_20131006&_r=0; "Abu Anas al Liby, al-Qaeda Suspect Nabbed in Libya, Pleads Not Guilty to Terrorism Charges," *CBS News,* October 15, 2013, http://www.cbsnews.com/news/abu-anas-al-liby-al-qaeda-suspect-nabbed-in-libya-pleads-not-guilty-to-terrorism-charges/. All accessed April 25, 2014.

33. George W. Bush, *Decision Points* (St. Ives: Virgin Books, 2010), 169–171.

34. Remarks by Richard B. Cheney, American Enterprise Institute, May 21, 2009, accessed May 19, 2014, http://www.aei.org/article/foreign-and-defense-policy/regional/india-pakistan-afghanistan/remarks-by-richard-b-cheney/.

35. Dick Cheney, *In My Time: A Personal and Political Memoir* (New York: Threshold Editions, 2011), 359–363.

36. Rodriguez with Harlow, *Hard Measures,* xiii

37. "Committee Study of the Central Intelligence Agency's Detention and Interrogation Program," Senate Select Committee on Intelligence, December 9, 2014, accessed December 15, 2015, https://web.archive.org/web/20141209165504/http://www.intelligence.senate.gov/study2014/sscistudy1.pdf.

38. "CIA Comments on the Senate Select Committee on Intelligence Report on the Rendition, Detention, and Interrogation Program," June 27, 2013, 4, accessed July 29, 2015, https://www.cia.gov/library/reports/CIAs_June2013_Response_to_the_SSCI_Study_on_the_Former_Detention_and_Interrogation_Program.pdf.

39. "CIA Fact Sheet Regarding the SSCI Study on the Former Detention and Interrogation Program," CIA, December 9, 2014, accessed July 29, 2015, https://www.cia.gov/news-information/press-releases-statements/2014-press-releases-statements/cia-fact-sheet-ssci-study-on-detention-interrogation-program.html.

40. "Statement from Director Brennan on the SSCI Study on the Former Detention and Interrogation Program," December 9, 2014, CIA, accessed December 11, 2014, https://www.cia.gov/news-information/press-releases-statements/2014-press-releases-statements/statement-from-director-brennan-on-ssci-study-on-detention-interrogation-program.html.

41. "CIA Comments on the Senate Select Committee on Intelligence Report," 3.

42. "CIA Fact Sheet Regarding the SSCI Study."

43. Clare Algar, "Torture Did Not Lead Us to Osama bin Laden," *Guardian,* May 5, 2011, http://www.theguardian.com/commentisfree/libertycentral/2011/may/05/torture-and-osama-bin-laden; John McCain, "Bin Laden's Death and the Debate over Torture," *Washington Post,* May 5, 2011, http://www.washingtonpost.com/opinions/bin-ladens-death-and-the-debate-over-torture/2011/05/11/AFd1mdsG_story.html; Peter Bergen, "'Zero Dark Thirty': Did Torture Really Net bin Laden?," CNN, December 11, 2012, http://www.cnn.com/2012/12/10/opinion/bergen-zero-dark-thirty/; "Fact Sheet: Torture and Obama bin Laden," Amnesty International, http://www.amnestyusa.org/pdfs/BinLadenAndTortureFactSheet.pdf. All accessed May 19, 2014.

44. Peter Bergen, *Manhunt: From 9/11 to Abbottabad—The Ten-Year Search for Osama bin Laden* (London: Bodley Head, 2012), 93.

45. Dana Priest, "NSA Growth Fuelled by Need to Target Terrorists," *Washington Post,* July 21, 2013, accessed May 19, 2014, http://www.washingtonpost.com/world/national-security/nsa-growth-fueled-by-need-to-target-terrorists/2013/07/21/24c93cf4-f0b1-11e2-bed3-b9b6fe264871_story.html.

46. Jeremy Scahill and Glenn Greenwald, "The NSA's Secret Role in the U.S. Assassination Program," *Intercept,* February 10, 2014, accessed May 19, 2014, https://firstlook.org/theintercept/article/2014/02/10/the-nsas-secret-role/.

47. Joshua Foust, "NSA Outrage Machine Misfires on Drone Kills," *War Is Boring,* accessed May 20, 2014, https://medium.com/war-is-boring/b63826b30825.

48. Greg Miller, Julie Tate, and Barton Gellman, "Documents Reveal NSA's Extensive Involvement in Targeted Killing Program," *Washington Post,* October 16, 2013, accessed May 19, 2014, http://www.washingtonpost.com/world/national-security/documents-reveal-nsas-extensive-involvement-in-targeted-killing-program/2013/10/16/29775278-3674-11e3-8a0e-4e2cf80831fc_story.html.

49. James Bamford, *The Shadow Factory: The Ultra-Secret NSA from 9/11 to the Eavesdropping on America* (London: Anchor Books, 2009), 135–136; David Axe, "That One Time the NSA Helped Kill a Guy," *War Is Boring,* accessed May 20, 2014, https://medium.com/war-is-boring/af4f6d6fec62.

50. "Boumediene v. Bush/Al Odah v. U.S.: The Supreme Court Decision," Center for Constitutional Rights, accessed May 20, 2014, http://ccrjustice.org/learn-more/faqs/factsheet-boumediene.

51. Remarks by the President Barack Obama at the Acceptance of the Nobel Peace Prize, Oslo City Hall, Oslo, Norway, December 10, 2009, whitehouse.gov, accessed May 19, 2014, http://www.whitehouse.gov/the-press-office/remarks-president-acceptance-nobel-peace-prize.

52. President George W. Bush Delivers Graduation Speech at West Point, June 1, 2002, United States Military Academy, West Point, New York, George W. Bush White House Archive, accessed May 19, 2014, http://georgewbush-whitehouse.archives.gov/news/releases/2002/06/20020601-3.html.

53. National Security Strategy of the United States, September, 2002, state.gov, accessed May 19, 2014, http://www.state.gov/documents/organization/63562.pdf, 15.

54. Russell Christopher, "Imminence in Justified Targeted Killing," in Claire Finkelstein, Jens David Ohlin, and Andrew Altman (eds.), *Targeted Killings: Law and Morality in an Asymmetrical World* (Oxford: Oxford University Press, 2012), 256. For a discussion on the fundamental distinction between status and conduct, see M. Maxwell, "Rebutting the Civilian Presumption: Playing Whack-A-Mole Without a Mallet?" in Finkelstein, Ohlin, and Altman (eds.), *Targeted Killings,* chapter 1, 31–59.

55. Jeffery McMahan, "Targeted Killing: Murder, Combat or Law Enforcement?" Finkelstein, Ohlin, and Altman (eds.), *Targeted Killing,* chapter 5, 135–155.

56. Greg Miller, "Plan for Hunting Terrorists Signals U.S. Intends to Keep Adding Names to Kill Lists," *Washington Post,* October 23, 2012, accessed May 19, 2014, http://www.washingtonpost.com/world/national-security/plan-for-hunting-terrorists-signals-us-intends-to-keep-adding-names-to-kill-lists/2012/10/23/4789b2ae-18b3-11e2-a55c-39408fbe6a4b_story.html.

57. David Von Drehle and R. Jeffrey Smith, "U.S. Strikes Iraq for Plot to Kill Bush," *Washington Post,* June 27, 1993, accessed May 19, 2014, http://www.washingtonpost.com/wp-srv/inatl/longterm/iraq/timeline/062793.htm; Richard Sale, *Clinton's Secret Wars: The Evolution of a Commander in Chief* (New York: Thomas Dunne, 2009), 6; Memo, "Plot against President Bush," May 13, 1993, in Richard Clarke, Transnational Threats, Terrorism, Box 7 of 13, National Security Council, Clinton Presidential Records, Clinton Presidential Library, Little Rock, Arkansas.

58. Seymour Hersh, "A Case Not Closed," *New Yorker*, November 1, 1993, accessed June 29, 2014, http://www.newyorker.com/archive/1993/11/01/1993_11_01_082_TNY_CARDS_000367232.

59. Sale, *Clinton's Secret Wars*, 8–9; Clarke, *Against All Enemies*, 83; Foreign Policy Contributors, "Clinton's Foreign Policy," *Foreign Policy* 121 (November–December 2000): 18–29.

60. Charles Arthur, "SkyGrabber: The $26 Software Used by Insurgents to Hack U.S. Drones," *Guardian*, December 17, 2009, http://www.theguardian.com/technology/2009/dec/17/skygrabber-software-drones-hacked; "Iraq Insurgents 'Hack into Video Feeds from U.S. Drones,'" *BBC News*, December 17, 2009, http://news.bbc.co.uk/1/hi/world/middle_east/8419147.stm; Siobhan Gorman, Yochi J. Dreazen, and August Cole, "Insurgents Hack U.S. Drones," *Wall Street Journal*, December 17, 2009, http://online.wsj.com/news/articles/SB126102247889095011. Both accessed May 19, 2014. The United States began encrypting the imagery shortly after this discovery was made.

61. Clarke, *Against All Enemies*, 83.

62. Sale, *Clinton's Secret Wars*, 9; Malcolm Lagauche, "The Forgotten Terrorist Attack," uruknet.info, accessed May 19, 2014, http://www.uruknet.info/?p=78864.

63. Sale, *Clinton's Secret Wars*, 9.

64. President Clinton's Address to the Nation, the Oval Office, Washington, D.C., June 26, 1993, YouTube.com, accessed May 19, 2014, https://www.youtube.com/watch?v=6mpWa7wNr5M; Drehle, Smith, "U.S. Strikes Iraq for Plot to Kill Bush," *Washington Post*.

65. Tom Engelhardt, *The American Way of War: How Bush's Wars Became Obama's* (Chicago: Haymarket Books, 2010), 144; Peter Van Buren, "The Divine Right of President Obama?" February 16, 2014, *TomDispatch.com*, accessed May 19, 2014, http://www.tomdispatch.com/blog/175807/.

66. Clarke, *Against All Enemies*, 204; Lawrence Wright, *The Looming Tower: Al-Qaeda's Road to 9/11* (New York: Penguin, 2007), 291–292.

67. Captain Gregory Ball, "Operation Desert Fox," Air Force Historical Studies Office, accessed May 20, 2014, http://www.afhso.af.mil/topics/factsheets/factsheet.asp?id=18632.

68. *The 9/11 Commission Report: Final Report of the National Commission on Terrorist Attacks upon the United States* (New York: W. W. Norton, 2004), 130–131.

69. Peter Bergen, *Holy War Inc.: Inside the Secret World of Osama bin Laden* (London: Phoenix, 2002), 123.

70. "Infinite Reach, 20 August 1998," Global Security.org, accessed April 16, 2014, http://www.globalsecurity.org/military/ops/infinite-reach_news.htm; Madeleine Albright, *Madame Secretary: A Memoir* (New York: Miramax Books, 2003), 468–469.

71. Coll, *Ghost Wars*, 290; William Maley, *The Afghanistan Wars* (New York: Palgrave Macmillan, 2009), 218–250; Ahmed Rashid, "The Taliban: Exporting Extremism," *Foreign Affairs* (November/December 1999); George Tenet, *At the Center of the Storm: My Years at the CIA* (New York: HarperCollins, 2007), 176–177.

72. Bergan, *Holy War Inc.*, 124.

73. "United States of America v, Usama bin Laden, et al.," Testimony of John Anticev, United States District Court, Southern District of New York, February 28, 2001, Cryptome.org, accessed May 20, 2014, http://cryptome.org/usa-v-ubl-12.htm.

74. "To Bomb Sudan Plant, or Not: A Year Later, Debates Rankle," *New York Times,* October 27, 1999, Internet Archive.org, accessed May 20, 2014, http://web.archive.org/web/20060909050523/http://www.library.cornell.edu/colldev/mideast/sudbous.htm; Bergen, *Holy War Inc.*, 126–127.

75. Richard Clarke, "Strategy for Eliminating the Threat from the Jihadist Networks of al-Qida," 7, National Security Archive, accessed May 20, 2014, http://www2.gwu.edu/~nsarchiv/NSAEBB/NSAEBB147/clarke%20attachment.pdf. For a comparison of Clarridge's problems in acquiring American eyes on the target, see chapter 3.

76. Ibid., 8.

77. Philip Alston, "Report of the Special Rapporteur on Extrajudicial, Summary or Arbitrary Executions," May 28, 2010, United Nations Human Rights Council, accessed May 19, 2014, http://www2.ohchr.org/english/bodies/hrcouncil/docs/14session/A.HRC.14.24.Add6.pdf.

78. "Lockheed Martin's Scorpion Successful in Test Flight," June 21, 2010, lockheedmartin.co.uk, http://www.lockheedmartin.co.uk/us/news/press-releases/2010/june/LockheedMartinsSCORPIONSu.html; "Year of the Drone," New American Foundation, http://counterterrorism.newamerica.net/drones. Both accessed June 16, 2014.

79. "Joint Special Operations Command," socom.mil, accessed May 22, 2014, http://www.socom.mil/pages/jointspecialoperationscommand.aspx.

80. Mark Mazzetti, *The Way of the Knife: The CIA, a Secret Army, and a War at the Ends of the Earth* (New York: Penguin Press, 2013), 63, 129.

81. Ibid., 129; Eric Schmitt and Mark Mazzetti, "Secret Order Lets U.S. Raid Al Qaeda," *New York Times,* November 9, 2009, accessed May 22, 2014, http://www.nytimes.com/2008/11/10/washington/10military.html?_r=0.

82. For more on the scale of JSOC's counterterrorist operations, see Jeremy Scahill, *Dirty Wars: The World Is a Battlefield* (London: Serpent's Tail, 2013); *Dirty Wars,* directed by Rick Rowley (Sundance Selects, 2013).

83. David E. Sanger, *Confront and Conceal: Obama's Secret Wars and Surprising Use of American Power* (New York: Broadway Paperbacks, 2013).

84. "Afghanistan," Office of the Secretary of Defense, Donald Rumsfeld to General Myers, Working Paper, October 17, 2001, National Security Archive, accessed May 22, 2014, http://www2.gwu.edu/~nsarchiv/NSAEBB/NSAEBB358a/doc19.pdf; http://www2.gwu.edu/~nsarchiv/NSAEBB/NSAEBB358a/#19.

85. "JITF-CT," Memo from Secretary of Defense Donald Rumsfeld to DCI George Tenet, September 26, 2001, Rumsfeld Papers Library, accessed May 22, 2014, http://library.rumsfeld.com/doclib/sp/1508/2001-09-26%20to%20George%20Tenet%20re%20JITF-CT.pdf.

86. "Strategic Thoughts," Secretary of Defense Donald, Memorandum for the President, September 30, 2001, National Security Archive, accessed May 22, 2014, http://www2.gwu.edu/~nsarchiv/NSAEBB/NSAEBB358a/doc13.pdf.

87. Ibid.

88. "CIA Support to Military Operations," CIA Library, cia.gov, accessed May 22, 2014, https://www.cia.gov/library/reports/archived-reports-1/Ann_Rpt_2001/smo.html.

89. See chapter 3 for details on the attack on Libya.

90. See chapter 1 for details on Operation Eagle Claw and its fallout.

91. President George H. W. Bush, "Address before a Joint Session of the Congress on the State of the Union," Washington, D.C., January 29, 1991, the American Presidency Project, accessed May 20, 2014, http://www.presidency.ucsb.edu/ws/?pid=19253.

92. Tudor A. Onea, *U.S. Foreign Policy in the Post–Cold War Era: Restraint versus Assertiveness from George H. W. Bush to Barack Obama* (New York: Palgrave Macmillan, 2013), 55–56; Brent Scowcroft quoted in Samantha Power, *A Problem from Hell: America and the Age of Genocide* (New York: Basic Books, 2002), 293.

93. Onea, *U.S. Foreign Policy in the Post–Cold War Era,* 56.

94. Madeleine Albright, "Yes, There Is a Reason to Be in Somalia," *New York Times,* August 10, 1993, accessed May 20, 2014, http://www.nytimes.com/1993/08/10/opinion/yes-there-is-a-reason-to-be-in-somalia.html.

95. Onea, *U.S. Foreign Policy in the Post–Cold War Era,* 57; John Dumbrell, *Clinton's Foreign Policy: Between the Bushes, 1992–2000* (New York: Routledge, 2009), 67.

96. Dumbrell, *Clinton's Foreign Policy,* 71.

97. Bill Clinton, *My Life* (New York: Knopf, 2004), 55. For the full story of the incident, see Mark Bowden, *Black Hawk Down* (New York: Atlantic Monthly Press, 1999).

98. Bill Roggio, "Shabaab leader Recounts al Qaeda's Role in Somalia in the 1990s," Long War Journal, December 31, 2011, accessed June 29, 2014, http://www.longwar-journal.org/archives/2011/12/shabaab_leader_recou.php#.

99. See chapter 4 for details on the aborted snatch operations against bin Laden in Afghanistan.

100. Clinton, *My Life,* 552.

101. Ibid., 935; Mark Knoller interview with William J. Clinton, CBS Radio, Dover, New Hampshire, January 11, 2001, Public Papers of the President, American Presidency Project, accessed June 15, 2014, http://www.presidency.ucsb.edu/ws/?pid=64800.

102. "Rachel Maddow Interview with Colin Powell," *Rachel Maddow Show, MSNBC News,* April 1, 2009, accessed June 22, 2014, http://www.nbcnews.com/id/30038761/ns/msnbc-rachel_maddow_show/t/full-text-transcript-colin-powell-talks-rachel-maddow/#.U6cCFLFwa7o; Jonathan Monten, Andrew Bennett, "Models of Crisis Decision Making and the 1990–91 Gulf War," *Security Studies* 19, no. 3 (2010): 486–520; Colin Powell, "U.S. Forces: Challenges Ahead," *Foreign Affairs* (Winter 1992/1993), accessed June 22, 2014, http://www.cfr.org/world/us-forces-challenges-ahead/p7508. For details on the Weinberger Doctrine, see chapter 1.

103. David Usborne, "Obituary: Les Aspin," *Independent,* May 23, 1995, accessed June 22, 2014, http://www.independent.co.uk/news/people/obituary—les-aspin-1620708.html.

104. Clarke, *Against All Enemies,* 190.

105. Ibid., 190.

106. Clarke, "Strategy for Eliminating the Threat from the Jihadist Networks of al-Qida," 7.

107. Albright, *Madam Secretary* (New York: Miramax, 2005), 468–469; Editorial, "Assessing the Blame For 9/11," *New York Times,* March 25, 2004, section A, 26.

108. Peter Bergen, *The Longest War: The Enduring Conflict between America and Al-Qaeda* (New York: Free Press, 2011), 54–55.

109. Nicholas Schmidle, "Getting Bin Laden," *New Yorker,* August 8, 2011, accessed May 22, 2014, http://www.newyorker.com/reporting/2011/08/08/110808fa_fact_schmidle?currentPage=all.

110. Bergen, *Manhunt,* 219–220.

111. President Barack Obama, interviewed in *Bin Laden: Shoot to Kill,* Channel 4, September 7, 2011, 10 P.M.

112. Owen Bowcott, "Pakistan says U.S. May Have Breached Sovereignty," *Guardian,* May 5, 2011, http://www.guardian.co.uk/world/2011/may/05/osama-bin-laden-pakistan-us-sovereignty; Harold Koh, "The Lawfulness of the U.S. Operation Against Osama bin Laden," *Opinio Juris,* http://opiniojuris.org/2011/05/19/the-lawfulness-of-the-us-operation-against-osama-bin-laden/. Both accessed May 22, 2014.

113. "Poll: Most Pakistanis Disapproved of U.S. Killing bin Laden," *CNN,* June 22, 2011, accessed May 22, 2014, http://edition.cnn.com/2011/WORLD/asiapcf/06/21/pakistan.bin.laden.poll/.

114. "U.S. Image in Pakistan Falls No Further Following bin Laden Killing," Pew Research Global Attitudes Project, June 21, 2011, accessed May 22, 2014, http://www.pewglobal.org/2011/06/21/u-s-image-in-pakistan-falls-no-further-following-bin-laden-killing/.

115. "Abbottabad Commission Report," Pakistan's Bin Laden Dossier, Al Jazeera, July 8, 2013, http://www.aljazeera.com/indepth/spotlight/binladen-files/2013/07/201378143927822246.html; Hannah Strange, "U.S. Raid That Killed bin Laden Was 'an Act of War,' says Pakistani Report," *Telegraph,* July 9, 2013, accessed May 22, 2014, http://www.telegraph.co.uk/news/worldnews/asia/paki-stan/10169655/US-raid-that-killed-bin-Laden-was-an-act-of-war-says-Pakistani-report.html.

116. "Pakistan Condemns Bin Laden Raid and US Drone Attacks," *BBC News,* May 14, 2011, accessed May 22, 2014, http://www.bbc.co.uk/news/world-south-asia-13398281.

117. "Abbottabad Commission Report."

118. "U.S. Image in Pakistan Falls No Further Following bin Laden Killing," Pew.

119. Ayman al-Zawahiri, letter quoted in Alan Cullison, "Inside Al-Qaeda's Hard Drive," *Atlantic,* September 1, 2004, accessed June 16, 2014, http://www.theatlantic.

com/magazine/archive/2004/09/inside-al-qaeda-s-hard-drive/303428/; Gilles Kepel and Jean-Pierre Milelli (eds.), *Al Qaeda in Its Own Words* (Cambridge, Mass.: Belknap Press of Harvard University Press, 2008), 53–56; Raymond Ibrahim (ed.), *The Al Qaeda Reader* (New York: Doubleday, 2007), 11–14; Bruce Lawrence (ed.), *Messages to the World: The Statements of Osama bin Laden* (London: Verso, 2005), 106–129, 133–144; Ayman al-Zawahiri, *Knights under the Prophet's Banner*, translated by Laura Mansfield, in *His Own Words: A Translation of the Writings of Dr. Ayman al-Zawahiri* (New York: TLG, 2006), 219–220.

120. "Al-Qaeda's Foreign Fighters in Iraq: A First Look at the Sinjar Records," West Point Counterterrorism Center, accessed June 16, 2014, http://www.isn.ethz.ch/isn/Digital-Library/Publications/Detail/?ots591=0c54e3b3-1e9c-be1e-2c24-a6a8c7060233&lng=en&id=45910.

121. Richard A. Serrano and David S. Cloud, "Times Square bomb suspect had ties to key Pakistani militants," *Los Angeles Times,* May 8, 2010, accessed June 16, 2014, http://articles.latimes.com/2010/may/08/nation/la-na-bomb-motive-20100508.

122. Mazzetti, *Way of the Knife,* 105–107.

123. For more on the Pakistani peace deal, see: Rohan Gunaratna and Syed Adnan Ali Shah Bukhari, "Making Peace with Pakistani Taliban to Isolate Al-Qaeda: Success and Failures," *Peace and Security Review* 1, no. 2 (Second Quarter, 2008): 1–25.

6. "The Only Game in Town"

Epigraph. "U.S. Airstrikes in Pakistan Called 'Very Effective,'" CNN, May 18, 2009, accessed June 16, 2014, http://articles.cnn.com/2009-05-18/politics/cia.pakistan.airstrikes_1_qaeda-pakistani-airstrikes?_s=PM:POLITICS.

1. Gregory D. Johnson, "The Untouchable John Brennan," *Buzzfeed,* April 24, 2015, accessed April 28, 2015, http://www.buzzfeed.com/gregorydjohnsen/how-cia-director-john-brennan-became-americas-spy-and-obamas#.ujM8lDRwv; James Mann, *The Obamians: The Struggle Inside the White House to Redefine American Power* (New York: Viking, 2012), 104–105, 113.

2. Senator Barack Obama, "Floor Statement on General Michael Hayden Nomination," May 25, 2006, accessed July 16, 2015, http://obamaspeeches.com/073-General-Michael-Hayden-Nomination-Obama-Speech.htm.

3. Jane Meyer, "The Secret History," *New Yorker,* June 22, 2009, accessed June 15, 2014, http://www.newyorker.com/magazine/2009/06/22/the-secret-history.

4. Mann, *The Obamians,* 104; Stephen Soldz, "Psychologists to Obama: Don't Name Torture Apologist John Brenner [*sic*] CIA Director," November 24, 2008, Alternet Civil Liberties, accessed June 15, 2014, http://www.alternet.org/story/108504psychologists_to_obama%3A_don%27t_name_torture_apologist_john_brenner_cia_director.

5. Johnson, "The Untouchable John Brennan."

6. Mann, *The Obamians,* 105.

7. John Rizzo quoted in Mark Mazzetti, *The Way of the Knife: The CIA, a Secret Army, and a War at the Ends of the Earth* (New York: Penguin Press, 2013), 90–91; Richard Whittle, *Predator: The Secret Origins of the Drone Revolution* (New York: Henry Holt, 2014), 221.

8. Leon Panetta, *Worthy Fights* (New York: Penguin Press, 2014), 198–201.

9. Karen DeYoung, "A CIA Veteran Transforms U.S. Counterterrorism Policy," *Washington Post*, May 29, 2014, accessed May 29, 2014, http://www.washingtonpost.com/world/national-security/cia-veteran-john-brennan-has-transformed-us-counterterrorism-policy/2012/10/24/318b8eec-1c7c-11e2-ad90-ba5920e56eb3_story.html.

10. Reid Cherlin, "Obama's Drone-Master," *GQ*, February 2013, accessed May 29, 2014, http://www.gq.com/story/john-brennan-cia-director-interview-drone-program.

11. Anne E. Kornblut, "Counterterrorism Advisor John Brennan: A Forceful Voice on Obama's Security Team," *Washington Post*, June 6, 2010, http://www.washingtonpost.com/wp-dyn/content/article/2010/06/05/AR2010060502969.html; Glenn Greenwald, "John Brennan's Extremism and Dishonesty Rewarded with CIA Director Nomination," *Guardian*, January 7, 2013. Both accessed March 29, 2014.

12. Johnson, "The Untouchable John Brennan." See chapter 3 for details on the formation of the CTC under Clarridge.

13. "Director of Terrorist Threat Integration Center Appointed," CIA Press Release Archive, March 11, 2003, accessed July 16, 2015, https://www.cia.gov/news-information/press-releases-statements/press-release-archive-2003/pro3112003.html.

14. Nancy Scola, "Obama, the 'Big Data' President," *Washington Post*, June 14, 2013, accessed June 4, 2014, http://www.washingtonpost.com/opinions/obama-the-big-data-president/2013/06/14/1d71fe2e-d391-11e2-b05f-3ea3f0e7bb5a_story.html.

15. Greg Miller, "Plan for Hunting Terrorists Signals U.S. Intends to Keep Adding Names to Kill Lists," *Washington Post*, October 23, 2012, accessed July 16, 2015, https://www.washingtonpost.com/world/national-security/plan-for-hunting-terrorists-signals-us-intends-to-keep-adding-names-to-kill-lists/2012/10/23/4789b2ae-18b3-11e2-a55c-39408fbe6a4b_story.html.

16. James Ball, "NSA Stores Metadata of Millions of Web Users for up to a Year, Secret Files Show," *Guardian*, September 30, 2013, accessed July 16, 2015, http://www.theguardian.com/world/2013/sep/30/nsa-americans-metadata-year-documents.

17. Chris Woods quoted in *Drone*, directed by Tonje Hessen Schei (Spectrum, 2015).

18. "Drone Wars Pakistan: Analysis," New America Foundation, accessed November 9, 2015, http://securitydata.newamerica.net/drones/pakistan-analysis.html. Despite collaborating with NAF, BIJ offers slightly different statistics, estimating that the administration undertook three hundred strikes in Pakistan during Obama's first term, resulting in the killing of 2,102 militants and 354 civilians (17 percent civilian casualties). Alice K. Ross, Chris Woods, and Sarah Leo, "The Reaper Presidency: Obama's 300th Drone Strike in Pakistan," Covert Drone War, Bureau of Investigative Journalism,

accessed June 22, 2014, http://www.thebureauinvestigates.com/2012/12/03/the-reaper-presidency-obamas-300th-drone-strike-in-pakistan/; "Summary of Information Regarding U.S. Counterterrorism Strikes Outside Area of Active Hostilities," Office of the Director of National Intelligence, July 1, 2016, https://www.dni.gov/files/documents/Newsroom/Press%20Releases/DNI+Release+on+CT+Strikes+Outside+Areas+of+Active+Hostilities.PDF; Charlie Savage and Scott Shane, "U.S. Reveals Death Toll from Airstrikes outside War Zones," *New York Times,* July 1, 2016, http://www.nytimes.com/2016/07/02/world/us-reveals-death-toll-from-airstrikes-outside-of-war-zones.html?_r=0.

19. In 2005, the creation of the post of director of national intelligence resulted in the role of director of central intelligence (DCI) changing to director of the Central Intelligence Agency (D/CIA).

20. Daniel Klaidman, *Kill or Capture: The War on Terror and the Soul of the Obama Presidency* (New York: Houghton Mifflin Harcourt, 2012), 41–42.

21. "Drone Wars Pakistan: Analysis"; "Summary of Information Regarding U.S. Counterterrorism Strikes outside Area of Active Hostilities."

22. John O. Brennan, Assistant to the President for Homeland Security and Counterterrorism, "Ensuring al-Qaeda's Demise," Paul H. Nitze School of Advanced International Studies, Washington D.C., June 29, 2011, accessed May 29, 2014, http://www.whitehouse.gov/the-press-office/2011/06/29/remarks-john-o-brennan-assistant-president-homeland-security-and-counter.

23. Bryan C. Price, "Targeting Top Terrorists: How Leadership Decapitation Contributes to Counterterrorism," *International Security* 36, no. 4 (Spring 2012): 9–46, 9–10.

24. Robert A. Pape, "The Strategic Logic of Suicide Terrorism," *American Political Science Review* 97, no. 3 (August 2003): 1–19, 14; Jenna Jordan, "When Heads Roll: Assessing the Effectiveness of Leadership Decapitation," *Security Studies* 18, no. 4 (December 2009): 719–755, 14; Steven R. David, "Fatal Choices: Israel's Policy of Targeted Killings," *Mideast Security and Policy Studies* 51 (September 2002): 1–26. For further studies concluding that leadership decapitation is an ineffective strategy, see: Robert A. Pape, *Bombing to Win: Air Power and Coercion in War* (Ithaca, N.Y.: Cornell University Press, 1996); Stephen T. Hosmer, *Operations against Enemy Leaders* (Santa Monica: Rand, 2001); Paul Staniland, "Defeating Transnational Insurgencies: The Best Offense Is a Good Fence," *Washington Quarterly* 29, no. 1 (Winter 2005/2006): 21–40; Mohammed M. Hafez and Joseph M. Hatfield, "Do Targeted Assassinations Work? A Multivariate Analysis of Israel's Controversial Tactic during Al-Aqsa Uprising," *Studies in Conflict and Terrorism* 29, no. 4 (April 2006): 359–382; Or Honig, "Explaining Israel's Misuse of Strategic Assassinations," *Studies in Conflict and Terrorism* 30, no. 6 (June 2007): 563–577. List compiled by Patrick B. Johnson, "Does Decapitation Work? Assessing the Effectiveness of Targeting in Counterinsurgency Campaigns," *International Security* 36, no. 4 (Spring 2012): 47–79, 47–48.

25. Audrey Kurth Cronin, *Ending Terrorism: Lessons for Defeating al-Qaeda* (London: Routledge, 2008), 27.

26. Johnson, "Does Decapitation Work?" 77.

27. Price, "Targeting Top Terrorists," 43–46.

28. "Letter from Osama bin Laden to Shaykh Mahmud ('Atiyya), October 21, 2010," Letters from Abbottabad, Combating Terrorism Center, West Point, May 3, 2012, SOCOM-2012-0000015, accessed July 21, 2015, https://www.ctc.usma.edu/posts/letters-from-abbottabad-bin-ladin-sidelined, 3.

29. Timothy Noah, "Al-Qaeda's Rule of Threes: Why Are We Always Killing Osama's 'No. 3' Operative?" *Slate*, June 1, 2010, accessed 29 May, 2014, http://www.slate.com/articles/news_and_politics/recycled/2010/06/alqaidas_rule_of_threes.html.

30. "U.S. Strikes 'Leadership Compound,'" CNN, November 27, 2001, http://edition.cnn.com/2001/WORLD/asiapcf/central/11/27/ret.omar.attack/index.html; John F. Burns, "A Nation Challenged: The Fugitives," *New York Times*, April 14, 2002, http://www.nytimes.com/2002/04/14/world/nation-challenged-fugitives-pakistans-interior-troubling-victory-hunt-for-al.html; Dan Eggen and Walter Pincus, "FBI, CIA Debate Significance of Terror Suspect," *Washington Post*, December 18, 2007, http://www.washingtonpost.com/wp-dyn/content/article/2007/12/17/AR2007121702151.html; Rehab el-Buri and Jonathan Karl, "Top Al-Qaeda Leader Killed," *ABC News*, January 31, 2008, http://abcnews.go.com/Blotter/story?id=4222911&page=1 Jeremy Kelly, "Al-Qaeda's Third in Command Killed in U.S. Drone Strike," *Telegraph*, June 1, 2010, http://www.telegraph.co.uk/news/worldnews/asia/pakistan/7792134/Al-Qaedas-third-in-command-killed-in-US-drone-strike.html. All accessed June 28, 2014.

31. Rohan Gunaratna, *Inside Al Qaeda: Global Network of Terror* (New York: Columbia University Press, 2002), 71–72.

32. Mia Bloom, "Dying to Kill: Motivations for Suicide Terrorism," in Ami Pedahzur (ed.), *Root Causes of Suicide Terrorism: The Globalization of Martyrdom* (New York: Routledge, 2006), 49.

33. Jordan, "When Heads Roll," 25.

34. Johnston, "Does Decapitation Work?" 77.

35. "U.S. Airstrikes in Pakistan Called 'Very Effective,'" CNN.

36. "White House Quarterly Report on Afghanistan and Pakistan," September 2011, accessed June 16, 2014, http://www.scribd.com/doc/66998459/WH-Report-on-Afghanistan-and-Pakistan-September-2011.

37. Michael Morell with Bill Harlow, *The Great War of Our Time: The CIA's Fight against Terrorism from Al 'Qa'ida to ISIS* (New York: Hachette Book Group, 2015), 137.

38. Abu Yahya al-Libi, "Guide to the Law Regarding Muslim Spies," June 2009, accessed April 23, 2014, http://xa.yimg.com/kq/groups/19377022/1660762922/name/%5Bthe+teacher+in+the+government+spy+Muslim+-+Sheikh+Abu+Yahya+al-Libi%5D+06.30.09+Archangel.pdf; Report No. 2438, July 9, 2009, Middle East Research Institute (MEMRI), Jihad and Terrorism Threat Monitor, accessed April 23, 2014, http://www.memrijttm.org/content/en/report.htm?report=3403&param=GJN; Jason Burke, "Abu Yahya al-Libi Obituary," *Guardian*, June 6,

2012, accessed April 23, 2014, http://www.theguardian.com/world/2012/jun/06/abu-yahya-al-libi.

39. Adam Rawnsley, "CIA Drone Targeting Tech Revealed, Qaeda Claims," *Wired,* August 7, 2009, http://www.wired.com/2009/07/infrared-beacons-guiding-cia-drone-strikes-qaeda-claims/; Declan Walsh, "Mysterious 'chip' is CIA's latest weapon against al-Qaeda targets hiding in Pakistan's tribal belt," *Guardian,* May 31, 2009, http://www.theguardian.com/world/2009/may/31/cia-drones-tribesmen-taliban-pakistan. Both accessed April 24, 2014.

40. Anthony Loyd, "US Drone Strikes in Pakistan Tribal Areas Boost Support for Taliban," *Times,* March 10, 2010, via Congressional Unmanned Systems Caucus, U.S. Congress, accessed April 24, 2014, http://unmannedsystemscaucus.mckeon.house.gov/pakistan/2010/03/us-drone-strikes-in-pakistan-tribal-areas-boost-support-for-taliban.shtml.

41. Zia Khan, "Taliban Create Cell to Hunt 'Spies' Assisting U.S. Drones," *Express Tribune,* March 28, 2011, accessed April 24, 2014, http://tribune.com.pk/story/138759/taliban-create-cell-to-hunt-spies-assisting-us-drones/.

42. Brian Glyn Williams, *Predators: The CIA's Drone War on Al-Qaeda* (Washington, D.C.: Potomac Books, 2013), 108.

43. "Letter from Osama bin Laden to Shaykh Mahmud ('Atiyya)," May 21, 2010, Letters from Abbottabad, Combating Terrorism Center, West Point, May 3, 2012, SOCOM-2012-0000019, 30, accessed 21 July, 2015, https://www.ctc.usma.edu/posts/letters-from-abbottabad-bin-ladin-sidelined; "Atiyah Abd al Rahman," Naming the Dead, Bureau of Investigative Journalism, accessed 21 July, 2015, https://www.thebureauinvestigates.com/namingthedead/people/nd451/?lang=en.

44. Price, "Targeting Top Terrorists," 9–10; Jason Burke, *The 9/11 Wars* (London: Allen Lane, 2011), 493.

45. Brian Fishman and Joseph Felter, "Al-Qa'ida's Foreign Fighters in Iraq: A First Look at the Sinjar Records," Combating Terrorism Center, West Point, January 2, 2007, 24, accessed July 20, 2015, https://www.ctc.usma.edu/posts/al-qaidas-foreign-fighters-in-iraq-a-first-look-at-the-sinjar-records; Brian Fishman (ed.), "Bombers, Bank Accounts, and Bleedout: Al-Qa'ida's Road in and Out of Iraq," Combating Terrorism Center, West Point, July 22, 2008, 62, accessed July 22, 2015, http://www.princeton.edu/~jns/publications/Sinjar_2_July_23.pdf.

46. Charles Lister, "Profiling the Islamic State," Brookings Doha Center, Number 13 (November 2014), 12, accessed 22 July, 2015, http://www.brookings.edu/~/media/Research/Files/Reports/2014/11/profiling%20islamic%20state%20lister/en_web_lister.pdf.

47. Daniel L. Byman and Jennifer R. Williams, "ISIS vs. Al Qaeda: Jihadism's Global Civil War," *National Interest* (March/April 2015), accessed March 5, 2015, http://nationalinterest.org/feature/isis-vs-al-qaeda-jihadism%E2%80%99s-global-civil-war-12304.

48. "Generation Kill: A Conversation with Stanley McChrystal," *Foreign Affairs* (March/April 2013), accessed July 22, 2015, https://www.foreignaffairs.com/

interviews/2013-02-11/generation-kill; Spencer Ackerman, "How Special Ops Copied Al-Qaida to Kill It," *Wired,* September 9, 2011, accessed August 11, 2015, http://www.wired.com/2011/09/mcchrystal-network/; Stanley McChrystal, *My Share of the Task: A Memoir* (New York: Portfolio, 2014), 242; Jeremy Scahill, *Dirty Wars: The World Is a Battlefield* (London: Serpent's Tail, 2013), 144–145.

49. Christopher Reuter, "The Terror Strategist: Secret Files Reveal the Structure of Islamic State," *Spiegel,* April 18, 2015, accessed June 22, 2015, http://www.spiegel.de/international/world/islamic-state-files-show-structure-of-islamist-terror-group-a-1029274.html; Liz Sly, "The Hidden Hand behind the Islamic State Militants? Saddam Hussein's," *Washington Post,* April 4, 2015, accessed April 6, 2015, http://www.washingtonpost.com/world/middle_east/the-hidden-hand-behind-the-islamic-state-militants-saddam-husseins/2015/04/04/aa97676c-cc32-11e4-8730-4f473416e759_story.html; Justine Drennan, "The Black-Market Battleground," *Foreign Policy,* October 17, 2014, accessed June 22, 2015, http://foreignpolicy.com/2014/10/17/the-black-market-battleground/.

50. Shiv Malik, Ali Younes, Spencer Ackerman, and Mustafa Khalili, "How ISIS Crippled al-Qaida," *Guardian,* June 10, 2015, accessed June 12, 2015, http://www.theguardian.com/world/2015/jun/10/how-isis-crippled-al-qaida.

51. Fawaz A. Gerges, *The Rise and Fall of Al-Qaeda* (New York: Oxford University Press, 2011), 33.

52. Basma Atassi, "Qaeda Chief Annuls Syrian-Iraqi Jihad Merger," Al Jazeera, June 9, 2013, accessed March 12, 2015, http://www.aljazeera.com/news/middleeast/2013/06/2013699425657882.html?utm=from_old_mobile.

53. Shiv Malik et al., "How ISIS Crippled al-Qaida."

54. "Sunni Rebels Declare New 'Islamic Caliphate,'" Al Jazeera, June 30, 2014, accessed July 23, 2015, http://www.aljazeera.com/news/middleeast/2014/06/isil-declares-new-islamic-caliphate-201462917326669749.html.

55. "This Is the Promise of Allah," Almayat Media Center, Islamic State, June 30, 2014, 5, accessed July 23, 2015, https://ia902505.us.archive.org/28/items/poa_25984/EN.pdf.

56. "Sunni Rebels Declare New 'Islamic Caliphate.'"

57. Daniel L. Byman, "Comparing Al Qaeda and ISIS: Different Goals, Different Targets," Testimony before the Subcommittee on Counterterrorism and Intelligence of the House Committee on Homeland Security, April 29, 2015, Brookings, accessed July 17, 2015, http://www.brookings.edu/research/testimony/2015/04/29-terrorism-in-africa-byman; Full translations of Zawahiri's communications regarding ISIS can be found here: "Ayman al-Zawahiri on Jihadist Infighting and the Islamic State of Iraq and al-Sham," jihadology.net, http://jihadology.net/category/individuals/ideologues/dr-ayman-al-zawahiri/. Also see: Ayman al-Zawahiri, translation of letter to the leaders of the two Jihadi groups, in http://s3.documentcloud.org/documents/710588/translation-of-ayman-al-zawahiris-letter.pdf. Translation by al-Jazeera, available in Basma Atassi, "Qaeda Chief Annuls Syrian-Iraqi Jihad Merger," Al Jazeera, June 9, 2013, http://www.aljazeera.com/news/middleeast/

2013/06/2013699425657882.html; Karen DeYoung, "Al-Qaeda Leader Zawahiri Seeks to End Infighting among Syrian Militants," *Washington Post,* January 23, 2014, http://www.washingtonpost.com/world/national-security/al-qaeda-leader-zawahiri-seeks-to-end-infighting-among-syrian-radicals/2014/01/23/05c80874-8451-11e3-8099-9181471f7aaf_story.html; Liz Sly, "Al-Qaeda Disavows Any Ties with Radical Islamist ISIS Group in Syria, Iraq," *Washington Post,* February 3, 2014, http://www.washingtonpost.com/world/middle_east/al-qaeda-disavows-any-ties-with-radical-islamist-isis-group-in-syria-iraq/2014/02/03/2c9afc3a-8cef-11e3-98ab-fe5228217bd1_story.html; Aryn Baker, "Why Al-Qaeda Kicked Out Its Deadly Syria Franchise," *Time,* February 3, 2014, http://time.com/3469/why-al-qaeda-kicked-out-its-deadly-syria-franchise/. All accessed June 16, 2014.

58. Hannah Fairfield, Tim Wallace, and Dereck Watkins, "How ISIS Expands," *New York Times,* May 21, 2015, accessed July 27, 2015, http://www.nytimes.com/interactive/2015/05/21/world/middleeast/how-isis-expands.html?_r=0.

59. Malik et al., "How ISIS Crippled al-Qaida."

60. Ibid.; Byman and Williams, "ISIS vs. Al Qaeda."

61. Brennan, "Ensuring al-Qaeda's Demise."

62. Sayyid Qutb, *Milestones* (Darya Ganj: Islamic Book Service, 2001), 21, 71.

63. Bruce Riedel, *The Search for Al Qaeda: Its Leadership, Ideology and Future* (Washington, D.C.: Brookings Institution Press, 2008), 40–41; "Profiles," Global Security.org, accessed June 13, 2014, http://www.globalsecurity.org/security/profiles/osama_bin_laden.htm.

64. Abdulla Azzam, 1987, quoted in Jason Burke, *Al-Qaeda: Casting a Shadow of Terror* (London: I. B. Tauris, 2003), 7–8.

65. Ayman al-Zawahiri, *Knights under the Prophet's Banner,* cited in David Aaron, *In Their Own Words: Voices of Jihad* (Santa Monica: Rand, 2008), 198.

66. Jacob N. Shapiro, *The Terrorist's Dilemma: Managing Violent Organizations* (Princeton: Princeton University Press, 2013), 10; Abdulkader H. Sinno, *Organizations at War in Afghanistan and Beyond* (Ithaca, N.Y.: Cornell University Press, 2008).

67. "Pakistan: CIA Drone Strikes, 2004 to Present," Bureau of Investigative Journalism Covert Drone War, accessed November 11, 2015, https://docs.google.com/spreadsheets/d/1NAfjFonM-Tn7fziqiv33HlGto9wgLZDSCP-BQaux51w/edit?pli=1#gid=477128060.

68. Rukmini Callimachi, "Mali Manual Suggests Al-Qaeda Has Feared Weapon," Associated Press, accessed June 13, 2014, http://bigstory.ap.org/article/mali-manual-suggests-al-qaida-has-feared-weapon.

69. "Pakistan: CIA Drone Strikes, 2004 to Present."

70. "Letter from Osama bin Laden to Shaykh Mahmud ('Atiyya)," October 21, 2010, Letters from Abbottabad, Combating Terrorism Center, West Point, May 3, 2012, SOCOM-2012-0000015, 2, accessed 21 July, 2015, https://www.ctc.usma.edu/posts/letters-from-abbottabad-bin-ladin-sidelined.

71. "Bin Laden to Shaykh Mahmud ('Atiyya)," October 21, 2010, 3.

72. Ibid., 2.

73. United States v. Abid Naseer, "Letter from Shaykh Mahmud to Osama bin Laden," U.S. Department of Justice, Government Exhibit 429 10-CR-019 (S-4) (RJD), February 15, 2015, 113, accessed March 25, 2015, http://kronosadvisory.com/Abid. Naseer.Trial_Abbottabad.Documents_Exhibits.403.404.405.420thru433.pdf.

74. United States v. Abid Naseer, Government Exhibit 421, 34–35; Owen Bennett-Jones, "Secret Cache of al Qaeda Messages to Osama bin Laden Corroborates Bureau Drone Strike Reports in Pakistan," Bureau of Investigative Journalism, March 23, 2015, accessed July 21, 2015, https://www.thebureauinvestigates.com/2015/03/23/ secret-cache-of-al-qaeda-messages-to-osama-bin-laden-corroborates-bureau-drone-strike-reports-in-pakistan/.

75. Lawrence Wright, *The Looming Tower: Al-Qaeda's Road to 9/11* (New York: Penguin, 2006), 167.

76. United States v. Abid Naseer, Government Exhibit 421, 35.

77. "Al-Qaeda's Operations Chief 'Killed in Pakistan,'" *BBC News,* August 27, 2011, accessed 21 July, 2015, http://www.bbc.co.uk/news/world-south-asia-14695569.

78. Bennett-Jones, "Secret Cache of al Qaeda Messages to Osama bin Laden Corroborates Bureau Drone Strike Reports in Pakistan," 113.

79. Morell, *The Great War of Our Time,* 139.

80. Justin McCarthy, "Confidence in U.S. Branches of Government Remains Low," Gallup, June 15, 2015, accessed 22 July, 2015, http://www.gallup.com/poll/183605/ confidence-branches-government-remains-low.aspx?utm_source=position1&utm_medium=related&utm_campaign=tiles.

81. General Counsel of the U.S. Department of Defense Jeh Charles Johnson, "The Conflict against Al Qaeda and Its Affiliates: How Will It End?" Oxford Union, University of Oxford, November 30, 2012, lawfareblog.com, accessed June 15, 2014, http://www.lawfareblog.com/2012/11/jeh-johnson-speech-at-the-oxford-union/.

82. Ibid.

83. "Remarks by Secretary of Defense Leon Panetta at the Center for a New American Security," Washington, D.C., November 20, 2012, defense.gov, accessed June 15, 2014, http://www.defense.gov/transcripts/transcript.aspx?transcriptid=5154; David Alexander, "U.S. Has Decimated al Qaeda Chiefs but Must Persist in Fight: Panetta," Reuters, November 20, 2012, accessed June 15, 2014, http://www.reuters.com/ article/2012/11/21/us-usa-qaeda-panetta-idUSBRE8AK03R20121121.

84. "Drone Wars," Bureau of Investigative Journalism, accessed July 23, 2015, https://www.thebureauinvestigates.com/category/projects/drones/drones-graphs/.

85. Seth G. Jones, "A Persistent Threat: The Evolution of al Qa'ida and Other Salafi Jihadists," Rand Corporation, June 2014, accessed June 16, 2014, http://www.rand. org/pubs/research_reports/RR637.html. For more information on al-Qaeda affiliates, see "Mapping Militant Organizations," stanford.edu, accessed June 16, 2014, http:// www.stanford.edu/group/mappingmilitants/cgi-bin/groups.

86. "Country Reports on Terrorism 2013," April 2014, Bureau of Counterterrorism, state.gov, accessed June 16, 2014, http://www.state.gov/j/ct/rls/crt/2013/.

87. "Pakistan Leaders Killed," New American Foundation, accessed November 11, 2015, http://securitydata.newamerica.net/drones/leaders-killed.html?country= Pakistan.

88. "Drone Wars Pakistan: Analysis," New American Foundation, accessed November 11, 2015, http://securitydata.newamerica.net/drones/pakistan-analysis.html.

89. "United Nations Security Council Resolution 1368," September 12, 2001, UN.org, accessed April 13, 2014, http://daccess-dds-ny.un.org/doc/UNDOC/GEN/N01/533/82/PDF/N0153382.pdf?OpenElement. For a full list of the ISAF troop numbers and nations contributing to the Afghanistan mission, see "Troop Numbers and Contributions," isaf.nato.int, accessed June 15, 2014, http://www.isaf.nato.int/troop-numbers-and-contributions/index.php.

90. "Transcript of First Presidential Debate between Senators Obama and McCain," September 26, 2008, CNN, accessed June 13, 2014, http://articles.cnn.com/2008-09-26/politics/debate.mississippi.transcript_1_mccain-and-barack-obama-first-presidential-debate-transcript?_s=PM:POLITICS.

91. Oliver Knox, "Obama Foes Blast Iraq Withdrawal amid Iran Worries," Agence France-Presse, October 21, 2011 http://www.google.com/hostednews/afp/article/ALeqM5j8kBusvTCpcuG4QYujbPyb6qU9ag?docId=CNG.372358000fff8eb3e0bd8527f1e55dc5.5b1; Frederick Kagan, "The Enterprise Blog," October 21, 2011, http://blog.american.com/author/fkagan/; "Status of Force Agreement between United States of America and Iraq," November 27, 2008, *New York Times,* http://graphics8.nytimes.com/packages/pdf/world/20081119_SOFA_FINAL_AGREED_TEXT.pdf; "American Servicemembers' Protection Act of 2002," Public Law 107–206, August 2, 2002, gpo.gov, http://www.gpo.gov/fdsys/pkg/PLAW-107publ206/pdf/PLAW-107publ206.pdf. All accessed June 13, 2014.

92. Remarks by President Barack Obama, "A New Strategy for Afghanistan and Pakistan," March 27, 2009, whitehouse.gov, accessed June 13, 2014, http://www.whitehouse.gov/blog/09/03/27/A-New-Strategy-for-Afghanistan-and-Pakistan/.

93. Ibid.

94. For an overview of the counterinsurgency approach developed by General David Petraeus, see General David Petraeus, Lt. General James F. Amos, Lt. Colonel John A. Nagl, and Sarah Sewall, *The U.S. Army and Marine Corp Counterinsurgency Field Manual* (Chicago: University of Chicago Press, 2007). Also see: David Kilcullen, *The Accidental Guerrilla: Fighting Small Wars in the Midst of a Big One* (London: Hurst, 2009), and David Galula, *Counterinsurgency Warfare: Theory and Practice* (Westport, Conn.: Praeger Security International, 2006).

95. Statistics provided by U.S. Department of Defense, "American Forces in Afghanistan and Iraq," *New York Times,* June 22, 2011, accessed June 13, 2014, http://www.nytimes.com/interactive/2011/06/22/world/asia/american-forces-in-afghanistan-and-iraq.html?ref=asia.

96. "Remarks by the President in Address to the Nation on the Way Forward in Afghanistan and Pakistan," West Point, December 1, 2009, accessed June 13, 2014, http://www.whitehouse.gov/blog/2009/12/01/new-way-forward-presidents-

address. For a full breakdown of the Obama administration's surge and breakdown plan, see Bob Woodward, *Obama's Wars: The Inside Story* (London: Simon and Schuster, 2010), 283.

97. "Best Practices in Counterinsurgency: Making High-Value Targeting Operations an Effective Counterinsurgency Tool," Directorate of Intelligence, Central Intelligence Agency, July 7, 2009, iii–1, Wikileaks, December 18, 2014, accessed December 20, 2014, https://wikileaks.org/cia-hvt-counterinsurgency/WikiLeaks_Secret_CIA_review_of_HVT_Operations.pdf.

98. Ibid, i.

99. "President Obama's Final Orders for Afghanistan Pakistan Strategy," November 29, 2009, in Woodward, *Obama's Wars*, 387–390.

100. "Bringing the U.S. War in Afghanistan to a Responsible End," whitehouse.gov, May 27, 2014, accessed June, 15, 2014, http://www.whitehouse.gov/the-press-office/2014/05/27/fact-sheet-bringing-us-war-afghanistan-responsible-end.

101. Greg Miller, "CIA Closing Bases in Afghanistan as It Shifts Focus amid Military Drawdown," *Washington Post,* July 23, 2013, accessed June 15, 2014, http://www.washingtonpost.com/world/national-security/cia-closing-bases-in-afghanistan-as-it-shifts-focus-amid-military-drawdown/2013/07/23/7771a8c2-f081-11e2-a1f9-ea873b7e0424_story.html.

102. Greg Miller, "Secret report raises alarms on intelligence blind spots because of AQ focus," *Washington Post,* March 20, 2013, http://www.washingtonpost.com/world/national-security/secret-report-raises-alarms-on-intelligence-blind-spots-because-of-aq-focus/2013/03/20/1f8f1834-90d6-11e2-9cfd-36d6c9b5d7ad_story.html; "Open Hearing on the Nomination of John O. Brennan to be Director of the Central Intelligence Agency," United States Senate Select Committee, Washington, D.C., February 7, 2013, intelligence.senate.gov, http://www.intelligence.senate.gov/130207/transcript.pdf, 69. Both accessed June 15, 2014.

103. Rhodri Jeffreys-Jones, "Why Was the CIA Established in 1947?" in Rhodri Jeffreys-Jones and Christopher Andrew (eds.), *Eternal Vigilance? 50 Years of the CIA* (London: Frank Cass, 1997), 33.

104. For more on the Phoenix Program, see: Dale Andradé, *Ashes to Ashes: The Phoenix Program and the Vietnam War—Cover for Assassination or Effective Counterinsurgency?* (New York: Lexington Books, 1990); Andrew R. Finlayson, "A Retrospective on Counterinsurgency Operations: The Tay Ninh Provincial Reconnaissance Unit and Its Role in the Phoenix Program, 1969–1970," *Studies in Intelligence* 51, no. 2 (June 2007), cia.gov/library, https://www.cia.gov/library/center-for-the-study-of-intelligence/csi-publications/csi-studies/studies/vol51no2/a--retrospective-on-counterinsurgency-operations.html; Seymour Hersh, "Moving Targets: Will the Counter-Insurgency Plan in Iraq Repeat the Mistakes of Vietnam?" *New Yorker,* December 15, 2003, http://www.newyorker.com/archive/2003/12/15/031215fa_fact?currentPage=all; Rhodri Jeffreys-Jones, *Cloak and Dollar: A History of American Secret Intelligence* (New Haven: Yale University Press, 2002), 203; Alfred McCoy, *A Question of Torture: CIA Interrogation, from the Cold*

War to the War on Terror (New York: Henry Holt, 2006), 68; Ken Tovo, "From the Ashes of the Phoenix: Lessons for Contemporary Counterinsurgency Operations," report for U.S. Army War College, Air University, au.af.mil, March 18, 2005, http://www.au.af.mil/au/awc/awcgate/army-usawc/ksil241.pdf. All accessed June 23, 2014.

105. For criticism of the CIA from the Obama administration, U.S. Senate, and media see: Ian Allen, "U.S. Intel Official Acknowledges Missed Arab Spring Signs," *intelnews.org*, July 23, 2012, http://intelnews.org/tag/0-us-intel-official-acknowledges-missed-arab-spring-signs/; Ken Dilanian, "U.S. Intelligence Official Acknowledges Missed Arab Spring Signs," *Los Angeles Times,* July 19, 2012, http://latimesblogs.latimes.com/world_now/2012/07/us-intelligence-official-acknowledges-missed-signs-ahead-of-arab-spring-.html; Kimberly Dozier, "Obama Criticizes Spy Agencies for Not Seeing Revolts," Associated Press, February 5, 2011 http://www.sfgate.com/world/article/Obama-criticizes-spy-agencies-for-not-seeing-2477011.php; Dylan Evans, "The Arab Spring and the CIA—One Year on," *Huffington Post,* February 2, 2012, http://www.huffingtonpost.co.uk/dylan-evans/the-arab-spring-and-the-cia_b_1246230.html; Charlotte Higgins, "Arab Spring Has Created 'Intelligence Disaster,' Warns Former CIA Boss," *Guardian,* August 28, 2011, http://www.theguardian.com/world/2011/aug/28/arab-spring-intelligence-disaster-scheuer; Mark Mazzetti, Eric Schmitt, and David D. Kirkpatrick, "Benghazi Attack Called Avoidable in Senate Report," *New York Times,* January 15, 2014, http://www.nytimes.com/2014/01/16/world/middleeast/senate-report-finds-benghazi-attack-was-preventable.html?_r=0, "Review of the Attacks on U.S. Facilities in Benghazi, Libya, September 11–12, 2012," Senate Select Committee on Intelligence, January 15, 2014, intelligence.senate.gov, http://www.intelligence.senate.gov/benghazi2014/benghazi.pdf; Shane Harris, "Jihadist Gains in Iraq Blindside American Spies," *Foreign Policy,* June 12, 2014, http://www.foreignpolicy.com/articles/2014/06/12/jihadist_gains_in_iraq_blindside_american_spies. All accessed June 15, 2014.

106. Eli Laek, Noah Shachtman, and Christopher Dickey, "Ex-CIA Chief: Why We Keep Getting Putin Wrong," *Daily Beast,* March 2, 2014, accessed June 15, 2014, http://www.thedailybeast.com/articles/2014/03/02/ex-cia-chief-why-we-get-putin-wrong.html.

107. Bob Herbert, "The Afghan Quagmire," *New York Times,* January 5, 2009, http://www.nytimes.com/2009/01/06/opinion/06herbert.html; John Barry, "Obama's Vietnam," *Newsweek,* January 30, 2009, http://www.thedailybeast.com/newsweek/2009/01/30/obama-s-vietnam.html; Hal Burnton, "For Troops, Afghanistan 'Like Vietnam without Napalm,'" *Seattle Times,* October 2, 2009, http://seattletimes.nwsource.com/html/nationworld/2009990698_usafghanvillage03.html, Robert Wright, "Worse Than Vietnam," *New York Times,* November 23, 2010, http://opinionator.blogs.nytimes.com/2010/11/23/afghanistan-and-vietnam/. All accessed June 15, 2014.

108. "Statistical Information about Fatal Casualties of the Vietnam War," archives.gov, accessed June 15, 2014, http://www.archives.gov/research/military/vietnam-war/casualty-statistics.html; "Operation Enduring Freedom casualties," iCasualties.org, accessed August 11, 2015 http://icasualties.org/.

109. Ahmed Rashid, "Taliban," in G. F. Krivosheev (ed.), *Soviet Casualties and Combat Losses in the Twentieth Century* (London: Greenhill Books and Stackpole Books, 1997), 18.

110. "Record Number Favors Removing U.S. Troops from Afghanistan," Pew Research, June 21, 2011, accessed June 29, 2014, http://www.people-press.org/2011/06/21/record-number-favors-removing-u-s-troops-from-afghanistan/.

111. "Iraq Versus Vietnam: A Comparison of Public Opinion," Gallup, August 24, 2005, http://www.gallup.com/poll/18097/iraq-versus-vietnam-comparison-public-opinion.aspx; Pew Research Center, June 21, 2011, http://www.people-press.org/2011/06/21/record-number-favors-removing-u-s-troops-from-afghanistan. Both accessed June 15, 2014; Joby Warwick, *The Triple Agent: The Al-Qaeda Mole Who Infiltrated the CIA* (New York: Doubleday, 2011). Also see D/CIA Leon Panetta, *Washington Post,* January 10, 2010, accessed June 15, 2014, https://www.cia.gov/news-information/press-releases-statements/press-release-2010/director-panettas-op-ed-on-terrorist-attack-in-afghanistan.html.

112. Wali Aslam, "A Critical Evaluation of American Drone Strikes in Pakistan: Legality, Legitimacy and Prudence," *Critical Studies on Terrorism* 4, no. 3 (2011): 313–329, 324.

113. Michael Gross, *Moral Dilemmas of Modern War: Torture, Assassination, and Blackmail in an Age of Asymmetric Conflict* (Cambridge: Cambridge University Press, 2009), 113.

114. Aslam, "A Critical Evaluation," 324.

115. Selig S. Harrison, "The Pashtun Time Bomb," *New York Times,* August 1, 2007, accessed April 24, 2014, http://www.nytimes.com/2007/08/01/opinion/01iht-edharrison.1.6936601.html?_r=0.

116. Syed Alam Mehsud quoted in Loyd, "US Drone Strikes."

117. Farhat Taj, "Drone Attacks—A survey," *The News,* March 5, 2009, accessed April 25, 2014, http://www.thenews.com.pk/TodaysPrintDetail.aspx?ID=165781&Cat=9&dt=3/4/2009.

118. Brian Glynn Williams, "The CIA's Covert Predator Drone War in Pakistan, 2004–2010: The History of an Assassination Campaign," *Studies in Conflict and Terrorism* 33, no. 10 (2010): 871–892, 884.

119. Clay Ramsay, Steven Kull, Stephen Weber, and Evan Lewis, "Pakistani Public Opinion on the Swat Conflict, Afghanistan, and the U.S.," July 1, 2009, World Public Opinion.org, http://www.worldpublicopinion.org/pipa/pdf/jul09/WPO_Pakistan_Jul09_rpt.pdf; "Pakistani Public Turns against Taliban, but Still Negative on U.S.," World Public Opinion.org, http://www.worldpublicopinion.org/pipa/articles/brasiapacificra/619.php?nid=&id=&pnt=619&lb=bras; C. Christine Fair, "Islamist Militancy in Pakistan: A View from the Provinces," 24 July, 2009, World Public Opinion.org, http://worldpublicopinion.org/pipa/articles/brasiapacificra/629.php?lb=bras&pnt=629&nid=&id. All accessed April 25, 2014.

120. "The Peshawar Declaration," reproduced in full in *Let Us Build Pakistan,* accessed April 24, 2014, http://lubpak.com/archives/47109; Williams, "The CIA's Covert Predator Drone War in Pakistan, 2004–2010," 891.

121. "The Peshawar Declaration." For more on the policy of strategic depth, see Charles G. Cogan, "Shawl of Lead," *Conflict* 10, no. 3 (1990): 189–204. For more on the link between the Pakistani government and the Taliban, see: William Maley, *The Afghanistan Wars* (New York: Palgrave Macmillan, 2002), 218–250, and Steve Coll, *Ghost Wars: The Secret History of the CIA, Afghanistan and bin Laden from the Soviet Invasion to September 10, 2001* (London: Penguin, 2005), 290.

122. "The Peshawar Declaration."

123. Massoumeh Torfeh, "The Pursuit of Peace in Afghanistan," Al Jazeera, July 16, 2015, http://www.aljazeera.com/indepth/opinion/2015/07/pursuit-peace-afghanistan-taliban-pakistan-150713141026071.html; May Jeong and Victor Mallet, "Taliban Slaughter Threatens Afghan Peace Talks," *Financial Times,* August 10, 2015, accessed August 11, 2015, http://www.ft.com/cms/s/0/19631474-3f61-11e5-b98b-87c7270955cf.html#axzz3iVoAOGyC; both accessed August 11, 2015.

124. Patrick Cockburn, *The Rise of Islamic State: ISIS and the New Sunni Revolution* (London: Verso, 2015), 52–55.

125. President Barack Obama, "Remarks by the President at the United States Military Academy Commencement Ceremony," U.S. Military Academy, West Point, New York, May 28, 2014, White House, accessed July 24, 2015, https://www.whitehouse.gov/the-press-office/2014/05/28/remarks-president-united-states-military-academy-commencement-ceremony.

126. Barton Gellman and Greg Miller, "'Black Budget' Summary Details U.S. Spy Network's Successes, Failures and Objectives," *Washington Post,* August 29, 2013, http://www.washingtonpost.com/world/national-security/black-budget-summary-details-us-spy-networks-successes-failures-and-objectives/2013/08/29/7e57bb78-10ab-11e3-8cdd-bcdc09410972_story.html; "Congressional Budget Justification FY2013," volume 1, National Intelligence Program Summary, February 2013, *Washington Post,* http://apps.washingtonpost.com/g/page/national/inside-the-2013-us-intelligence-black-budget/420/. Both accessed June 26, 2014.

127. "Congressional Budget Justification FY2013," *Washington Post.*

128. Rhodri Jeffreys-Jones, *The CIA and American Democracy* (New Haven: Yale University Press, 2003), xxii.

129. Dana Priest and William M. Arkin, "A hidden world, growing beyond control," *Washington Post,* July 19, 2010, accessed August 11, 2015, http://www.pulitzer.org/files/entryforms/WashPost_TSA_Item1.pdf.

130. Ken Dilanian, "FBI Chief: Islamic State Group Bigger Threat Than al-Qaida," *Washington Post,* July 23, 2015, accessed July 24, 2015, http://www.washingtonpost.com/national/fbi-chief-islamic-state-group-bigger-threat-than-al-qaida/2015/07/22/79bcaf90-30db-11e5-a879-213078d03dd3_story.html?postshare=4001437617888286&utm_source=Sailthru&utm_medium=email&utm_term=%2ASituation%20Report&utm_campaign=SitRep0723.

131. Lister, "Profiling the Islamic State," 16, 24.

132. Audrey Kurth Cronin, "Why Counterterrorism Won't Stop the Latest Jihadist Threat," *Foreign Affairs* 94, no. 2 (March 2015): 87–98.

133. Missy Ryan, "U.S. Drone Believed Shot Down in Syria Ventured into New Area, Official Says," *Washington Post,* March 19, 2015, accessed August 11, 2015, https://www.washingtonpost.com/world/national-security/us-drone-believed-shot-down-in-syria-ventured-into-new-area-official-says/2015/03/19/891a3d08-ce5d-11e4-a2a7-9517a3a70506_story.html.

134. "Operation Inherent Resolve: Targeted Operations against ISIL Terrorists," U.S. Department of Defense, accessed August 11, 2015, http://www.defense.gov/home/features/2014/0814_iraq/.

135. Ibid.

136. Aaron Mehta, "Odierno: ISIS Fight Will Last '10 to 20 Years,'" *Defense News,* http://www.defensenews.com/story/defense/2015/07/17/odierno-isis-fight-last-10-20-years/30295949/.

137. Greg Miller, "U.S. Launches Secret Drone Campaign to Hunt Islamic State Leaders in Syria," *Washington Post,* September 1, 2015, accessed 2 September, 2015, https://www.washingtonpost.com/world/national-security/us-launches-secret-drone-campaign-to-hunt-islamic-state-leaders-in-syria/2015/09/01/723b3e04-5033-11e5-933e-7d06c647a395_story.html.

138. Michael Scheuer, "Damn Iraq, Start Caring for America First," *Michael Scheuer's Non-Intervention.com,* June 18, 2014, accessed July 23, 2015, http://non-intervention.com/1223/damn-iraq-start-caring-for-america-first/.

Conclusion

Epigraph. Barack Obama, "Remarks by the President at the National Defense University," Fort McNair, Washington, D.C., May 23, 2013, whitehouse.gov, accessed July 5, 2016, https://www.whitehouse.gov/the-press-office/2013/05/23/remarks-president-national-defense-university.

1. Gregory D. Johnson, "The Untouchable John Brennan," *Buzzfeed,* April 24, 2015, accessed April 28, 2015, http://www.buzzfeed.com/gregorydjohnsen/how-cia-director-john-brennan-became-americas-spy-and-obamas#.ujM8lDRwv.

2. Richard Whittle, *Predator: The Secret Origins of the Drone Revolution* (New York: Henry Holt, 2014), 156.

3. "FY 2011 Military Construction Project Data: Ramstein Air Base, Germany," Department of the Air Force, February 2010, accessed May 27, 2015, https://www.ndr.de/geheimer_krieg/satcom101.pdf%20; Jeremy Scahill, "Germany Is the Tell-Tale Heart of America's Drone War," *Intercept,* April 17, 2015, accessed 2 May, 2015, https://firstlook.org/theintercept/2015/04/17/ramstein/.

4. "Military Construction Program Fiscal Year 2012 Budget Estimates," Department of the Air Force, February 2011, accessed May 27, 2015, http://www.saffm.hq.af.mil/shared/media/document/AFD-110210-037.

5. Ian G. R. Shaw, "Predator Empire: The Geopolitics of U.S. Drone Warfare," *Geopolitics* 18, no. 3 (2013): 536–559, at 537.

6. Charles S. Maier, *Among Empires: American Ascendancy and Its Predecessors* (Cambridge: Harvard University Press, 2006), 109.

7. C. Collin Davies, *The Problem of the North-West Frontier, 1890–1908,* 2nd ed. (New York: Cambridge University Press, 1932), 2.

8. Joby Warrick and Greg Miller, "U.S. Intelligence Gains in Iran Seen as Boost to Confidence," *Washington Post,* April 8, 2012, accessed August 27, 2014, http://www.washingtonpost.com/world/national-security/us-sees-intelligence-surge-as-boost-to-confidence/2012/04/07/gIQAlCha2S_print.html.

9. Gordon Lubold, "Exclusive: Iraq, in a Major Shift, Might Want Some U.S. Troops Back," *Foreign Policy,* May 8, 2014, http://complex.foreignpolicy.com/posts/2014/05/08/exclusive_iraq_in_a_major_shift_might_want_some_us_troops_back#correction; Eri Lake, "Iraq Wants America Back to Fight Insurgents with Air Strikes," *Daily Beast,* June 11, 2014, http://www.thedailybeast.com/articles/2014/06/11/iraq-wants-america-back-to-fight-al-qaeda-with-air-strikes.html; Chelsea J. Carter, Arwa Damon, and Raja Razek, "U.S. Has Armed Drones over Baghdad, Official Says," CNN, http://www.cnn.com/2014/06/27/world/meast/iraq-crisis/; Associated Press, "US flying armed drones in Iraq," *Guardian,* June 28, 2014, http://www.theguardian.com/world/2014/jun/28/us-flying-armed-drones-in-iraq; Mark Thompson, "Armed U.S. Drones Flying over Baghdad," *Time,* June 27, 2014, accessed June 29, 2014, http://time.com/2933508/iraq-drones-isis-obama/.

10. Michael R. Gordon and Eric Schmitt, "Iran Secretly Sending Drones and Supplies into Iraq, U.S. Officials Say," *New York Times,* June 25, 2014, http://www.nytimes.com/2014/06/26/world/middleeast/iran-iraq.html?_r=0; Elias Groll, "Iran Is Deploying Drones in Iraq: Wait, What? Iran Has Drones?" *Foreign Policy,* June 25, 2014, accessed June 29, 2014, http://blog.foreignpolicy.com/posts/2014/06/25/iran_is_deploying_drones_in_iraq_wait_what_iran_has_drones.

11. Ed Lopez, "Innovative Technology Gains New Potential," *Army Technology* (July/August), accessed August 29, 2015, http://armytechnology.armylive.dodlive.mil/index.php/2015/07/01/4-4/.

12. "The National Military Strategy of the United States of America 2015," Joint Chiefs of Staff, June 2015, accessed July 24, 2015, http://www.jcs.mil/Portals/36/Documents/Publications/National_Military_Strategy_2015.pdf; Joseph Trevithick, "The U.S. Army Plans to Obliterate Russian and Chinese Drones with a Huge Chain Gun," *War Is Boring,* July 23, 2015, accessed July 24, 2015, http://theweek.com/articles/567635/army-plans-obliterate-russian-chinese-drones-huge-chain-gun.

13. Friederike Heine, "Merkel Buzzed by Mini-Drone at Campaign Event," *Spiegel,* September 16, 2013, accessed July 26, 2015, http://www.spiegel.de/international/germany/merkel-campaign-event-visited-by-mini-drone-a-922495.html; Dan Bilefsky, "France Arrests 3 with Drones by Power Plant," *New York Times,* November 6, 2014, accessed August 14, 2015, http://www.nytimes.com/2014/11/07/world/europe/3-found-with-drones-near-nuclear-plant-are-questioned-in-france.html?_r=0; Michael D. Shear and Michael S. Schmidt, "White House Drone Crash Described as a U.S. Worker's Drunken Lark," *New York Times,* January 27, 2015, http://www.nytimes.com/2015/01/28/us/white-house-drone.html?_r=0; "Japan Radioactive Drone: Tokyo Police Arrest Man," *BBC News,* April 25, 2015, http://www.bbc.co.uk/news/

world-asia-32465624; Hogwit, "Flying Gun," YouTube, July 10, 2015, accessed July 26, 2015, https://www.youtube.com/watch?v=xqHrTtvFFIs.

14. Matt McFarland, "American Red Cross Takes Serious Look at Using Drones for Disaster Relief, Holds Off for Now," April 21, 2015, http://www.washingtonpost.com/blogs/innovations/wp/2015/04/21/american-red-cross-takes-serious-look-at-using-drones-for-disaster-relief-holds-off-for-now/; Adam Clark Estes, "Some Good Things Drones Can (Actually) Do," *Gizmodo,* December 3, 2013, accessed July 26, 2015, http://gizmodo.com/some-good-things-drones-can-actually-do-1475717696; Mark Zuckerberg, Facebook post, March 27, 2014, accessed August 14, 2015, https://www.facebook.com/zuck/posts/10101322049893211.

INDEX

1972 Munich Massacre, 160, 168.
  See also Black September
1983 barracks bombing, 31, 35, 106:
  failure to retaliate for, 54. *See also*
  Beirut; Hezbollah
1993 Beirut embassy bombing, 33, 35,
  61. *See also* Beirut; Hezbollah
1998 U.S. embassy bombings, 185, 194,
  202
9/11, 10–11, 78–79, 88, 116, 148, 189–190:
  "changed everything," 62, 146; CIA/
  FBI rivalry prior to, 181; Commission,
  55, 69–70, 154, 156, 160, 163, 168;
  drone deployed after, 119, 120, 196,
  239; empowered by, 179; post, 22, 59,
  102, 131, 144, 173, 175, 198; preceding,
  21, 157, 198; target those responsible,
  209; war footing after, 196
2008 economic crash, 18, 207

Abbas, Abdul, 77, 82–83; captured by
  JSOC, 77
Abbottabad, 9: bin Laden's compound
  in, 7, 81; Commission, 204–205;
  documents taken from, 219; JSOC
  raid of, 197, 203; political fallout,
  203–204

Abramowitz, Morton, 76
Abu Bakr al Sidiq Brigade, 226,
  227
Achille Lauro hijacking, 63–64, 74–78,
  84–85, 89, 181
Adnani, Abu Mohamed al-, 222
aerial reconnaissance, 102, 104, 106, 110,
  114–115, 244: over Afghanistan, 148,
  195, 202, 248
"Afghan Eyes," 104, 116, 118–119,
  148
Afghanistan, 2–7, 10, 55, 64–65, 69–71,
  79, 194, 202; breakup of, 237; Bush's
  neglect of, 86; CIA reducing bases in,
  233; civil war in, 71; cost of war in,
  207; jihad in, 70, 131, 155, 159, 223;
  JSOC operating in, 197; lack of U.S.
  intel in, 195; mujahideen in, 70, 131;
  negotiation with Taliban, 239;
  Northern Alliance in, 88, 159;
  quagmire, 235; regime change in, 198;
  support for war in, 236; Tarnak Farm
  in, 104; a terrorism "incubator," 224;
  training camps in, 139; U.S. backed
  government in, 239; U.S. invasion of,
  87, 203, 206, 210, 232–236. *See also*
  "Afghan Eyes"; AfPak

AfPak, 72, 94, 196, 207, 224–225; 228, 239: border region, 10, 86, 184; campaign aim in, 215; central front of war, 231; definition of, x; safe haven in, 20, 207, 229; stats on drone strikes in, 225

Aideed, Mohamed Farah, 200. *See also* Somalia

Air Combat Command, 148

air piracy, 82, 180, 181

Albania, 112, 113

Albright, Madeleine, 166–167: on Somalia, 200

Alec Station: *See* Bin Laden station

Aleppo, 222

Allen, Charles, 104: interest in drones, 116–117

al-Qaeda, 14, 15, 16, 86, 114, 120, 121, 207: actionable intel on, 117; assassination plots by, 169; Bhutto killed by, 79; bureaucratic structure, 217, 240; camps, 194, 202; core leadership, 218–219, 220, 223–224, 228, 229–230; cruise missiles used on, 101, 191–194; crusader narrative of, 205–206, 217–218; decimated, 218–219, 228, 229, 239, 242; drones used on, 196, 220, 237; efforts to hack drones, 225; efforts to neutralize, 138, 143; embassy bombings by, 104, 155, 158, 185, 194; fear of drones, 70; fighters detained, 179, 186–187, 195; finances, 223; focus on "far enemy," 154; franchise, 209–210, 221, 229, 230; identified as threat, 129–130; jihad against U.S., 143; killing hostages, 125; Lackawanna cell, 152; leadership targeting of, 20, 60, 82, 152–153, 215–216, 218, 230–231; leaders isolated, 220–221; legal sanction to hunt, 144; lethal force against, 165–167, 174; limited support to attack, 166; new kind of threat, 173; Northern Alliance and, 88; opportunity for CIA, 143–144; Pakistanis killed by, 79; poor morale, 225–226; recruitment into, 205, 217; resilience of, 217–218; Resolution 1368 on, 145; safe house, 2, 227; split with ISIS, 220–224; state of armed conflict with, 59, 164, 189, 190, 196, 228; strengthened by CIA, 71; third-in-command, 217; tracked by NSA, 188; in tribal areas, 69–70; as unlawful combatants, 146, 149; use of JSOC against, 203–204. *See also* bin Laden, Osama; Jabat al-Nusra (JAN); safe haven; Sinjar Records; terrorism; unlawful combatants

al-Shabaab, 200

Alston, Philip, 16, 19, 81

Amber drone, 104–105, 123. *See also* Karem, Abraham; Leading Systems Inc.

American Civil Liberties Union (ACLU), 15

American Enterprise Institute, 186, 231

Ames, Aldrich, 140

Ames, Robert: death of, 33, 35

Andrews Air Force Base, 181

anti-Americanism, 18, 58, 71, 73, 77, 85, 207–208: in Pakistan, 204, 237–238; resulting from torture, 189

Antiterrorism Act, 180

AQ in the Arabian Peninsula, 125, 176: loyal to Zawahiri, 222; significant threat to U.S., 230

Aquilla drone, 106, 108: scrapped, 107

Arab-Israel peace process, 78

Arab League, 181

Arab Spring, 20, 229, 234–235, 240–241, 246

Arafat, Yasser, 78: links to terrorism, 78. *See also* Palestinian Liberation Front; Palestinian Liberation Organization

Armitage, Richard, 55, 80, 96

Army Rangers, 200, 203

Article 51, 16, 44, 67, 81, 145, 176, 191. *See also* United Nations

Aspin, Les, 191, 197, 201–202: resignation of, 202

Assad, Hafaz al-, 82

assassination, 14, 42–44, 90–91, 237: ban on, 55, 61, 66, 99, 101, 145–146, 158, 165, 175; ban superseded, 146; Bush's EO 13355 on, 146; CIA in breach of ban on, 66; Ford's EO 11905 on, 14, 134; Reagan's EO 12333 on, 42, 55, 61, 90; language of, 159, 164; NSC in breach of ban on, 99; plot against president, 169, 191–192; as U.S. policy, 134, 165; by CIA, 118, 120, 134, 147. *See also* Church Committee

assertive nationalists, 169–170

asymmetric warfare, 48

Attar, Layla al-, 192

Aum Shinrikyo, 183

AUMF, 15, 59, 146, 163–164, 209: a "blank check," 144–145, 149: exploitation of, 176; passed after 9/11, 163, 209

Awlaki, Anwar al-, 16, 133: Justice Department paper on, 151–152

"Axis of Evil," 209

Azzam, Abdullah, 224

Baader-Meinhof Gang, 26

Ba'ath Party, 220–221

Baer, Robert, 121: criticism of CIA, 162, 164, 177

Baghdadi, Abu Bakr al-, 220: caliph, 222; "quiet and uncharismatic," 221. *See also* Islamic State (ISIS)

Bagram Air Base, 1, 6, 226

Baker, James, 109

Balawi, Khalil al-, 3, 236

Balkans: F-16 shot down over, 110; F-117 shot down over, 115; "Lofty View," 112; proving ground for drones, 116; weather over, 109–110

Ball, George: criticism of Schultz, 32–33; root causes of terrorism, 32

Bashir, Salman, 204

Bay of Pigs, 165, 201

Beirut, 25, 65, 103, 117, 181: bombing of U.S. embassy in, 33, 35, 61; bombing of Marine barracks in, 31, 35, 65, 89; car bomb in, 65; G-2 agents in, 65–68. *See also* hostages

Bellinger III, John, 145

Beltway politics, 164

Benjamin, Medea, 100

Bergdahl, Bowdrie, 125–126

Berger, Sandy, 104, 118: press briefing on al-Qaeda, 167

Berntsen, Gary, 162, 164, 177

Bhutto, Benazir, 71: killed by al-Qaeda, 79

big data, 212

Big Safari, 114–116, 119–120, 126: create WILD Predator, 115; weaponize Predator drone, 116

Bilateral Extradition Treaty, 80

bin Laden, Osama, 31, 55, 114, 116, 117: aided by ISI, 155; authority to pursue, 144; Azzam mentor to, 224; based in Sudan, 155; call to kill Americans, 59–60; "dead or alive," 87; declaration of jihad, 104, 129, 143, 158, 169; efforts to kill, 143, 194; efforts to provoke U.S., 205; failure to neutralize, 138, 163, 201; filmed at Tarnak Farm, 104, 118, 195; focus on "far enemy," 154; introduced to Taliban, 155; JSOC raid against, 197; links to Somalia, 200; no *bay'a* from ISI, 220; no indictment against, 155; planned raid against, 156–163; prosecuting, 156; retaliation against, 166; return to Afghanistan, 155; role of EITs in locating, 186; "a star," 223; use of safe haven, 158; warnings about drones, 217, 225–226. *See also* Abbottabad; al-Qaeda; Bin Laden station

Bin Laden station 154: snatch plan, 156–163; TRODPINT team, 156, 159; unclear on authority, 163. *See also* Central Intelligence Agency (CIA); Counterterrorism Center (CTC); Massoud, Ahmed Shah; Scheuer, Michael

Black, Cofer, 104, 118, 146: support for armed Predator, 117, 119; "take the gloves off," 120

Black Hawk helicopter, 200: stealth, 203

Black September, 168

black sites, 176, 179–180, 207, 209. *See also* Central Intelligence Agency (CIA); enhanced interrogation (EITs); GREYSTONE

Blue brothers, 107–108, 110, 123: "Predator" drone, 108, 113

Boko Haram, 222

Bourguiba, Habib, 82

Boutros-Ghali, Boutros, 199

Brennan, John, 17: advisor in 2008 election, 210; on CIA's use of EITs, 186–187; drone campaign architect, 210–213, 215, 247; "deputy president," 212; echoed North's ideas, 60: focus on AfPak region, 210; focus on espionage, 233–234, 235; 211; shared office with Panetta, 212

Brown, Senator Frank, 169

Brugger, Fred, 141–142

Buckley, William, 102: torture and death of, 35

Bush, George H. W., 85: assassination plot, 191; assertive nationalists, 169–170; and Boutros-Ghali, 199–200; defense cuts by, 107; Iran-Contra pardons, 136, 138; new world order, 199; and NSDD 207, 92; policy on Balkans, 109. *See also* Bush administration (G. H. W.)

Bush, George W., 10–11: approval rating of, 87; attacks Clinton's policies, 172; "Axis of Evil," 209; Doctrine, 190; empowered by 9/11, 179; endorsement of preemption, 59, 189–190; enhanced interrogation, 179; EO 13355, 146; fail to capture bin Laden, 150, 232; favorability ratings, 85; Finding on HVT signed by, 15, 120, 145–146; motivation for capture, 185–186; support for EITs, 185; transition with Obama, 131, 144, 149, 150, 184; unilateral action, 68; use of rendition, 179; use of targeted killing, 61, 146, 152–153. *See also* Bush administration (W.).

Bush administration (W): abuses of the, 211; AQ Network Executive Order, 197; armed Predator ready for, 196; drone casualties under, 215; military deployments, 205, 208; NSS 2002, 59, 190; neglect of Afghanistan, 86; reject lib. Int., 172; status of forces agreement, 231; targeting Americans, 152–153; transition to Obama, 131; unilateral targeted killing, 171. *See also* Bush, George W.

Bush administration (G. H. W.), 139: democracy promotion, 170; hegemony, 170; new world order, 169–170; no new CT policy, 139

Camp Chapman, 2, 236

Cannistraro, Vincent, 122–123

Carter, Jimmy, 10, 125: Eagle Claw backlash, 45–46; Iranian hostages and, 23. *See also* Operation Eagle Claw

Casey, William, 11, 12, 22, 57, 143: on assassination, 91; attitude toward Congress, 137; Bob Woodward on, 65–66; brain tumor, 123; and

Clarridge, 89–91; close relations with North, 38, 41; common ground with Schultz, 34–37; criticism of Reagan policy, 36; criticism of Weinberger, 36, 47; deal with Saudis, 66; death of, 89, 123, 135; death of Buckley, 35; ending G-2 program, 69; Fadlallah affair and, 65–67, 72; farsighted, 122; fascinated by CTC drone, 102; handpicked Sporkin, 44; hostage policy adopted, 125; impact of Iran-Contra on, 122; link to Blue brothers, 108; in OSS, 37, 43; on penetrating Hezbollah, 103; political alliance with Haig, 34; "Reagan revival" of CIA, 34, 43, 135; Soviet funding of terror, 34; U.S. at war with terrorism, 36. *See also* counterterrorism hardliners

CENTCOM, 120

Central Intelligence Agency (CIA), 3, 114, 117: aid for Contras, 108; attack upon, 153; authorized to kill, 120; backing of mujahideen, 70; in Balkans, 112; beyond the reach of, 230; blamed for Bay of Pigs, 165; blowback, 71, 147; budget concerns, 118, 139–140, 142, 156, 162; budget increased for, 139–140, 241–242; control of operations, 68–71; counterterrorism role, 26, 94; casualties, 236; covert action, 10, 42, 134, 175; defensive mentality of, 91–92; "dirty assets," 141; domestic spying, 42, 134; drone development by, 96, 111–112; 121; drone operators, 147–149; effort to locate bin Laden, 104; emphasis on HUMINT, 117, 195, 207; enhanced interrogation, 179–180; fear of prosecution at, 163; focused on USSR, 93; foreign agents, 69–70, 86; GNAT-750-45 transferred to, 111–112, 114; hybrid program with JSOC, 10; interrogation by, 186–187; a killing machine, 174; lead in counterterror, 62; links with FBI, 181, 182; links to GA-ASI, 112; militarized, 88; Memorandums for, 118; morale at, 142, 162, 175; oversight of, 42–44, 91, 129, 177, 186, 234; paramilitary force, 12, 93, 153, 157, 163, 233–234, 241; politics and, 162–163; post-9/11 role of, 13; "Reagan revival" of, 34; resistance to drones, 120; risk aversion, 15, 118, 129, 138, 142, 144, 162–163; "rogue operatives," 66, 156; secured basing rights, 202; in Syria, 10; targeted by Hezbollah, 35; targeted killing by, 163; Title 50, 15; use of drones, 9–10, 18, 216; withdrawal from Afghanistan, 69. *See also* Baer, Robert; Berntsen, Gary; Bin Laden station; black sites; Casey, William; Church Committee; Clarridge, Duane; Counterterrorism Center; drone campaign/program; drone strikes; enhanced interrogation (EITs); Gates, Robert; GREYSTONE; Phoenix program; Pike Committee; Scheuer, Michael; Tenet, George

Cheney, Dick, 62, 146: assertive nationalist, 169–170, 172; support for EITs, 185–186

Christopher, Warren, 191

Church Committee, 14, 42–43, 90–91, 122, 133–134, 175

Church, Frank, 42, 122

civilian casualties, 64, 73, 161: efforts to minimize, 44, 59, 66, 67, 99, 120, 133; from drone strikes, 6, 17–18, 227, 238; post-9/11 attitude to, 226. *See also* collateral damage

Clapper, James R., 214: on strike methodology, 215

Clark, Ramsey, 100

Clarke, Richard, 21, 117, 130, 191, 196, 202, 211: "Afghan Eyes," 104, 118, 171, 195; calls to arm drone, 119, 195; criticism of CIA, 160; on INF Treaty, 171–172; on Lewinsky scandal, 165–166; "see it/shoot it option," 21, 119, 171, 195; strategy to eliminate AQ, 195

Clarridge, Duane, 12, 99, 112, 121–124, 126–127, 129, 177, 195: aligned with hardliners, 90–91; on assassination ban, 101; Beirut hostage plan, 103, 110; Casey impressed by, 90; co-located with Allen, 117; and Contras, 90; counterterrorism strategy, 91–92; CTC established by, 89, 91–92, 95; on Eagle Program, 102; farsighted, 122; father of drone campaign, 128; forced to retire, 123, 136; impact of Iran-Contra on, 122; link to Blue brothers, 108; long-term impact of, 212–213; on preemption, 91; a risk taker, 121–122; vision for CTC, 154–155; on Webster, 136–137. See also Casey, William; Central Intelligence Agency (CIA); Counterterrorism Center (CTC)

"clash of civilizations," 205

Clinton, Hillary, 19

Clinton, William, J., 16, 55: assassination ban concerns, 164; avoid cycle of violence, 160; before 9/11 Commission, 160; CIA budget cut by, 139–140; CIA budget increased by, 140–141; discussed bin Laden threat, 167; edited bin Laden Memo, 159; failed to build consensus, 166; first inaugural, 170; focus upon economy, 139; globalization president, 191; Lewinsky scandal, 165–166; liberal internationalist, 170; Memorandums for CIA, 118, 145–146, 156–163, 174; no retaliation for Cole, 201; plans to use JSOC, 202; policy on Balkans, 109–110; refused to authorize JSOC, 157; transition with Bush, 201; unilateral stance, 170–171; unsupported by Reno, 167–168; Wag the Dog, 166

Clinton administration: efforts to kill bin Laden, 130, 158; identified nonstate terrorism, 129–130; introduced PDD 39, 182–183; lack of authorization, 162–163; no war paradigm for terror, 164; strategic outlook of, 169–172; use of cruise missiles, 178–179, 191–194; use of rendition, 178. See also Clinton, William J.

CNN, 192, 204

Cohen, William, 197

Cold War, 27, 57, 65, 93, 198: end of, 108, 139; consensus post, 109; MAD during, 250; outdated thinking of, 122

collateral damage, 7, 45, 60, 101, 128, 192, 226: definition of, x; estimate, 4; from signature strikes, 215; in Libya, 198–199; prevention of, 128, 132, 157, 193, 196, 213; U.S. attitude toward, 161, 226–227. See also civilian casualties

combatant's privilege, 147

Comey, James, 243

Comprehensive Crime Act, 180

Congress, 180, 202: control over foreign policy, 42; counterterror funding, 243; criticism of CIA, 234–235; "Era of Trust," 134; on EITs, 186–187; House Intelligence Committee, 114; lack of security clearance, 13; lack of support, 166, 203; oversight from, 42–44, 91, 129, 131–134, 228; public trust in, 228; Senate Intelligence Committee, 36, 43, 132, 139, 159; Tenet hearing before, 142–143 connectedness, 213

containment, 166–167

Contra rebels, 38, 64, 133: U.S. aiding: 64, 108; 131–132

Counterterrorism Center (CTC), 2, 12–13, 16, 20, 86, 144, 210: Afghan assets, 104; "Afghan Eyes," 104, 195, 248; Air Force pilots and, 148–149; Alec Station, 154; armed drone concept from, 114; Charles Allen's role in, 117; Clarridge director of, 91–92; concerns over drones, 176; created Eagle Program, 89; criticized by Rumsfeld, 198; data-driven approach, 213; drone feed to, 119; drone operators, 147–149; and DST, 93, 95, 101; emphasis on HUMINT, 117, 184, 207; establishment of, 11, 72, 131, 156, 181; expansion of, 93–94, 121; Finding required by, 144; foreign agents, 101, 225; a fusion center, 89, 93, 128; GPS locators for agents, 70, 183–184, 219; hub for WoT in AfPak, 72, 88; hunt for bin Laden, 13, 104; impact of Iran-Contra on, 122, 123, 136–138; interdisciplinary, 155; interrogation run by, 186–187; lack of legal sanction, 130, 155; lack of lethal capability, 155; like a "war room," 121–122; links to DoD, 93, 234; NSA support for, 188; oversight of, 177; paramilitary force, 93, 153, 157; "Radio Shack" approach, 95–96, 101, 112, 126; Reagan Finding, 92; reluctance to kill, 146; a revolution, 93; secret action teams, 92; signature strikes by, 215; Special Operations Group, 93; strike outside warzone, 146; supported by Casey, 89; tight control of drones, 69; use of drones, 195–196, 207–208, 210, 213–214, 239. *See also* Bin Laden station; black sites; Central Intelligence Agency (CIA);

Clarridge, Duane; drone campaign/ program; drone strikes; enhanced interrogation (EITs); Scheuer, Michael; TRODPINT team

counterterrorism hardliners, 11, 22, 37, 65, 86, 95, 97, 153: alternative plans of, 101; building upon ideas of, 131; bypass Weinberger, 63; changes since days of, 156; Cold War mindset of, 122; destroyed by Iran-Contra, 123; farsighted, 122; influence on CIA, 154; long-term impact of, 59, 212, 248; policies pushed by, 62; rejection of, 182; unsuccessful, 41; views in NSDD 138, 39. *See also* Casey, William; North, Oliver, NSDD 138; Schultz; George

counterterrorism policy, 7, 26, 178, 207–208, 228, 239, 247: controversy of, 189; defensive measures, 178; evolution of, 39, 62, 63, 74, 86–87, 101, 133, 228; failure to implement, 35, 40–41, 54, 57–58, 71, 86; focused on al-Qaeda, 150; lethal force in, 29, 33, 35, 130–131, 245. *See also* NSDD 138

Counterterrorism Pursuit Teams, 184. *See also* GPS locators

covert action, 10, 91, 94, 138, 140, 156: Al Gore on, 170; legal sanction for, 143 149, 158, 171; lethal, 134, 138, 163; negative impact of, 122, 143; pendulum, 15, 130–135, 140, 175–176; seductive nature of, 165

Crazi, Bettino, 83

Creech Air Force Base, 4, 148

Crimea: Russian invasion of, 235

Crowe, Admiral William J., 48, 101

cruise missiles, 18, 55, 101, 117, 192, 202: adopted by Clinton, 101, 171, 191–194; against al-Qaeda, 117, 166, 171, 191, 193–195; drone as a, 105; guided by Predator, 118–119; limitations of, 196, 202.

DARPA, 105–106, 114: funds GA-ASI, 112–113; funds Karem, 105

decapitation strategy, 207, 210, 215, 231, 239, 242: increases mortality rate, 216; misguided, 216; unintended consequences, 242. *See also* drone strikes, precision strikes, strikes on terrorists, targeted killing

Defense Intelligence Agency, 27, 103, 117: created JITF-CT, 198

Delta Force, 23, 199: *Achille Lauro* raid, 75, 77; Battle of Mogadishu, 200, 203; links with CIA, 93; rendition from Libya, 185. *See also* Operation Eagle Claw

democracy promotion, 150, 170

Department of Defense: expanded CT role, 196–199; ill equipped for CT, 196, 197; JITF-CT created within, 198; links to CIA, 234; no Afghan plan, 203; Weinberger Doctrine, 46–47. *See also* JSOC; Pentagon; Rumsfeld, Donald

Department of Justice, 59, 86, 186: on Awlaki killing, 151–152; EITs legal, 186; on Geneva Conventions, 68; memo on preemptive strikes, 14; white paper on preemption, 14, 59–60

Derwish, Kemal, 152

Deutch, John, 140–142: "Deutch rules," 141–142; removing "stovepipes," 141

Diplomatic Security Act, 180

Diplomatic Security Service, 182

Disposition Matrix, 2, 174, 212–213, 248

Djibouti, 188

Donilon, Tom, 214–215

drone campaign/program, 44, 83, 86, 91, 92, 161: accuracy of, 214; ad-hoc nature of, 132; Alston report on, 81; AUMF and, 145; Brennan architect of, 212; chain of command for, 120–121, 147; CIA concerns over, 176; denial of safe haven, 224, 229; DI's role in promoting, 232; dismantle al-Qaeda, 60; a distraction, 235; drying up recruitment, 74; efficacy of, 18, 20, 242; escalated, 74, 207–208; expansion of, 213; infrastructure for, 118–119; intel collection for, 184; lack of accurate info on, 227; legal architecture of, 130, 176; lessons from *Achille Lauro,* 77–80; limitations of, 229; minimal oversight of, 174; oversight of, 132, 227–228; own legal architecture, 174; in Pakistan, 68–69, 237–238, 239; Pakistani complicity in, 80–81; persistence of, 220; preemption central to, 190; primary CT tool, 198; primary goals of, 215–236; propaganda against, 238; protests against, 84; public opinion on, 83–85; replaced rendition, 190, 191; safeguards for, 215; scaled down, 214; seductive nature of, 132, 152; signature strikes, 215; surge in the, 232; targeted killings by, 164; territorial limitations, 228–229; transfer to DoD, 132; transparency of, 15, 133, 214; UNSC Resolution 748 and, 81; terrorism and, 73; tight CIA control of, 69; tool for preemption, 60. *See also* civilian casualties; drone strikes; precision strikes; strikes on terrorists; targeted killing

drone(s), 207: Air Force pilots flying, 4, 148; Aquilla, 106; bases, 94; "Big Bird," 111; budget scandal over, 106–107; CIA development of, 96; definition of, x; hacked by insurgents, 192, 225; impact in Balkans, 114; importance of intel for, 184; launch recovery element, 118; legality of, 171; proliferation of, 250–253; proving ground for, 116; psychological impact of, 224; refusal to give Pakistan, 70; surveillance with, 230, 244; tech in infancy, 126; to get intel on terrorists,

103; UAV JPO for, 107; used for terrorism, 251–253; weaponized, 116. *See also* Eagle Program

drone strikes: against al-Qaeda, 185, 188, 196, 224; against Taliban, 185; analysis behind, 95; endurance of, 220; first lethal, 120–121; information gap on, 227; killed hostages, 125; legality of, 60–61, 147; methodology of, 215; militants killed by, 230–231; as national self-defense, 60–61, 149; official statistics on, 214; outside a warzone, 121, 146, 151; "PlayStation mentality," 19; president authorizes, 192–193; proportionate, 81–82; signature, 215; very effective, 218–224. *See also* civilian casualties

Eagle Program, 13, 104–105, 106, 111, 117, 121, 123, 127–128: adopted by DoD, 105; came full circle, 146; created by CTC, 89; loaded with C-4, 102; prototype drone, 89, 101–103

Egypt: extradition treaty, 78; major non-NATO ally, 85; radical Islamists in, 78–79; response to *Achille Lauro*, 78–80; U.S. relations with, 82–83

Eisenhower, Dwight D., 32

Emmerson, Ben, 79–80

enhanced interrogation (EITs), 131, 147, 176, 179, 185–187: certified legal, 186; defended by Brennan, 211; Obama's executive order on, 185; political controversy of, 187, 188–189, 207; as torture, 18, 189, 207; unconstitutional, 176. *See also* black sites; rendition

Fadlallah affair, 63–74, 89, 157: Bir al Abed massacre, 73; death of Robert Stethem, 73

Fadlallah, Mohammed Hussayn, 12, 63–64: assassination attempt, 64–65; links to Hezbollah, 65

Fallon, William J., 80

Farouq, Ahmed, 125

FATA, 17, 74, 184, 206, 208, 215, 219, 250: impact of drones on, 17, 236–238. *See also* North-West Frontier Province; Pakistan

Federal Bureau of Investigation (FBI), 93, 135, 228, 242–243: arrest AQ operative, 195; closer links to CIA, 141, 181–182; competition with CIA, 181; informant for, 154; Joint Terrorism Taskforce, 182, 242; jurisdiction over terror, 180–181; prosecuting terrorists, 153; role in finding bin Laden, 187

Feinstein, Dianne, 132

Ford, Gerald R., 14, 165

Fry, Scott, 116–117

Gaddafi, Mu'ammar, 10, 95, 127, 199; captured by rebels, 127; daughter Hana, 100; Lockerbie bombing, 98; reputation damaged, 98; sponsorship of terror, 13, 25, 95, 96; targeting of, 13, 25, 98–101. *See also La Belle* bombing, 97; Libya; Operation El Dorado Canyon

Garrison, William F., 201

Gates, Robert, 135: on Clarridge, 90; criticism of Reagan's NSC, 40; support for CTC, 92; on Webster, 137–138

GCHQ, 97

General Atomics (GA-ASI), 111, 118, 148: advances in drone tech, 213; awarded Predator contract, 114; bought out LSI, 107, liaised with Charles Allen, 117; worked with Big Safari, 119. *See also* Blue brothers

Geneva Conventions, 68, 146, 189. *See also* international law

George, Clair, 136
Gilani, Yousaf: request for drones, 69, 204–205
Gjader airbase, 112
GNAT-750, 107, 111, 114, 117, 244-45; CIA/USAF use of, 148. *See also* Karem, Abraham
Gore, Al, 170
GPS locators, 70, 183–184, 219. *See also* Counterterrorism Pursuit Teams
GREYSTONE, 179–180, 185: political fallout from, 191. *See also* black sites; enhanced interrogation (EITs)
ground control station, 4, 148. *See also* drone(s)
G-2 agents, 65–68, 72: "rogue operatives," 66; program shut down by Casey, 69
Guantánamo Bay, 9, 18, 149, 180, 189, 207. *See also* enhanced interrogation (EITs)
Guatemala, 141
Gulf War, 116, 201

Haig, Alexander, M., 26: assassination attempt on, 26; criticism of Reagan, 52; exit from cabinet, 28; hardline on terrorism, 26–28; political alliance with Casey, 34
Haq, Mohammed Zia-al-, 70
Harthi, Ali Qaed Senyan al-, 121, 188: bombing of USS *Cole*, 152
Hasenfus, Eugene, 133
Hass, Kenneth: death of, 35
Hawkins, Charles, 105, 106
Hayden, Michael, 210: on CIA overstretch, 235; on signature strikes, 214–215
Hekmatyar, Gulbuddin, 70
Hellfire missile, 4, 6, 152, 213: chosen for Predator drone, 116; Griffin missile rival to, 213; successful launch of, 119.
Hezbollah, 65, 117, 246: bombing of

Marine barracks, 31, 35, 106; difficulty in penetrating, 103; Islamic Jihad, 31, 73; hijacking of TWA 847, 73; kidnapping Americans, 102; kidnapping of Buckley, 35, 102–103. *See also* Beirut; hostages
hijacking, 63–64, 180, 249: of TWA 847, 73. See also *Achille Lauro* hijacking
Holder, Eric, 151–152: criticism of torture, 176
hostages, 124–125: in Beirut, 13, 25, 122, 124–125, 135, 195; death of, 125; held by Iran, 23, 180; killed in drone strike, 125; located by drone, 124–125; Locating Task Force, 103, 116, 117; negotiations over, 124; rescue failed, 165; U.S. public attitude toward, 126. *See also* Iran-Contra
Hughes Aircraft Company, 107, 114
human rights, 9, 16, 27, 32, 147: advocates, 12, 83, 133, 147, 214; drones proportionate under, 81–82; U.N. rapporteur on, 79; watch, 15
Hussein, Saddam, 110, 162, 199, 220: Bush assassination plot, 191. *See also* Iraq
hybrid warfare, 247–248

illegal enemy combatant, 15, 190
imperial presidency, 42, 193
INF Treaty: armed drone in breach of, 171
Inman, Robert, 90
Intelligence Oversight Board, 90
International Criminal Court, 100, 171
international law, 68, 77, 144, 148: bin Laden beyond reach of, 156: contradictions with U.S. aims, 170, 173; lack of consensus within, 190; terrorists within, 173; U.S. efforts to stay within, 171, 209
Inter-Services Intelligence, 70: aid bin Laden, 155; cooperation with FBI,

182; and Gulbuddin Hekmatyar, 70; Mumbai massacre and, 71; support for Taliban, 71, 81, 194

Iran, 54–55: American hostages, 23, 180; lifting of trade sanctions, 24; militia backed by, 25; put on notice, 97–98; U.S. arms shipments to, 122. *See also* Iran-Contra

Iran-Contra, 13, 38, 89, 108, 129, 131–132, 147: Boland Amendment, 38; Charles Allen involved in, 117; impact on CTC, 122, 123, 135, 156, 175; NSC's role in, 57; participants pardoned, 136, 138; Tower Commission on, 57; years preceding, 182

Iraq, 86: American forces in, 205; boost to jihadists, 86, 205; cost of war in, 207; cruise missiles fired at, 192; democracy promotion in, 150; expansion of ISIS in, 235; lack of WMD, 190; mobile Scuds, 110; Sinjar records from, 205; status of forces agreement, 231; U.S. invasion of, 150, 220; U.S. withdrawal from, 231. *See also* Hussein, Saddam; Islamic State (ISIS)

Islamabad, 182–183, 206, 237: U.S. embassy in, 69, 80, 169

Islamic State (ISIS), 9–10, 20, 173, 176, 233, 242–245: announces caliphate, 222; Ba'athists recruited, 220–221; capture of Mosul, 235; declares caliphate, 222; expansion into Syria, 220; leaders not pledged *bay'a*, 220; shot down F-16, 244; significant threat to U.S., 230; split with al-Qaeda, 220–224; split with Jabat al-Nusra, 221; use of black markets, 221

Israel, 25, 38, 73, 78: aggressive counterterrorism, 29, 168; airstrike on PLO by, 75; in cycle of violence, 160; Karem from, 105, 111; rescue in Entebbe, 29; Scuds fired at, 110. *See also* Mossad

Israel, Kenneth, 114

Italy, 28, 248: asserts jurisdiction, 77; US relations with, 82–83: collapse of govt., 83

Jabat al-Nusra (JAN): endorsed by Zawahiri, 220; progress in Syria, 221; significant threat to U.S., 230. *See also* al-Qaeda; Islamic State (ISIS); Syria

jihadist movement: evolution of, 229–230; expanding, 240–241; expansion disguised as, 238; safe haven for, 224; Salafist, 229–230; splits within, 220–224, 246; warfighting, 173

Johnson, Jeh, 228

Joint Chiefs of Staff, 48, 76, 96, 101: chairman of, 201, 202; reject CT mission, 198–199; reject snatch plan, 137

Jonathan Institute, 29

Jowlani, Abu Muhammad al-, 220–223: *bay'a* to Zawahiri, 221 *See also* Islamic State (ISIS); Jabat al-Nusra (JAN)

JSOC, 93, 103, 117, 207, 208, 243: Abbottabad raid, 197, 203; *Achille Lauro* raid, 75; drone program, 10, 176, 245, 248; errors by, 132; expansion of, 196–199; at Fort Bragg, 197; kill Zarqawi, 220; links with CIA, 10, 93, 197; "mowing the lawn," 203; purge ISI leadership, 220–221; Quick Reaction Force, 200

Jumper, John P., 116

just war, 149, 172: *jus ad bellum*, 16, 67–68, 86, 100–101.

Karem, Abraham, 104, 123: bankrupt, 107; support from CIA, 112; Woolsey impressed by, 111. *See also* Leading Systems Inc.

Kasi, Mir Ami, 182–184: prosecuted in US court, 153

Kels, Charles, 147

Kennedy, John, F., 165, 201

Khan, Imran, 84: anti-drone rallies, 84

Khan, Malik Gulistan, 215

Khanabad airfield, 119

Khomeini, Ayatollah, 23

Kilcullen, David, 231

Klinghoffer, Leon, 74–75: return of body, 82

Koch, Noel: criticism of Reagan NSC, 40–41

Kosovo, 115, 116. See also Balkans

Kostelnikm, Michael C., 116

Ku-band SATCOM link, 113

Kuwait, 188, 191, 199, 206

La Belle bombing, 97. See also Gaddafi, Mu'ammar; Operation El Dorado Canyon

Lake, Anthony, 154, 192: recommended Brennan, 210

leadership targeting, 20, 207, 215–216: evidence of impact, 220–221. See also decapitation strategy; drone strikes; precision strikes; strikes on terrorists; targeted killing

Leading Systems Inc. (LSI), 105, 107, 110, 114, 123: GNAT-750 developed by, 107; purchase by GA, 109. See also Amber drone; General Atomics (GA-ASI); GNAT-750; Karem, Abraham

Lebanon, 25, 31, 48, 61, 65–66, 69, 72, 122: anti-Americanism in, 73. See also Beirut; hostages

Lee, Barbara, 144–145

legal architecture, 15, 62, 129, 146, 209, 243: backbone of War on Terror, 130: drone campaign's own, 174; roots from Reagan admin, 153; unstable foundations of, 176

lethal force, 22: against al-Qaeda, 165–167, 174; as deterrent, 58; large-scale, 161; legal sanction for, 164–165; overreliance upon, 232; unsupported by Reno, 167–168; U.S.'s preferred approach, 191. See also preemptive force

Lewinsky, Monica, 165

liberal internationalist, 170, 200

liberal interventionist, 109

Libi, Abu Anas al-, 185

Libi, Abu Yahya al-, 70, 217: on drone strikes, 218–219

Libya, 10, 55, 96, 127: Benghazi attack in, 235; bombing of, 239; bombing of Pan Am 103, 81, 98; drone casualties in, 214; preemptive strike against, 96; rendition from, 185; uprising in, 229. See also Gaddafi, Mu'ammar; La Belle bombing; Operation El Dorado Canyon

Long Commission, 38–39: criticism of Weinberger, 39; on terrorism as warfare, 39. See also Beirut

machine-learning, 213

Maqdisi, Abu Muhammad al-, 222–223

Massoud, Ahmed Shah, 159–160

McCain, John, 150, 231

McChrystal, Stanley, 221

McFarlane, Robert, 53, 89: Achille Lauro raid, 76; lack of access to Reagan, 40; NSDD 109 violated, 54–55; pardoned for Iran-Contra, 136

McMahon, Bernie, 66

McMahon, John: criticism of NSDD 138, 41–43; criticism of Fadlallah plan, 66, 72

Meese, Edwin, 182

Mohammed, Khalid Sheikh, 59–60, 217: waterboarded, 186

Morell, Michael, 218, 227–228
Mossad, 168
Mubarak, Hosni, 75, 79, 82, 83, 85, 97:
    assassination attempt on, 155;
    overthrown, 234–235; a "U.S. agent,"
    82
mujahideen, 70, 79, 227, 240: CIA
    support to Afghan, 131; TRODPINT
    team from, 156, 159, 183–184
Mullen, Mike, 81
Mumbai massacre, 71
Musharraf, Perez, 80

National CT Center, 212
National Geospatial Agency, 213
National Intelligence Estimate, 27
National Security Act of 1947, 12,
    149
National Security Agency (NSA), 1–2,
    103, 117, 128, 207, 210: budget
    doubled, 242; Fort Meade, 1;
    improved capabilities, 187–188, 213;
    intercepted Libya cable, 97; Snowden
    leaks on, 133, 150; surveillance by,
    220; use of metadata, 188; work with
    CTC, 188
National Security Council (NSC),
    40; Achille Lauro plan, 76; insert
    self-defense in MON, 158;
    paralysis of, 53; targeted Gaddafi,
    98–99.
National Security Strategy: 2002, 59,
    190; 2010, 172
NATO, 127, 226: convoys blocked, 204;
    intervention in Balkans, 109, 113, 115;
    intervention in Libya, 10; mission in
    Afghanistan, 210, 231, 233; Obama
    working with, 172; operations against
    Serbs, 13; relations with, 82; Sigonella
    base, 76–77; status of forces
    agreement, 233
Navy SEALS, 7, 9: Abbottabad raid,
    203, 219; Achille Lauro raid, 75; free

Captain Phillips, 124; kill hostage,
    124–125; links with CIA, 93
neoconservative, 59, 109, 169, 172:
    PNAC, 59. See also Wolfowitz,
    Paul
Netanyahu, Benjamin, 29
neutralization, 65, 90: controversy of
    term, 41–44; definition of, xi: of
    terrorists, 21, 41, 77, 86; term
    changed, 44. See also assassination
New American Foundation, 17,
    213–214
New American Security, 229
Nicaragua, 38: Blue brothers' business
    in, 108; Sandinista coup in, 108; U.S.
    support for Contras in, 38, 64, 90,
    108, 122
North, Oliver, 11, 22, 66, 129: Achille
    Lauro raid, 76; close relations with
    Casey, 38, 41; dismissed by Reagan,
    123; drafting of NSDD 138, 39–44;
    Fadlallah affair and, 72; farsighted,
    122; ideas echoed by Brennan, 60;
    Iran-Contra, 37–38, 122; link to Blue
    brothers, 108; man of action, 38; plan
    to use cruise missiles, 101; plan to use
    SEALs, 101; precedent for drone
    campaign, 37. See also
    counterterrorism hardliners
Northern Alliance, 88, 159. See also
    Massoud, Ahmed Shah
North-West Frontier Province, 206:
    impact of drones on, 236–238. See also
    FATA; Pakistan
NSDD 138, 33, 39–44, 65–67, 82, 144,
    146: active defense measures, 39;
    controversy over, 90; failure to
    implement, 63; influence on WoT,
    58–62, 245, 249; preemption in, 69,
    92, 95; use of foreign agents, 69–70,
    72
NSDD 207, 92, 94: terrorism as crime
    and war, 94

Obama, Barack H., 17: 2008 campaign, 150, 189, 209, 212, 231; abandons unlawful combatant, 149; Abbottabad raid, 203; advised by Brennan, 210–213; AfPak strategy, 206, 210, 225, 231–234, 239; "assassin-in-chief," 193; authorizes strikes, 192–193; bans torture, 176, 185, 189; call to impeach, 100; compared to Reagan, 85–86; counterterror options, 207–208; criticism of Iraq war, 150; democracy promotion, 150; "Drone President," 11, 14, 86; embracing drone warfare, 86–87, 179; expands drone campaign, 151, 176, 207–208, 229; expands oversight, 132–133, 151; failure to close Gitmo, 151; first year in office, 213–214; focus to al-Qaeda, 150–151, 196–197, 209–210; Iraq a "strategic mistake," 231; just war used by, 149; on military force, 150; Nobel Peace Prize, 68, 149, 172, 189; opposed to Hayden, 210; on Pakistan, 151, 206; reframing War on Terror, 149, 209–210; second-term, 214; transition with Bush, 184; use of big data, 212; use of int. law, 172, 176; use of JSOC, 203; use of rendition, 179, 185; use of targeted killing, 61, 179; weak economy, 207; withdrawal from Iraq, 231. See also legal architecture; Obama administration
Obama administration: adoption of drone strikes, 18, 239; alignment with NSDD 138, 61; conviction drones work, 228; counterterrorism strategy, 59; criticism of CIA, 234–235; drone strikes effective, 218; on end of WoT, 228–229; escalated drone campaign, 74, 247; first term, 218; formalized drone network, 119; hypocrisy of, 68; legal case for drones, 60–61; military deployments, 205; National Security Strategy, 172; Pakistan unwilling/unable, 81; rejection of Bush policies, 13, 188; releases strike data, 214; shuts down GREYSTONE, 185, 188; transition from Bush, 131; use of JSOC, 197; use of Sporkin argument, 60–61, 68; use of UNSC Resolution 748, 81; views on targeted killing, 164; war with al-Qaeda, 59, 172. See also Obama, President Barack H.
O'Connell, Geoff, 157: warned by Reno, 168
October surprise, 24
Odeh, Mohammed, 195
Omar, Mullah: targeted by Predator drone, 120
Operation: Allied Force, 115; Cyclone, 131; Deliberate Force, 113; Desert Storm, 110; Enduring Freedom, 120; Flower, 96–97; Goldenrod, 181–182; Infinite Reach, 55, 166, 194; Inherent Resolve, 10, 244; Iraqi Freedom, 77; Nomad Vigil, 113. See also Operation Eagle Claw; Operation El Dorado Canyon
Operation Eagle Claw, 103, 202, 203: backlash against, 45–46; failure of, 23–24, 199
Operation El Dorado Canyon, 97–101: civilian casualties from, 99–100; FIII malfunctioned during, 99; FIII downed during, 100; targeted Gaddafi, 98–99
OSS, 37, 91

Pakistan, 68–69, 78, 149, 194, 198: attitude toward drones, 84, 238; breakup of, 237; campaign against Taliban, 206; complicity in drones, 80–81; drone casualties in, 214; impact of drones on, 236–238; Lal Masjid mosque, 206; links to Taliban, 79, 81, 194; loss of Bangladesh, 204;

Mumbai massacre and, 71; nuclear program, 183; peace treaty with Taliban, 206; Peoples Party, 204; Peshawar Declaration, 238; radical Islamists in, 79; request for drones, 69, 237–238; response to bin Laden raid, 204; rivalry with India, 70, 79, 194, 238; sanctions on, 183; sovereignty of, 79–80, 237; Tehrik-e-Insaff party, 84; training camps in, 139; tribal regions of, 237. *See also* FATA; Gilani, Yousaf; Haq, Mohammed Zia-al-; North-West Frontier Province; Waziristan; Zadari, Asif Ali

Pakistan-Afghanistan Department (PAD), 2, 94

Palestinian Liberation Front, 74–75; 77, 80, 122: stateless terrorist group, 93

Palestinian Liberation Organization, 74–75: Israeli raid on, 75, Unit 17, 75

Pan Am, 103, 81, 98

Panetta, Leon, 211, 234: on Brennan, 212; criticism of torture, 176; drone strikes effective, 218; drone strike limitations, 228–229

PATRIOT Act, 9

pattern of life, 213

Pavitt, James, 117, 118

Pax Americana, 109

Paul, Rand: filibuster over drone policy, 13

"peace dividend," 107

Peacekeeper missiles, 111

Pentagon, 198–199, 244, 245: attacked on 9/11, 61; impact of Somalia on, 200–203; need to diversify forces, 39; on danger of drones, 251; Panetta moved to, 234. *See also* Department of Defense

Perry, William, 197, 202

Peters, Whit, 118

Petraeus, David, 231, 234

Phoenix program, 175, 234. *See also* Central Intelligence Agency (CIA)

Pike Committee, 42

Pillar, Paul, 162–163, 164

Poindexter, John, 41: *Achille Lauro* raid, 76; Beirut hostages, 103, 117; plan to use SEALs, 101; on Reagan's Alzheimer's, 56; targeted Gaddafi, 98–99. *See also* National Security Council (NSC)

Policy Standards and Procedures, 14

post-Cold War, 139, 169–170, 173, 200

Powell, Colin, 51: Doctrine, 201–203

precision strikes, 13, 174, 208, 213: Pakistan's lack of, 206. *See also* decapitation strategy; leadership targeting; targeted killing

Predator drone, 2, 5, 80, 114, 146, 152: accident prone, 118, 126–127; adoption of, 138; "Afghan Eyes" success, 104, 118, 195; attacked Gaddafi, 127; bureaucratic deadlock over, 118; compared to Reaper, 213; custom made for CT, 196; delayed deployment of, 120, 171–172; deployment over Afghanistan, 13; derived from GNAT-750, 113; first lethal action by, 120–121; guide cruise missiles with, 118; Hellfire missile for, 116, 195; intel on terrorists, 103–104; joint use with USAF, 13; Langley's resistance to, 117; link to Eagle Program, 102; link to Karem, 112; loss to Serb AA fire, 113; SATCOM link for, 113, 115, 119; sensor ball on, 114; technical problems with, 113; transferred to USAF, 114; use over Balkans, 13; weaponized, 116, 171; WILD variant of, 114; a work in progress, 126. *See also* Reaper drone

preemptive force, 11, 22, 41, 65–67, 128, 209: against terrorists, 69, 86, 95, 189, 212, 239; as deterrent, 58–60; established in NSS 2002, 59, 190; seen as heavy-handed, 217; as self-defense, 36, 39–40, 44, 66, 92; in U.S. policy, 190; U.S. public support for, 87. *See also* NSDD 138
proxy agents, 18, 70

Qur'an, 221
Qutb, Sayidd, 224

Rahman, Atiyah Abd al-, 219
Rahman, Omar Abdel-, 154
Ralston, Joe, 194
Ramstein Air Base, 5, 119, 248
Rand Corporation, 31, 43, 216, 229–230
Rashid, Mohammed, 136
Raytheon, 114, 119, 213
Reagan, Ronald, W., 49, 82, 111, 125: on *Achille Lauro,* 75, 76; Alzheimer's, 56; on bombing Libya, 97; call for impeachment, 100; cannot let terrorists go, 87, 89; compared to Eisenhower, 51; counterterrorism Finding, 92; doctrine, 64–66; "evil empire," 45; harmed by Iran-Contra, 165; leadership of, 24, 49–50; on Libyan terrorism, 96; novice in foreign policy, 50–51; obsession with hostages, 89–90; order violated, 54–55; perception of, 98; pressure on Casey, 89–90; "Reaganism," 51; rhetoric of, 24, 98; signed EO 12333, 42; signed NSDD 138, 239; State of the Union 1985, 64; underestimated, 52–53; "unleashing" the CIA, 138, 175. *See also* counterterrorism hardliners; Reagan administration
Reagan administration, 16, 144, 207, 249: ad-hocism, 50; on assassination, 164; backlash for Libya bombing, 100,

199; censure of Italy, 82–83; CIA for counterterrorism, 143; collusion with Iran, 24; counterterrorism policy, 25, 57; criticism of military buildup; 39, 44–45; dysfunction within, 51–58, 71–72, 180–181; established WoT principles, 58; extradition request, 80, 83; fragmented, 26; funding for Contras, 38, 64, 131–132; lack of decisive action, 30; neo-Reaganite, 59; obsession with hostages, 102; paralysis of NSC, 53; supplying arms to Iran, 25; weakness of NSC, 40–41. *See also* counterterrorism hardliners
Reaper drone, 2, 18, 102, 213, 248: link to Eagle Program, 102; a "true hunter-killer," 127
Regan, Donald, 51
rendition, 9–10, 178, 189, 207, 211: first successful, 181; politically poisonous, 188–189, 191; reaffirmed in PDD 39, 183. *See also* black sites; GREYSTONE
Reno, Janet, 167–168: 9/11 Commission testimony, 168; illegal to kill bin Laden, 168; underestimated al-Qaeda, 169
Reprive, 6
Rice, Condoleezza, 145, 172
Riedel, Bruce, 81, 155
Rizzo, John, 120, 141, 153: on CIA's weakness, 140; on Clinton Findings, 156–163; criticism of Reno, 167–168; drafted Reagan Finding, 156; on Panetta, 211; on Webster, 136
Rockefeller Commission, 42
Rodriguez, Jose: defense of EITs, 186. *See also* black sites; GREYSTONE
Romney, Mitt, 19
rules of engagement, 4, 45, 76, 99, 121
Rumsfeld, Donald: AQ Network Executive Order, 197; expands DoD CT capability, 196–199; plans for

regime change, 198; rivalry with CIA, 197–198. *See also* Bush administration; Department of Defense; JSOC

safe haven, 92, 117, 155–156, 196, 217: denial of, 20, 149–150, 207, 210, 224–225, 239; rendition from, 183; terrorists use of, 180, 183; value of, 225
Sarajevo, 109. *See also* Balkans
satellite: Sputnik, 105; spy, 110, 125, 141, 183
satellite phone, 188
Saudi Arabia, 66, 94, 206: drone base in, 119; targeted by Scuds, 110
ScanEagle drone, 124. *See also* drone(s)
Scheuer, Michael, 157: criticism of managers, 161–162, 164; on defensive mentality, 161; on Islamic State, 245–246. *See also* Bin Laden station
Schroen, Garry, 159–160
Schultz, George, 11, 22, 26, 52, 57, 153: *Achille Lauro* raid, 76; call for active defense, 29–30; common ground with Casey, 34–35; criticism of, 32–33; criticism of Weinberger, 31–32, 47, 53; Fadlallah affair and, 72; forefather of War on Terror, 28–37; prior military experience, 32; realist, 32. *See also* counterterrorism hardliners
Scowcroft, Brent, 199
Scud missiles, 110
self-defense against terrorism, 11, 158–159, 174, 190. *See also* Article 51; Sporkin, Stanley
September 11 attacks, 9. *See also* 9/11
Serbs: counter surveillance by, 110. *See also* Balkans; Sarajevo; Yugoslavia
Shahzad, Faisal, 73
Sharia Law, 240
Sheikh Sa'id, 2
Shelton, Hugh, 202
signature strikes, 215

Sinjar Records, 205–206
Situation Room, 89, 115, 211
SkyGrabber, 192
Smith, Jeffrey, 141
Snowden, Edward, 133, 150, 187–188, 241. *See also* National Security Agency (NSA)
Somalia, 149: al-Shabaab in, 200; American soldiers killed in, 200; Battle of Mogadishu, 200; civil war, 199; drone casualties in, 214; JSOC operating in, 197, 229; UN mission in, 199–200; U.S. withdrawal from, 31. *See also* Aideed, Mohamed Farah
sovereignty, 16, 77, 80–81: violating Pakistan's, 204–205, 237
Soviet Union, 64, 93: collapse of, 138–139, 169, 171, 199; jihad against, 71, 131; KGB, 28; sponsorship of terrorism, 26–28; withdrawal from Afghanistan, 69, 70, 236. *See also* Cold War
Spadolini, Giovanni, 83
Speakes, Larry, 83
Sporkin, Stanley, 136, 144, 146: argument used by Obama, 60–61, 68; on assassination ban, 44, 158, 164; Casey's right hand man, 44; self-defense rationale, 66–67, 99
Stark, James, 76
State Department, 25, 57, 155: attacked by terrorists, 33, 35; Bureau of Counterterrorism, 230; criticism of, 32; drone export licenses, 107; on legality of armed drone, 171; links to CIA, 35; criticism of G-2 agents, 67–68; Patterns of Global Terrorism, 154; on safe havens, 155. *See also* 1993 Beirut embassy bombing; 1998 U.S. embassy bombings
Stethem, Robert, 73
Strickland, Frank, 111: order violated, 54–55

strikes on terrorists, 41: as national
self-defense, 60–61. *See also*
decapitation strategy; drone strikes;
leadership targeting; precision strikes;
targeted killing
Studeman, William, 192
Sudan, 155
Sullivan, Kevin, 116
Supreme Court, 151, 189
Suri, Abu Khalid al-, 222
surveillance state, 174, 209
Syria, 54–55, 150, 233: civil war in, 20,
220, 221, 229, 240–241; expansion of
ISIS in, 235; intervention in Iraq, 220;
port of Latakia, 244; put on notice,
97–98

Taliban, 20, 79, 86, 121: bin Laden a
guest of, 155, 194; degrade the, 210;
deny safe haven for, 208, 210, 225,
236; fear of spies, 219; fighters
detained, 179; foreign agents and the,
69; Haqqani network, 81; links to ISI,
70–71, 79, 81, 194; mistaken for, 215;
momentum reduced, 208, 235, 236;
Northern Alliance and, 88, 159;
Pakistani, 219; Pakistanis killed by, 79;
peace negotiations with, 239; prisoner
exchange with, 125; state of armed
conflict, 189, 190, 228; targeted by
drones, 60; tracked by NSA, 188; as
unlawful combatants, 146; UNSC
resolution on, 78, 145; U.S. relations
sour with, 155–156; U.S. at war with,
189, 196; use of JSOC against, 203.
See also unlawful combatants
targeted killing, 23, 87, 138, 213, 243: of
al-Qaeda, 60; as assassination, 14,
164–165; authorized by Bush, 120;
CIA report on, 232, 246; definition
of, xi; by drone strike, 164, 174; a
euphemism, 164; extrajudicial, 14;
high-value targets, 4, 215, 221; Israel's

policy of, 168; of militants, 230–231;
negative effects of, 232; roots of
policy of, 153; for self-defense, 164; of
unlawful combatants, 146. *See also*
decapitation strategy; drone strikes;
leadership targeting
targeters, 2, 95, 128, 245
Tarnak Farm, 104, 156
Taszar airfield, 113–114
Teicher, Howard, 76, 97
Tenet, George, 155, 198: before 9/11
Commission, 69–70, 154; on bin
Laden snatch plans, 157–158; on CIA
and terror mission, 143; concern over
covert action, 118, 120, 142–143;
criticism of Reno, 168; declaration of
war on AQ, 162; on Lewinsky scandal,
165–166, 167; recommended
Brennan, 210; reduce CIA covert
action, 143, 158; resistance to drones,
117–118, 120; and Rumsfeld, 198;
sworn in as DCI, 142–143; warning
Clinton, 155. *See also* Central
Intelligence Agency (CIA)
terrorism: act of war, 58–59, 182, 228;
ad-hoc, 154; bureaucratic structure,
217; as Cold War issue, 26–28; as
criminal activity, 46, 180–182, 228;
cycle of violence, 75; definition of, xi;
drones used for, 251–253;
international, 45, 93, 181; "lone-
wolf," 243; as national security threat,
92, 239; nonstate, 129, 139, 173, 175,
250; offensive action against, 29;
political costs of, 23; recruitment, 74,
217; root causes of, 32; Soviet
sponsorship of, 26–28, 34; state-
sponsorship of, 31, 45, 96, 97–98;
threat to America's image, 61. *See also*
al-Qaeda; terrorist(s)
terrorist(s): between criminal and war,
173; neutralize, 21, 44, 58, 121, 125,
131, 155, 212, 239; pattern of life of,

119; Threat Integration Center, 212
The Terror Network, 27–28
Townes Commission, 111
TRODPINT team, 156, 159, 183–184: blueprint for CTPT, 184. *See also* Counterterrorism Center (CTC)
Truman, Harry S., 166–167, 234
Turabi, Hassan, 155
Turco, Frederick, 92

U-2 spy plane, 110
UAV Joint Program Office, 107, 111
unilateral action, 61, 68, 77, 169, 239: adopted by Clinton, 170, 183, 191–192; Bush's authorization of, 172; under Obama, 133, 151, 172, 209; rejected by Clinton, 170, 171; risks of, 168
United Nations, 15, 155, 169: Article 2, 81; Article 51, 16, 44, 67, 81, 145, 176, 191; authorize Afghan invasion, 203; Charter, 16, 67, 204; to deal with terrorism, 46; Security Council, 78, 172; in Somalia, 199–200; Resolution 748, 81; Resolution 1368, 78–79, 145. *See also* international law
unlawful combatants, 146–147
U.S. Air Force, 114, 118: drone program, 9, 248; personnel undertaking strikes, 7; weaponizing UAVs, 116. *See also* Big Safari
U.S. Code: Title 10, 12, 149; Title 50, 15, 149
U.S. General Accounting Office, 106
USS: *Bainbridge,* 124; *Butte,* 181; *Cole,* 121, 152, 188, 201; *San Antonio,* 185; *Saratoga,* 77, 181
Uzbekistan, 119, 202

Vessey, John, 48
Vietnam War, 45, 46, 65: compared to Afghan, 235–236; Phoenix program in, 175; syndrome, 31

voice recognition, 2. *See also* Disposition Matrix; National Security Agency (NSA)

Walsh, Lawrence E., 136
Ward, Terry, 141–142
War on Terror, 9–13, 16, 72, 86, 123, 205: Bush's declaration of, 22, 88; CIA spearheading, 88, 144; cost of, 207; detainees, 180; end of, 228, 239–240; importance of AUMF in, 145; JSOC raids common in, 203; legal architecture of, 130; links to Reagan policy, 58–62; NSA's role in, 188; reframed by Obama, 149, 150, 209–210. *See also* AUMF
Watkins, James D., 48: terrorism as a war, 48
Waziristan, 1–6, 73–74, 80, 206, 219, 223: al-Qaeda in, 226–227. *See also* FATA; North-West Frontier Province; Pakistan
weapons of mass destruction (WMD), 9, 182–183, 209: Iraq's lack of, 190
Webster, William, 135–138
Weinberger, Caspar, 44, 52, 57, 82, 111: Arabist inclinations, 48, 82; criticism from Casey, 36; criticism of military buildup, 39; criticism from Schultz, 31–32, 53; defense of Reagan, 52–53; deterrence of USSR, 56; *Fighting for Peace,* 44; on Gaddafi's daughter, 100; impact of Eagle Claw on, 24; Libya and, 96–97; military buildup, 12, 44–45, 65; NSDD 109 violated by, 54–55; opposed *Achille Lauro* raid, 76; opposition to hardliners, 22; pardoned for Iran-Contra, 136; rejects counterterrorism, 49, 197; terrorism as criminal, 46; Doctrine, 46–47, 201–202. *See also* Reagan administration
West Point, 189, 231, 240

WikiLeaks, 232
Williams, Bob, 106
Wing Operations Center, 4
Wolfowitz, Paul, 169–170
Woolsey, James: advocate for Karem, 111, 139–140
World Trade Center, 153: 1993 bombing of, 182

Yazid, Mustafa Abu al-, 1–6, 217: killing of, 226–227; on killing Bhutto, 79
Yemen, 16, 146, 149, 225: drone casualties in, 214; first drone strike in, 152; hostage rescue in, 125; JSOC operating in, 197, 229
Yoo, John, 14
Younis, Fawaz, 181–182
Yousef, Ramzi, 153–154, 182–183: prosecuted in US court, 153

Yugoslavia: post-Cold war break up of, 109. See also Balkans; Sarajevo

Zadari, Asif Ali, 84, 204–205
Zarqawi, Abu Musab al-, 221: founder of AQI, 223; killed by JSOC, 220. See also al-Qaeda; Iraq; Islamic State (ISIS)
Zawahiri, Ayman al-, 59, 155: Afghan "incubator," 224; *bay'a* from Jowlani; contained by drones, 222, 223; development of narrative, 217; on efficacy of drones, 70; efforts to provoke U.S., 205; intervention in dispute, 222; lack of authority, 223; no *bay'a* from ISI, 220; order to dissolve ISIS, 222; training in Afghanistan, 70. See also al-Qaeda; bin Laden, Osama
"Zionist Crusaders," 205, 221, 240